Symmetries in Science XVI 2013

Journal of Physics: Conference Series Volume 538

Bregenz, Austria
21-26 July 2013

ISBN: 978-1-63439-463-5
ISSN: 1742-6588

Printed from e-media with permission by:

Curran Associates, Inc.
57 Morehouse Lane
Red Hook, NY 12571

Some format issues inherent in the e-media version may also appear in this print version.

Publisher supplied ISSN: 1742-6588

Printed by Curran Associates, Inc. (2014)

For permission requests, please contact the Institute of Physics
at the address below.

Institute of Physics
Dirac House, Temple Back
Bristol BS1 6BE UK

Phone: 44 1 17 929 7481
Fax: 44 1 17 920 0979

techtracking@iop.org

Additional copies of this publication are available from:

Curran Associates, Inc.
57 Morehouse Lane
Red Hook, NY 12571 USA
Phone: 845-758-0400
Fax: 845-758-2634
Email: curran@proceedings.com
Web: www.proceedings.com

Table of contents

Volume 538

Symmetries in Science XVI
21–26 July 2013, Bregenz, Austria

Accepted papers received: 20 August 2014
Published online: 10 October 2014

Preface

Papers

Journal of Physics: Conference Series **538** (2014) 011001 doi:10.1088/1742-6596/538/1/011001

Preface

This volume of the proceedings "Symmetries in Science XVI" is dedicated to the memory of Miguel Lorente and Allan Solomon who both participated several times in these Symposia. We lost not only two great scientists and colleagues, but also two wonderful persons of high esteem whom we will always remember.

Dieter Schuch, Michael Ramek

There is a German saying "all good things come in threes" and "Symmetries in Science XVI", convened July 20-26, 2013 at the Mehrerau Monastery, was our third in the sequel of these symposia since taking it over from founder Bruno Gruber who instigated it in 1988 (then in Lochau).

Not only the time seemed to have been perfect (one week of beautiful sunshine), but also the medley of participants could hardly have been better. This time, 34 scientists from 16 countries (more than half outside the European Union) came together to report and discuss their latest results in various fields of science, all related to symmetries. The now customary grouping of renowned experts and talented newcomers was very rewarding and stimulating for all. The informal, yet intense, discussions at "Gasthof Lamm" occurred (progressively later) each evening till well after midnight and finally till almost daybreak!

However, prior to the opening ceremony and during the conference, respectively, we were informed that Miguel Lorente and Allan Solomon had recently passed away. Both attended the SIS Symposia several times and had many friends among present and former participants. Professor Peter Kramer, himself a long-standing participant and whose 80th birthday commemoration prevented him from attending SIS XVI, kindly agreed to write the obituary for Miguel Lorente. Professors Richard Kerner and Carol Penson (both also former attendees) penned, at very short notice, the tribute to Allan Solomon. The obituaries are included in these Proceedings and further tributes have been posted to our conference website.

In 28 lectures and an evening poster-session, topics ranging from theoretical chemistry and molecular physics via fundamental problems in quantum theory to thermodynamics, nonlinear dynamics, soliton theory and finally cosmology, were examined and lively discussed. Nearly all the talks can also be viewed on the conference website. The majority of participants contributed to these Proceedings but some were unable to do so as their results were either previously submitted or published elsewhere. We refer to:

- Quesne C 2013, J. Math. Phys. **54**, 102102.
- Spera M 2013, Nankai Series in Pure, Applied Mathematics and Theoretical Physics: **11** Symmetries and Groups in Contemporary Physics: pp. 593-598; Proceedings of the XXIX International Colloquium on Group-Theoretical Methods in Physics Tianjin, China, 20–26 August 2012 (World Scientific, Singapore)
- Snobl L and Winternitz P 2014, Classification and Identification of Lie Algebras, CRM Monograph Series **33** (Montreal) ISBN-10: 0-8218-4355-9, ISBN-13: 978-0-8218-4355-0 http://www.ams.org/bookstore?fn=20&arg1=crmmseries&ikey=CRMM-33.

Our personal thanks to Daniel and family! Endless support from the Schenk Family who, among other things, sponsored (yet again) the entire conference dinner (including wines and banquet hall) meant that some costs could be alleviated. We could therefore assist various colleagues from economically-weak countries, despite the lack of external funding.

A financial deficit meant we would have had to forego the Conference Proceedings, published in previous years by IOP. After long deliberations, and with donations from Gerhard Berssenbrügge, Dr. Dr. Stephan Hauk and Dr. Volker Weisswange, this could be facilitated. We are very grateful to these private donors for their generous and wholehearted support. The staff of Collegium Mehrerau is also to be thanked for their hospitality.

Finally, our sincere thanks to Yvette not only for her preparatory work and support during the conference, but also for her persistent interest and help in producing the Proceedings within a reasonable time.

Dieter Schuch, Frankfurt am Main, Germany
Michael Ramek, Graz, Austria
August 2014

Peer review statement

All papers published in this volume of *Journal of Physics: Conference Series* have been peer reviewed through processes administered by the proceedings Editors. Reviews were conducted by expert referees to the professional and scientific standards expected of a proceedings journal published by IOP Publishing.

Journal of Physics: Conference Series **538** (2014) 012001 doi:10.1088/1742-6596/538/1/012001

In Memory of Alexander (Allan) Isaac Solomon (1936–2013)

Richard Kerner and Karol Penson
Université Pierre et Marie Curie (Paris-VI)
Paris, France

Our colleague and friend Allan I. Solomon passed away in the University College Hospital London, UK, on April 3, 2013, after a long and brave fight with a terrible cancer.

Allan's scientific activity, rich and original, can be characterized as a deep exploration of physical consequences of various symmetries in physics, especially in quantum phenomena. He participated many times in the conference series "Symmetries in Science", in the series of workshops "Quantum Theory and Symmetries", and above all, he was one of the most active members of the Standing Committee of the ICGTMP meetings ("International Colloquia on Group Theoretical Methods in Physics") until 2009, and Honorary Member until his last days.

Alexander Isaac Solomon was born in Glasgow on December 11, 1936. Soon after, his family moved to Belfast where his father worked with Allan's uncle in the family music business; Allan attended primary school in Belfast. Later on, his family moved to Dublin where Allan continued his scholarship at Wesley College and got his first degree. He studied science at Trinity College of Dublin University and later on in Cambridge, England.

For his postgraduate studies Allan went to Peterhouse, Cambridge where he worked on elementary particles and then to France (Orsay) where he obtained a doctorate in Theoretical Physics in 1963. The same year he married Paulette Baroukh whom he met in Paris while studying at Sorbonne.

Paulette was always a support to Allan in good times and bad and was fascinated by her husband's academic work. She accompanied him all over the world on his scientific trips while continuing her own career with people affected by learning difficulties. She passed away in June 2014, less than two years after Allan's death.

It was 1963—the time when the Americans were coming to Europe and causing a true brain drain hiring young graduates in scientific and medical fields. Allan was then at the Dublin Institute for Advanced Studies and was offered a job in the States to work with "Republic Aviation" industry. He also worked for NASA studying cosmic radiation and its effects on supersonic travel. Although Allan very much enjoyed being in New York and his area of research, he pined for academia and accepted a considerably lower-paid job teaching at NYU Polytechnic in Brooklyn. During that period he made part of his living as a professional jazz trumpet player.

Academic research was confirmed to be Allan's passion and when Yuval Ne'eman invited him to work at the recently established Tel Aviv University, he gladly accepted, despite a move to the Levant and a further drop in salary. Then he and Paulette returned to the Dublin Institute for Advanced Studies with their baby daughter Nathalie. Here he spent two fruitful years in Dublin working alongside and forging life-long friendships with Professor J. L. Synge and Professor C. Lanczos.

His next job was in England, at the newly established Open University and it was there where Allan found his *raison d'être.* After writing numerous textbooks for the Open University, Allan went on to make over 150 radio and television broadcasts with the BBC, lecturing to Open University students around the world. Allan's position at the Open University gave him a lot of freedom to travel and, during his life, he was able to pursue research in collaboration with colleagues in Italy, Israel, China, Singapore and France as well as giving lectures in many other places. His second daughter Annabel was born in that period too.

In the late seventies Allan joined an international group of scientists, headed by Joe Birman and several other eminent Western scholars, and travelled to the former USSR to visit Soviet scientists who were *"refuseniks"* and had been kicked out of their universities. Allan and other visiting scientists gave them lectures in their homes, often huddled in small kitchens, to keep their Soviet counterparts up-to-date with the latest research in their fields.

His scientific interests concentrated around the influence of symmetries on various phenomena in quantum physics, including magnetism, superconductivity, ferroelectricity and, later on, coherent and entangled states, quantum information and cryptography. Among other contributions he developed in 1971 a pioneering application of group theory to superfluid systems which remains a classic to these days.

In the eighties he collaborated with Jacob Katriel and Mario Rasetti on quantum optics and, above all, with Joseph Birman with whom he co-authored 15 publications, including their "magnum opus" on an SU(8) grand unification of superconductivity and various charge and spin density waves. Starting from mid-nineties, Allan developed an intense collaboration with a Parisian group headed by Karol Penson, working on coherent states, combinatorics and quantum control theory. His co-authors of this period included Sonia Schirmer, Hongchen Fu, Gérard Duchamp, Pawel Blasiak, Kasia Gorska, Andrzej Horzela and Apostolos Vourdas. In all his works Allan combined mathematical mastery with extraordinary physical intuition and curiosity which always led him to address challenging new problems.

Allan's life was successful and exemplary in many ways:

–he founded a beautiful family and was a beloved husband, father and grandfather;

–as a scientist, he was successful as an outstanding researcher in both applied and theoretical physics;

–he displayed pedagogical skills in a novel form of popularizing science on radio and television;

–he was a respected member of the British Jewish community and a great friend and defender of Israel;

–he was a staunch promotor of academic freedom, international collaboration and tolerance.

He will be missed and remembered by his family, friends and by all colleagues who had the privilege to meet him and appreciate his deep knowledge, openness and humanity, inimitable affability and his wonderful sense of humor.

Symmetries in Science XVI IOP Publishing
Journal of Physics: Conference Series **538** (2014) 012002 doi:10.1088/1742-6596/538/1/012002

In Memory of Miguel Lorente (1931–2013)

Peter Kramer
Institut für theoretische Physik, Universität Tübingen
72076 Tübingen, Germany

In his scientific work since 1972, Miguel Lorente applied group and representation theory to a broad range of problems in theoretical physics. He followed E. P. Wigner's 1949 idea that elementary systems correspond to irreducible representations of relativistic groups. In 1972–74 Lorente, with Paul Roman, explored position and spin operators in relativistic systems [1]. In relation to these systems, he worked particularly on Bargmann–Wigner equations of massless and spin S=2 fields, on Cayley parametrization, conformal invariance and on group contraction. In cooperation with Bruno Gruber, he analyzed topics in representation theory as unitarization, induced scalar products and indecomposable representations [2]. From the beginning, Lorente's interest turned beyond continuous to discrete or lattice versions of Poincare transformations of space-time, including lattice translations and discrete Lorentz transformations. He was convinced that, on Planck's scale of space-time at 10^{-33} cm, quantum physics must become discrete and theoretical physics must develop the corresponding tools for that scale. Due to the lattice assumption, space and time translations become discrete steps and the generators of space and time translations are to be replaced by difference operators. Lorente analyzed discrete Lorentz transformations and formulated the Dirac and other relativistic equations and their solutions on relativistic lattices [3–5].

Lorente published several studies on the discrete counterparts of the classical polynomials. He reconstructed, for the discrete case, all the tools and techniques known from continuous polynomials as lowering and raising operators and factorization [6].

Since 1990, Lorente took up ideas of discrete quantum gravity, also called spin-foam theories. In contrast to the continuous fields underlying quantum as well as superstring theories, it is based on a discretization of space-time on Planck's scale. In the discrete theory, space-time is geometrically triangulated. To the geometric elements like vertices, edges, faces and tetrahedra of this triangulation one attaches quantum fields interpreted as probability amplitudes. This way towards unification of general relativity and quantum mechanics was opened by T. Regge [7] in the early sixties. Regge introduced some triangulations in a Riemann manifold. He constructed a local curvature at the vertices and, out of it, a finite action which, in a continuous limit, becomes the Hilbert–Einstein action underlying classical gravity. This calculus was made more precise by Ponzano and Regge [8]. They associate to every tetrahedron in the triangulation a $6j$-symbol depending on six spin quantum numbers j. This approach is linked to a discretization of space and time proposed by R. Penrose [9]. Given some graph representing the interaction of elementary units satisfying the rules of angular momentum without an underlying space, Penrose constructed out of this spin network the properties of total angular momentum as a derived concept.

In his own work, Lorente was well-prepared to take the step from lattices to networks. In spin foams, irreducible representations of a relativistic group are attached as fields to the edges of a network such that vertices of the network imply couplings of representations. In many visits to Tübingen since 1993 we analyzed together topological properties, duality, embedding and representation theory on networks [10].

Miguel Lorente had also studied theology and philosophy and their relations to physics [11]. It was always a pleasure to work with Lorente and share his deep understanding and broad knowledge of theoretical physics. He was an active participant of the series of Conferences on Symmetry, initiated by Bruno Gruber since the year 1988 in Lochau and Bregenz, Austria. Lorente was a gifted musician; he played by heart on the pianoforte the classical pieces of music from medieval Spanish to romantic and modern composers. At the Universities of Oviedo and Comillas Madrid, Lorente lectured and directed theses as professor of physics. Dr. Olaya Fernandez Herrero, who got her doctor's degree in physics of music with Lorente, told me that he was one of the few physics professors in Spain who also lectured on acoustics, including tonality and consonance.

With his passing, our community of friends of symmetry and physics has lost a gifted scientist, a charming colleague and a dear friend.

References

[1] Lorente M and Roman P 1974 General expressions for the position and spin operators of relativistic systems *J. Math. Phys.* **15** 70–4

[2] Lorente M, Gruber B and Lenczewski R 1990 Induced Scalar Products and Unitarization *J. Math. Phys.* **31** 587–93

[3] Lorente M 1976 Basis for a discrete special relativity *Int. J. Theor. Phys.* **15** 927–47

[4] Lorente M 1993 A new Scheme for the Klein–Gordon and Dirac Fields on the Lattice with Axial Anomaly *Int. J. Group Theor. in Phys.* **1** 105–21

[5] Lorente M and Kramer P 1999 Representations of the discrete inhomogeneous Lorentz group and Dirac wave equation on the lattice *J. Phys. Math. Gen.* **A 32** 2481–97

[6] Lorente M 2003 Raising and lowering operators and their factorization for generalized classical orthogonal polynomials of discrete variable on homogeneous and non-homogeneous lattices *J. Phys. Math. Gen.* **A 36** 183–98

[7] Regge T 1961 General Relativity without coordinates *Il Nuovo Cimento* **19** 558–71

[8] Ponzano G and Regge T 1968 Semiclassical limit of Racah coefficients *Spectroscopy and group theoretical methods in Physics* ed F Bloch *et al.* (Amsterdam: North Holland)

[9] Penrose R 1971 Angular momentum: an approach to combinatorial space-time *Quantum theory and Beyond* ed T Bastin (Cambridge: C.U.P.)

[10] Lorente M and Kramer P 2011 Discrete Quantum Gravity *De las Ciencias a la Teología, Ensayos interdisciplinares, Homenaje a Manuel García Doncel* ed Jesús Romero Moñivas (Estella: Editorial Verbo Divino) pp 133–56

[11] Lorente M www.newton.ac.uk/programmes/DIS/seminars/032616306.html

Journal of Physics: Conference Series **538** (2014) 012003 doi:10.1088/1742-6596/538/1/012003

A framework for dynamical generation of flavor mixing

M. Blasone[†], P. Jizba[‡], G. Lambiase[†] and N.E. Mavromatos[♯]

[†] INFN and Università di Salerno, Via Giovanni Paolo II, 132 – 84084 Fisciano (SA), Italy
[‡] FNSPE, Czech Technical University in Prague, Břehová 7, 115 19 Praha 1, Czech Republic
[♯] Theoretical Particle Physics and Cosmology Group, Physics Department, King's College London, Strand, London WC2R 2LS, UK

Abstract. We present a dynamical mechanism à la Nambu–Jona-Lasinio [1] for the generation of masses and mixing for two interacting fermion fields. The analysis is carried out in the framework introduced long ago by Umezawa et al. [2], in which mass generation is achieved via inequivalent representations, and that we generalize to the case of two generations. The method allows a clear identification of the vacuum structure for each physical phase, confirming previous results [3] about the distinct physical nature of the vacuum for fields with definite mass and fields with definite flavor. Implications for the leptonic sector of the Standard Model are briefly discussed.

1. Introduction

Particle mixing, and neutrino oscillations in particular [4], has undergone rapid development, both theoretically and experimentally. Neutrino oscillations are nowadays firmly established by many experiments [5] involving solar, atmospheric or reactor neutrinos and their basic properties are resonably well understood[1]. On the theoretical side this has, in turn, produced a large number of theoretical ideas [8] trying to work out possible extensions of the original Standard Model which does not accommodate non-zero neutrino masses and mixings. In spite of this, the true origin of the mixing is still rather elusive, though it is generally believed that it is the result of physics occurring at much higher energies than the electroweak scale.

In the context of quantum field theory (QFT), a complex vacuum structure has been found to arise in connection with flavor mixing [3]. In this approach, flavor states for mixed particles are consistently defined as eigenstates of the flavor charges: from this, several results have been derived, including exact oscillation formulas which exhibit corrections with respect to the usual ones[11]. However, one of the limitations of this approach resides in the fact that only free fields have been considered. This is not a problem when discussing single-particle properties like neutrino oscillations, however, it is inadequate if one wants to explore salient dynamical aspects such as energy considerations related to flavor vacuum or the dynamical generation mechanisms for such a condensate.

In this paper, we perform a first step towards this goal by addressing a more complete treatment of flavor mixing and its ensuing vacuum structure, in the context of a model with interacting fields: we consider a simple model with two fields and dynamical symmetry breaking

[1] An exception could be the puzzling phenomenon known as GSI-Oscillation-Anomaly [6, 7].

à la Nambu–Jona-Lasinio (NJL) [1], which allows for the generation of (unequal) masses and of the corresponding mixing at one stroke[2]. In this study, we take advantage of the reformulation of NJL mechanism given by Umezawa et al. in Ref. [2], in which mass generation is achieved via inequivalent representations, a built-in property of QFT.

At this stage we should remark that a dynamical generation of flavor mixing has been considered in a related but not identical context in Ref. [12]. There, we have identified, within a string-inspired framework, a microscopic mechanism for *dynamical generation of mixing* within the concept of the "flavor vacuum" of [3] by identifying the latter with a ground state that was populated by space-time point-like brane defects ("D-particles" in the brane-theory terminology [13]). The Lorentz invariance breaking induced by the recoil of the defects, during their interaction with the neutrino states in such a set up, was fully consistent with the corresponding violation of the symmetry by the flavor vacuum [14]. At an effective field theory level, the interactions of the neutrinos with the defects gave rise to effective contact four fermion interactions, which lead to dynamical formation of flavor-mixing condensates à la NJL model. It is the purpose of this paper to discuss a more general situation, beyond specific models, where such a dynamical formation of flavor-vacuum condensates and the resulting mixing can be discussed based on the inequivalent representation properties of the QFT flavor vacuum of [3].

The present paper is organized as follows: we first review, closely following Ref. [2], the issue of inequivalent representations in QFT and the V-limit procedure, which are then used to discuss the dynamical mass generation for the NJL model. In Section 3 we extend the treatment to the case of two fermion fields with an interaction term that allows both for unequal masses and flavor mixing generation. Results are discussed together with open issues. Section 4 is devoted to further speculations and conclusions.

2. Inequivalent representations and dynamical mass generation

In this section we review the dynamical mass generation mechanism by Nambu and Jona-Lasinio [1], as reformulated by Umezawa, Takahashi and Kamefuchi [2], in terms of inequivalent representations.

2.1. Inequivalent representations in QFT

It is well known that in QFT the vacuum is not a trivial object: far from being "empty", it can have a rich condensate structure with non-trivial topological properties and non-equivalent quantum mechanical sectors (or phases). This complexity is due to the fact that QFT possesses an infinite number of degrees of freedom and this allows for the existence of different (unitarily inequivalent) representations of the field algebra. So, in particular, for a given dynamics one can have several Hilbert spaces, built on inequivalent vacua and representing different phases of the system with generally very different physical properties and distinct elementary excitations (see, e.g., Ref. [15, 16, 17]). This situation is drastically different from that of Quantum Mechanics, which deals with systems with a finite number of degrees of freedom, and where typically only one Hilbert space is admitted due to Stone–von Neumann's theorem [18, 19].

In order to arrive at the concept of inequivalent representations let us consider a system of *Fermi* fields enclosed in a finite-volume (volume V) box. Let $|0\rangle$ be a fiducial reference vacuum state with the corresponding set of creation and annihilation operators, $a^r_{\mathbf{k}}$ and $b^r_{\mathbf{k}}$, respectively. These satisfy the usual Clifford algebra

$$[a^r_{\mathbf{k}}, a^{s\dagger}_{\mathbf{l}}]_+ \;=\; [b^r_{\mathbf{k}}, b^{s\dagger}_{\mathbf{l}}]_+ \;=\; \delta_{\mathbf{k},\mathbf{l}}\delta_{rs}\,, \tag{1}$$

[2] An early attempt in this direction can be found in Refs.[9]. See also the general fomulation of the NJL model given in Ref.[10].

with other anticommutators being zero. Here $r = 1,2$ is the helicity index and

$$\mathbf{k} \;=\; \frac{2\pi}{V^{1/3}}\,\mathbf{n}, \qquad n_1, n_2, n_3 \;\; \text{integers}\,. \tag{2}$$

The expansion for the field is:

$$\psi(x) \;=\; \frac{1}{\sqrt{V}}\sum_{\mathbf{k},r}\left[u^r_{\mathbf{k}}\,a^r_{\mathbf{k}}\,e^{i\mathbf{k}\cdot\mathbf{x}} + v^r_{\mathbf{k}}\,b^{r\dagger}_{\mathbf{k}}\,e^{-i\mathbf{k}\cdot\mathbf{x}}\right], \tag{3}$$

where the spinor wavefunctions $u^r_{\mathbf{k}}$, $v^r_{\mathbf{k}}$ carry the time dependence through the factors $e^{-i\omega_k t}$ and $e^{i\omega_k t}$, respectively, with $\omega_{\mathbf{k}} = \sqrt{\mathbf{k}^2 + m^2}$.

Our interest lies in finding all possible unitary transformations of the vacuum state $|0\rangle$ that satisfy simple (physically motivated) consistency criteria. To this end we assume that the unitary transformation has the form

$$G \;=\; e^{iF}\,, \tag{4}$$

where F is some self-adjoint operator which itself is some functional of creation and annihilation operators. Invariance of the vacuum state under translations and rotations (vacuum is homogeneous and isotropic) implies that G must satisfy the commutation relations

$$[\mathbf{P}, G] \;=\; [\mathbf{J}, G] \;=\; 0\,, \tag{5}$$

where $\mathbf{P}$ and $\mathbf{J}$ are total momentum and total angular momentum operators, respectively. Assuming further that the vacuum preserves the total fermion charge we should also require that

$$[\mathcal{Q}, G] \;=\; 0\,, \tag{6}$$

with the charge $\mathcal{Q}$

$$\mathcal{Q} \;=\; \sum_{\mathbf{k},r}\left(a^{r\dagger}_{\mathbf{k}}a^r_{\mathbf{k}} - b^{r\dagger}_{\mathbf{k}}b^r_{\mathbf{k}}\right). \tag{7}$$

Since the vacuum states typically refer to asymptotic fields (in-fields) which have linear field equations, one can restrict the attention to F's that are only quadratic in the creation and annihilation operators that constitute the asymptotic fields. With this the above conditions (5) and (6) imply that the unitary transformation G can be parametrized with only two parameters $\vartheta^r_{\mathbf{k}}$ and $\varphi^r_{\mathbf{k}}$, namely (see, e.g. Ref. [15])

$$G(\vartheta, \varphi) \;=\; \exp\left[\sum_{\mathbf{k},r}\vartheta^r_{\mathbf{k}}\left(b^r_{-\mathbf{k}}a^r_{\mathbf{k}}e^{-i\varphi^r_{\mathbf{k}}} - a^{r\dagger}_{\mathbf{k}}b^{r\dagger}_{-\mathbf{k}}e^{i\varphi^r_{\mathbf{k}}}\right)\right]. \tag{8}$$

The invariance of G under rotation ensures that ϑ and φ depend only on $k \equiv |\mathbf{k}|$. In addition, it can be argued [2] that ϑ is independent on r and $\varphi^r_k = -(-1)^r\varphi_k$. The explicit form of G allows now to define new (quasi)particle annihilation and creations operators as

$$\alpha^r_{\mathbf{k}} \;=\; G(\vartheta,\varphi)a^r_{\mathbf{k}}G^\dagger(\vartheta,\varphi) \tag{9}$$

$$\;=\; \cos\vartheta_k\,a^r_{\mathbf{k}} + e^{i\varphi^r_k}\sin\vartheta_k\,b^{r\dagger}_{-\mathbf{k}}\,, \tag{10}$$

$$\beta^r_{\mathbf{k}} \;=\; G(\vartheta,\varphi)b^r_{\mathbf{k}}G^\dagger(\vartheta,\varphi) \tag{11}$$

$$\;=\; \cos\vartheta_k\,b^r_{\mathbf{k}} - e^{i\varphi^r_k}\sin\vartheta_k\,a^{r\dagger}_{-\mathbf{k}}\,. \tag{12}$$

The corresponding inverse transformation can be easily deduced:

$$\begin{aligned} a^r_{\mathbf{k}} &= G^\dagger(\vartheta,\varphi)\alpha^r_{\mathbf{k}}G(\vartheta,\varphi) = \cos\vartheta_k\,\alpha^r_{\mathbf{k}} - e^{i\varphi^r_k}\sin\vartheta_k\,\beta^{r\dagger}_{-\mathbf{k}}\,, \\ b^r_{\mathbf{k}} &= G^\dagger(\vartheta,\varphi)\beta^r_{\mathbf{k}}G(\vartheta,\varphi) = \cos\vartheta_k\,\beta^r_{\mathbf{k}} + e^{i\varphi^r_k}\sin\vartheta_k\,\alpha^{r\dagger}_{-\mathbf{k}}\,. \end{aligned} \tag{13}$$

The transformations (9)-(13) preserve the commutation relations, and represent thus a Bogoliubov transformation in the usual sense [15]. In this respect the labels $\{\vartheta_k, \varphi^r_k\}$ yield the most general parametrization for the Bogoliubov transformation of creation and annihilation operators. A physical picture behind the Bogoliubov transformation (12) is that $a^{r\dagger}_{\mathbf{k}}$ (and $b^{r\dagger}_{\mathbf{k}}$) create above the vacuum state $|0\rangle$ particle (anti-particle) quanta with momentum $\mathbf{k}$ and helicity r, whereas the quasiparticles (and anti-quasiparticles) created by $\alpha^{r\dagger}_{\mathbf{k}}$ (and $\beta^{r\dagger}_{\mathbf{k}}$) are the elementary excitations above the vacuum state

$$|0(\vartheta,\varphi)\rangle = G(\vartheta,\varphi)|0\rangle = \prod_{\mathbf{k},r}\left(\cos\vartheta_k - e^{i\varphi^r_k}\sin\vartheta_k\,a^{r\dagger}_{\mathbf{k}}b^{r\dagger}_{-\mathbf{k}}\right)|0\rangle\,, \tag{14}$$

which is annihilated both by $\alpha^r_{\mathbf{k}}$ and $\beta^r_{\mathbf{k}}$.

In the finite volume limit all vacuum states $|0(\vartheta,\varphi)\rangle$ are equivalent (i.e., they describe the same unique physical ground-state). In the infinite-volume limit the situation is drastically different. This can be seen by noticing that from (14) we have (for $V \to \infty$):

$$\langle 0|0(\vartheta,\varphi)\rangle = \exp\left[\sum_{\mathbf{k},r}\log(\sin\vartheta_k)\right] = \exp\left[\frac{V}{(2\pi)^3}\int d^3\mathbf{k}\,\log(\sin^2\vartheta_k)\right] \to 0\,. \tag{15}$$

Since as the fiducial vacuum one can chose any of the infinitely many vacuum states $|0(\vartheta,\varphi)\rangle$, the previous result implies that in the infinite-volume limit all the vacua with different ϑ's and φ's are orthogonal, i.e.,

$$\langle 0(\vartheta,\varphi)|0(\vartheta',\varphi')\rangle \to 0\,, \qquad \vartheta',\varphi' \neq \vartheta,\varphi\,. \tag{16}$$

The representation (9) (or (11)) of the Bogoliubov transformation loses its meaning for an (infinite) QFT system, in as much as the operator $\exp(iF)$ occurring in it has no domain on the representation space involved [16]. This fact, however, has no direct bearing on QFT which uses the well-defined form (10) (or (12)). It just states that the operator-algebra representations which are used in QFT yield unitarily inequivalent Fock-space representations. In other words, the vacuum states $|0(\vartheta',\varphi')\rangle$ and $|0(\vartheta,\varphi)\rangle$ for different ϑ's and φ's do not belong to the same Hilbert space.

Broken symmetry is a typical framework in which the above inequivalent-representations picture is of a particular importance. This is because the different vacuum states describing the broken symmetry phases cannot be connected by unitary representatives of the symmetry group in question [20]. What happens there is that each vacuum state induces a truly different representation of the operator algebra in each broken phase [15, 16]. Such a multiple vacuum structure was used in Ref. [2] to reformulate mass generation via dynamical breakdown of symmetry in the Nambu–Jona-Lasinio model. We will review this in Section 2.3 and further extend in Section 3 to the case of two generations in order to accomodate for flavor mixing.

2.2. V-limit procedure

To proceed, we review now the so-called V-limit procedure introduced by Umezawa *et al.* in Ref. [2]. Let us consider matrix elements of QFT operators, say Q, between states $|\Phi_i(\vartheta,\varphi)\rangle$, generated from the vacuum state $|0(\vartheta,\varphi)\rangle$ by a suitable action of creation and annihilation operators. The index "i" is a multi-index distinguishing various states, and the two real parameters ϑ and φ label the different (unitarily inequivalent) vacuum states. In particular the V-limit of Q with respect to a representation characterized by the parameters $\{\vartheta,\varphi\}$ is defined as

$$\langle\Phi_i(\vartheta,\varphi)|V\text{-lim}[Q]|\Phi_j(\vartheta,\varphi)\rangle \equiv \lim_{V\to\infty}\langle\Phi_i(\vartheta,\varphi)|Q|\Phi_j(\vartheta,\varphi)\rangle\,, \tag{17}$$

for all i and j. It should be noted that the V-limit is not the same as the week limit because the basis of the representation in which the limit is carried out may depend (and as a rule it does) on the volume V. The matrix element on the right-hand side of (17) is operationally calculated by phrasing the full (Heisenberg-picture) fields ψ present in Q in terms of free fields ψ_{in} enclosed in a finite-volume (volume V) box. The mapping between ψ and ψ_{in} is known as the Yang–Feldman equation or also Haag's map [15, 21, 22, 23]. Formally it can be written in the form [15, 22]; $\psi(x) = S^\dagger T(S\psi_{\text{in}}(x))$, where S and T are the S-matrix and time-ordering symbol, respectively.

The free field in the representation $\{\vartheta,\varphi\}$ can be obtained from the free field expansion (3) via the Bogoliubov transformation (9)-(12). As a result one has, for the *same* field operator:

$$\psi(x) = \frac{1}{\sqrt{V}}\sum_{\mathbf{k},r}\left[u^r_{\mathbf{k}}(\vartheta,\varphi)\alpha^r_{\mathbf{k}}\,e^{i\mathbf{k}\cdot\mathbf{x}} + v^r_{\mathbf{k}}(\vartheta,\varphi)\beta^{r\dagger}_{\mathbf{k}}\,e^{-i\mathbf{k}\cdot\mathbf{x}}\right], \tag{18}$$

with

$$\alpha^r_{\mathbf{k}}\,|0(\vartheta,\varphi)\rangle = \beta^r_{\mathbf{k}}\,|0(\vartheta,\varphi)\rangle = 0\,. \tag{19}$$

The Dirac spinors $u^r_{\mathbf{k}}(\vartheta,\varphi)$ and $v^r_{\mathbf{k}}(\vartheta,\varphi)$ are related with the fiducial representation spinors via the relation

$$\begin{aligned} u^r_{\mathbf{k}}(\vartheta,\varphi) &= u^r_{\mathbf{k}}\cos\vartheta_k + v^r_{-\mathbf{k}}\,e^{-i\varphi^r_k}\sin\vartheta_k\,,\\ v^r_{\mathbf{k}}(\vartheta,\varphi) &= v^r_{\mathbf{k}}\cos\vartheta_k - u^r_{-\mathbf{k}}\,e^{i\varphi^r_k}\sin\vartheta_k\,. \end{aligned} \tag{20}$$

By employing the operatorial Wick theorem [24], it is a simple exercise to show that for *free* fields in the $\{\vartheta,\varphi\}$ representation we have

$$\begin{aligned} V\text{-lim}\left[\int d^3\boldsymbol{x}\,\bar\psi_\alpha(x)\psi_\beta(x)\right] &= \int d^3\boldsymbol{x}\,:\bar\psi_\alpha(x)\psi_\beta(x): + \int d^3\boldsymbol{x}\,iS^-_{\alpha\beta}(\vartheta,\varphi)\,,\\ V\text{-lim}\left[\int d^3\boldsymbol{x}\,\bar\psi_\alpha(x)\psi_\beta(x)\bar\psi_\gamma(x)\psi_\delta(x)\right] &= \\ = iS^-_{\alpha\beta}(\vartheta,\varphi)\int d^3\boldsymbol{x}\,&:\bar\psi_\gamma(x)\psi_\delta(x): + iS^+_{\gamma\delta}(\vartheta,\varphi)\int d^3\boldsymbol{x}\,:\bar\psi_\alpha(x)\psi_\beta(x):\\ + \; iS^-_{\alpha\delta}(\vartheta,\varphi)\int d^3\boldsymbol{x}\,&:\bar\psi_\gamma(x)\psi_\beta(x): + iS^+_{\gamma\beta}(\vartheta,\varphi)\int d^3\boldsymbol{x}\,:\bar\psi_\alpha(x)\psi_\delta(x):\\ + \int d^3\boldsymbol{x}\sum_{\text{contractions}}&S^+(\vartheta,\varphi)S^+(\vartheta,\varphi)\,. \end{aligned} \tag{21}$$

where we have introduced the two-point Wightman functions evaluated with respect to the $|0(\vartheta,\varphi)\rangle$ vacuum:

$$iS^+_{\alpha\beta}(\vartheta,\varphi) = \langle 0(\vartheta,\varphi)|\bar\psi_\alpha(x)\psi_\beta(x)|0(\vartheta,\varphi)\rangle, \tag{22}$$

$$iS^-_{\beta\alpha}(\vartheta,\varphi) = \langle 0(\vartheta,\varphi)|\psi_\beta(x)\bar\psi_\alpha(x)|0(\vartheta,\varphi)\rangle\,. \tag{23}$$

Note that, due to translational invariance of the vacuum state $|0(\vartheta,\varphi)\rangle$, the two-point Wightman function is x-independent. This implies that the last terms in both equations in (21) are the c-numbers proportional to the volume V.

It is useful to consider the explicit form of the following quantities (see Ref.[2]):

$$C_p \equiv i \lim_{V\to\infty} \langle 0(\vartheta,\varphi)|\bar\psi(x)\gamma_5\psi(x)|0(\vartheta,\varphi)\rangle = \frac{2}{(2\pi)^3}\int d^3\mathbf{k}\, \sin 2\vartheta_k\ \sin\varphi_k \tag{24}$$
$$C_s \equiv \lim_{V\to\infty} \langle 0(\vartheta,\varphi)|\bar\psi(x)\psi(x)|0(\vartheta,\varphi)\rangle = -\frac{2}{(2\pi)^3}\int d^3\mathbf{k} \left[\frac{m}{\omega_k}\cos 2\vartheta_k - \frac{k}{\omega_k}\sin 2\vartheta_k \cos\varphi_k\right].$$

2.3. Gap equation

Following Ref.[2], we now apply the above developed concepts to the study of mass generation in the NJL model. This is described by the following Hamiltonian[3]

$$H = H_0 + H_{\rm int}\,, \tag{25}$$
$$H_0 = \int d^3\mathbf{x}\, \bar\psi\,(-i\boldsymbol{\gamma}\cdot\boldsymbol{\nabla} + m)\,\psi\,, \tag{26}$$
$$H_{\rm int} = \lambda \int d^3\mathbf{x} \left[(\bar\psi\psi)^2 - \left(\bar\psi\gamma^5\psi\right)^2\right]. \tag{27}$$

Considering the lowest order in the Yang–Feldman expansion, the V-limit of H gives [2]:

$$V\text{-lim}\,[H] = \bar H_0 + \text{c-number}\,, \tag{28}$$

with

$$\bar H_0 = H_0 + \delta H_0\,, \qquad \delta H_0 = \int d^3\mathbf{x}\ \{f\,\bar\psi\psi + ig\,\bar\psi\gamma_5\psi\}\,. \tag{29}$$

and

$$f = \lambda C_s\,, \qquad g = \lambda C_p\,. \tag{30}$$

So far we have not specified the vacuum state $|0(\vartheta,\varphi)\rangle$ of interest. In fact, most of the vacuum states and ensuing representations are not physically acceptable. The physically admissible representations are only those which satisfy appropriate renormalization condition, namely that the V-limit of the full Hamiltonian H should describe the quasiparticle (i.e., diagonal) Hamiltonian with the correct relativistic dispersion condition. Thus one requires that [2]:

$$\bar H_0 = \sum_r E_k\left(\alpha_{\mathbf{k}}^{r\dagger}\alpha_{\mathbf{k}}^r + \beta_{\mathbf{k}}^{r\dagger}\beta_{\mathbf{k}}^r\right) + W_0\,, \tag{31}$$

with the free particle dispersion relation $E_k = \sqrt{k^2+M^2}$. The mass M corresponds to the mass of elementary excitations (or quasiparticles) over the physical vacuum. The vacuum energy (or condensate density W_0) is fixed by setting the values of parameters ϑ and φ of the physical representation into the expression [2]:

$$W_0 = \sum_r \int d^3\mathbf{k}\ \Big[2\omega_k \sin^2\vartheta_k - (-1)^r g \sin 2\vartheta_k \sin\varphi_k^r$$
$$-\frac{f}{\omega_k}\left(m\cos 2\vartheta_k - k\sin 2\vartheta_k\cos\varphi_k^r\right) - \omega_k\Big] = -2\int d^3\mathbf{k}\, E_k\,. \tag{32}$$

[3] Here, as in Ref.[2], we consider the general case where $m \neq 0$. Results for $m = 0$ are then obtained as a special case.

After tedious but straightforward calculations one finds that the condition (31) is satisfied[4] when the following conditions hold [2]

$$\cos(2\vartheta_k) = \frac{1}{E_k}\left[\frac{m}{\omega_k}f(\vartheta,\varphi)+\omega_k\right], \tag{33}$$

$$\sin(\varphi_k^r) = g(\vartheta,\varphi)(-1)^r\left[g^2(\vartheta,\varphi)+\frac{k^2}{\omega_k^2}f^2(\vartheta,\varphi)\right]^{-\frac{1}{2}}, \tag{34}$$

$$M^2(\vartheta,\varphi) = (m\ +\ f(\vartheta,\varphi))^2+g^2(\vartheta,\varphi) = (m\ +\ \lambda C_s)^2+\lambda^2 C_p^2\,, \tag{35}$$

Since f and g depend on the parameters $\{\vartheta,\varphi\}$ via Eqs.(30) and (24), the above solutions give rise to two non-linear equations

$$f = f(\vartheta(f,g),\varphi(f,g)) \quad \text{and} \quad g = g(\vartheta(f,g),\varphi(f,g))\,, \tag{36}$$

which can be recasted as

$$C_p\left(1+\frac{2\lambda}{(2\pi)^3}\int\frac{d^3\mathbf{k}}{E_k}\right) = 0\,, \tag{37}$$

$$C_s\left(1+\frac{2\lambda}{(2\pi)^3}\int\frac{d^3\mathbf{k}}{E_k}\right) = -\frac{2m}{(2\pi)^3}\int\frac{d^3\mathbf{k}}{E_k}\,. \tag{38}$$

These equations determine the mass M.

In Ref. [2] two possibilities are discussed

$$C_p=0, \qquad M=m-\frac{2\lambda}{(2\pi)^3}M\int\frac{d^3\mathbf{k}}{E_k}, \tag{39}$$

$$m=0, \qquad 1+\frac{2\lambda}{(2\pi)^3}\int\frac{d^3\mathbf{k}}{E_k}=0. \tag{40}$$

The second case, Eq. (40), is only allowed for $\lambda<0$. Eq. (39) for $m=0$ and $M\neq 0$ is a special case of Eq. (40).

For $m\neq 0$, Eq. (39) gives perturbative corrections to the mass:

$$M = m-\frac{2\lambda}{(2\pi)^3}m\int\frac{d^3\mathbf{k}}{\omega_k}+\cdots \tag{41}$$

On the other hand, the solution Eq. (40) has a non-perturbative character and expresses the dynamical breakdown of (chiral) symmetry.

3. Two-flavor mixing

We now consider the dynamical symmetry breaking for the case of two fermion fields, for which in general a non-diagonal mass matrix will be obtained, thus generating flavor mixing in addition to nonzero masses. Here we mainly intend to present some qualitative aspects of this extension, relegating a complete discussion to a future paper. The notation is over-simplified: we omit

[4] The condition $E_k>0$ is also enforced.

spacetime dependence and drop momentum and helicity indices. Let us consider a fermion field doublet $\boldsymbol{\psi}$ whose Hamiltonian density is given as

$$\mathcal{H} = \mathcal{H}_0 + \mathcal{H}_{\rm int}\,, \tag{42}$$

$$\mathcal{H}_0 = \bar{\boldsymbol{\psi}}\,(-i\boldsymbol{\gamma}\cdot\boldsymbol{\nabla} + \boldsymbol{M}_0)\,\boldsymbol{\psi}\,, \tag{43}$$

with $\boldsymbol{\gamma}$ being a shorthand for $\mathbb{I}\otimes\boldsymbol{\gamma}$ with $\mathbb{I}$ being the 2×2 identity matrix and

$$\boldsymbol{\psi} = \begin{pmatrix} \psi_{\rm I} \\ \psi_{\rm II} \end{pmatrix} \quad \text{and} \quad \boldsymbol{M}_0 = \begin{pmatrix} m_{\rm I} & 0 \\ 0 & m_{\rm II} \end{pmatrix}. \tag{44}$$

The interaction Hamiltonian $H_{\rm int}$ can be assumed in the generic form

$$\mathcal{H}_{\rm int} = (\bar{\boldsymbol{\psi}}\,\Gamma\,\boldsymbol{\psi})\,(\bar{\boldsymbol{\psi}}\,\Gamma'\,\boldsymbol{\psi})\,, \tag{45}$$

where Γ and Γ' are some doublet spinor matrices. For simplicity we will, in the following, consider only the scalar counterterms, i.e., we will put $g_{\rm I} = g_{\rm II} = 0$. Recalling Eq.(34), this in turn implies that $\varphi_{\rm I} = \varphi_{\rm II} = 0$ in the Bogoliubov transformations for fields $\psi_{\rm I}$ and $\psi_{\rm II}$. This assumption simplifies considerably the following treatment, without altering the results of our analysis. The case including also pseudoscalar potential and related counterterms will be discussed elsewhere.

In this case the V-limit renormalization term $\delta\mathcal{H}_0$ has the generic structure

$$\delta\mathcal{H}_0 = \delta\mathcal{H}_0^{\rm I} + \delta\mathcal{H}_0^{\rm II} + \delta\mathcal{H}_{\rm mix} \tag{46}$$

$$= f_{\rm I}\,\bar{\psi}_{\rm I}\psi_{\rm I} + f_{\rm II}\,\bar{\psi}_{\rm II}\psi_{\rm II} + h\,(\bar{\psi}_{\rm I}\psi_{\rm II} + \bar{\psi}_{\rm II}\psi_{\rm I})\,. \tag{47}$$

Instead of the Bogoliubov transformation (13), inequivalent representations are now defined through a 4×4 canonical transformation, which can be conveniently parameterized as:

$$\begin{pmatrix} \alpha_A \\ \beta_A^\dagger \\ \alpha_B \\ \beta_B^\dagger \end{pmatrix} = \begin{pmatrix} c_\theta\,\rho_{A\rm I} & c_\theta\,\lambda_{A\rm I} & s_\theta\,\rho_{A\rm II} & s_\theta\,\lambda_{A\rm II} \\ -c_\theta\,\lambda_{A\rm I} & c_\theta\,\rho_{A\rm I} & -s_\theta\,\lambda_{A\rm II} & s_\theta\,\rho_{A\rm II} \\ -s_\theta\,\rho_{B\rm I} & -s_\theta\,\lambda_{B\rm I} & c_\theta\,\rho_{B\rm II} & c_\theta\,\lambda_{B\rm II} \\ s_\theta\,\lambda_{B\rm I} & -s_\theta\,\rho_{B\rm I} & -c_\theta\,\lambda_{B\rm II} & c_\theta\,\rho_{B\rm II} \end{pmatrix} \begin{pmatrix} a_{\rm I} \\ b_{\rm I}^\dagger \\ a_{\rm II} \\ b_{\rm II}^\dagger \end{pmatrix}. \tag{48}$$

where $c_\theta \equiv \cos\theta$, $s_\theta \equiv \sin\theta$ and

$$\rho_{ab} \equiv \cos\frac{\chi_a - \chi_b}{2}, \quad \lambda_{ab} \equiv \sin\frac{\chi_a - \chi_b}{2}, \quad \chi_a \equiv \cot^{-1}\left[\frac{k}{m_a}\right], \qquad a,b = {\rm I}, {\rm II}, A, B\,. \tag{49}$$

The transformation (48) contains thus three parameters (θ, m_A, m_B) to be fixed in terms of the quantities $(f_{\rm I}, f_{\rm II}, h)$ in order to diagonalize the Hamiltonian.

Let us start by considering the case in which no mixing arises after the V-limit. In such a situation, the Hamiltonian reduces into the sum of two singlet-field Hamiltonians, each being the same as the one studied in the previous section, i.e.

$$\bar{\mathcal{H}}_0 = \sum_{i={\rm I},{\rm II}} \left(\mathcal{H}_0^i + \delta\mathcal{H}_0^i\right). \tag{50}$$

The Bogoliubov matrix (48) that describes this situation must be block diagonal, i.e.

$$\begin{pmatrix} \alpha_A \\ \beta_A^\dagger \\ \alpha_B \\ \beta_B^\dagger \end{pmatrix} = \begin{pmatrix} \rho_{A\rm I} & \lambda_{A\rm I} & 0 & 0 \\ -\lambda_{A\rm I} & \rho_{A\rm I} & 0 & 0 \\ 0 & 0 & \rho_{B\rm II} & \lambda_{B\rm II} \\ 0 & 0 & -\lambda_{B\rm II} & \rho_{B\rm II} \end{pmatrix} \begin{pmatrix} a_{\rm I} \\ b_{\rm I}^\dagger \\ a_{\rm II} \\ b_{\rm II}^\dagger \end{pmatrix}, \tag{51}$$

and the diagonalization condition reads (cf Eq.(35)):

$$m_A \;=\; m_{\rm I} + f_{\rm I}\,, \quad m_B \;=\; m_{\rm II} + f_{\rm II}\,. \tag{52}$$

This is the same as the condition derived in Section 2.3, if we make the identification

$$\vartheta_i = \frac{1}{2}\left(\cot^{-1}\left[\frac{k}{m_a}\right] - \cot^{-1}\left[\frac{k}{m_i}\right]\right), \qquad (a,i) = (A,{\rm I}),(B,{\rm II})\,. \tag{53}$$

The resulting Hamiltonian (50) is now expressed in terms of the A, B modes.

Let us now come back to the full Hamiltonian (42). After the V-limit, in general we obtain an Hamiltonian density of the form:

$$\bar{\mathcal{H}}_0 \;=\; \sum_{i={\rm I,II}} \left(\mathcal{H}_0^i \;+\; \delta\mathcal{H}_0^i\right) \;+\; \delta\mathcal{H}_{\rm mix}\,. \tag{54}$$

In order to select among the inequivalent representations, we have to impose an appropriate renormalization condition on the form of the Hamiltonian (54). With respect to the simple case described in Section 2.3, where only one field was present, we have now two distinct possibilities:

- One possibility is to impose the condition that the Hamiltonian (54) becomes fully diagonal in two fermion fields, ψ_1 and ψ_2, with masses m_1 and m_2:

$$\bar{\mathcal{H}}_0 \;=\; \sum_{j=1,2} \bar{\psi}_j\,(-i\gamma\cdot\nabla \;+\; m_j)\,\psi_j\,. \tag{55}$$

The condition for the complete diagonalization of (54) is found to be:

$$\theta \;\to\; \bar{\theta} \;\equiv\; \frac{1}{2}\,\tan^{-1}\left[\frac{2h}{m_\mu - m_e}\right], \tag{56}$$

$$m_{_A} \;\to\; m_1 \;\equiv\; \frac{1}{2}\left(m_e + m_\mu - \sqrt{(m_\mu - m_e)^2 + 4h^2}\right), \tag{57}$$

$$m_{_B} \;\to\; m_2 \;\equiv\; \frac{1}{2}\left(m_e + m_\mu + \sqrt{(m_\mu - m_e)^2 + 4h^2}\right). \tag{58}$$

where we introduced the notation $m_e = m_{\rm I} + f_{\rm I}$, $m_\mu = m_{\rm II} + f_{\rm II}$. In passing we might observe that (56)-(58) imply the useful mass relations

$$\begin{aligned} m_e \;&=\; m_1\cos^2\bar{\theta} \;+\; m_2\sin^2\bar{\theta} \\ m_\mu \;&=\; m_2\cos^2\bar{\theta} \;+\; m_1\sin^2\bar{\theta}\,. \end{aligned} \tag{59}$$

In the following we will denote the vacuum state associated with such a representation as

$$|0(\bar{\theta}, m_1, m_2)\rangle \;\equiv\; |0\rangle_{1,2}\,, \tag{60}$$

since it is simply the tensor product states of the vacua for the free fields ψ_1 and ψ_2. In addition, the vacuum expectation value ${}_{1,2}\langle 0|\cdots|0\rangle_{1,2}$ will be denoted for short as $\langle\cdots\rangle_{1,2}$. With this the vacuum expectation value of the Hamiltonian in this representation can be shown to have the form:

$$\langle\bar{H}_0\rangle_{1,2} \;=\; -2\int d^3\mathbf{k}\,\left(\sqrt{k^2 \,+\, m_1^2} \;+\; \sqrt{k^2 \,+\, m_2^2}\right). \tag{61}$$

- Another possible representation is obtained by a partial diagonalization of (54), leaving untouched $\delta\mathcal{H}_{\rm mix}$. This will lead to the Hamiltonian density

$$\bar{\mathcal{H}}_0 \;=\; \sum_{\sigma=e,\mu} \bar{\psi}_\sigma\left(-i\gamma\cdot\boldsymbol{\nabla} \;+\; m_\sigma\right)\psi_\sigma \;+\; h\left(\bar{\psi}_e\psi_\mu \;+\; \bar{\psi}_\mu\psi_e\right). \tag{62}$$

Such a representation is obtained by setting

$$\theta \;\to\; 0\,, \tag{63}$$

$$m_A \;\to\; m_e \;\equiv\; m_{\rm I} + f_{\rm I}\,, \tag{64}$$

$$m_B \;\to\; m_\mu \;\equiv\; m_{\rm II} + f_{\rm II}\,. \tag{65}$$

The vacuum in this representation is denoted as

$$|0(\theta = 0, m_e, m_\mu)\rangle \;\equiv\; |0\rangle_{e\mu}\,, \tag{66}$$

and will be called the flavor vacuum. An important point to be noticed is that the mixing term in Eq.(62) is form-invariant under the transformation (48), provided $\theta = 0$.

Denoting ${}_{e,\mu}\langle 0|\cdots|0\rangle_{e,\mu}$ as $\langle\cdots\rangle_{e,\mu}$, the vacuum expectation value of the Hamiltonian in this representation is given as:

$$\langle \bar{H}_0\rangle_{e,\mu} \;=\; -2\int d^3\mathbf{k}\,\left(\sqrt{k^2 \;+\; m_e^2} \;+\; \sqrt{k^2 \;+\; m_\mu^2}\right), \tag{67}$$

since $\langle\delta\mathcal{H}_{\rm mix}\rangle_{e,\mu} \;=\; 0$.

In passing it should be stressed that the expressions for vacuum energies (61) and (67) should have some ultraviolet momentum cutoff in order to ensure meaningful stable vacua.

3.1. Physical motivations for the choice of the representation

The choice between the representations $|0\rangle_{1,2}$ and $|0\rangle_{e,\mu}$ has to be motivated on physical grounds. In this sense, the requirement that the Hamiltonian is only partially diagonalized, cf. Eq.(62), which corresponds to the representation built on the flavor vacuum $|0\rangle_{e,\mu}$, seems to be the one which better fits the situation present in the Standard Model, where the flavor fields describe the physical particles, and do not have in general a diagonal mass matrix[5]. In this representation, mixing can be seen as the effect of an external field, as discussed in Refs.[25].

The difference between the two above representations can be also seen via the respective gap equations. In particular, in the representation $|0\rangle_{1,2}$ the gap equation will be formally written as a set of 3 non-linear equations for $f_{\rm I}$, $f_{\rm II}$ and h in the form

$$\begin{aligned}
f_{\rm I} \;&=\; f_{\rm I}(\bar{\theta}, m_1, m_2) \;=\; f_{\rm I}(\bar{\theta}(f_{\rm I}, f_{\rm II}, h), m_1(f_{\rm I}, f_{\rm II}, h), m_2(f_{\rm I}, f_{\rm II}, h))\,,\\
f_{\rm II} \;&=\; f_{\rm II}(\bar{\theta}, m_1, m_2) \;=\; f_{\rm II}(\bar{\theta}(f_{\rm I}, f_{\rm II}, h), m_1(f_{\rm I}, f_{\rm II}, h), m_2(f_{\rm I}, f_{\rm II}, h))\,,\\
h \;&=\; h(\bar{\theta}, m_1, m_2) \;=\; h(\bar{\theta}(f_{\rm I}, f_{\rm II}, h), m_1(f_{\rm I}, f_{\rm II}, h), m_2(f_{\rm I}, f_{\rm II}, h))\,.
\end{aligned} \tag{68}$$

Here the explicit forms of $f_{\rm I}$, $f_{\rm II}$ and h are determined in terms of the vacuum expectation values $\langle\bar{\psi}_{\rm I}\psi_{\rm I}\rangle_{1,2}$, $\langle\bar{\psi}_{\rm I}\psi_{\rm II}\rangle_{1,2}$ and $\langle\bar{\psi}_{\rm II}\psi_{\rm II}\rangle_{1,2}$. This is a direct generalization of Eq.(36).

[5] The fact that charged leptons, for example, are not mixed, while neutrinos are, is just a matter of convention, since the generation of masses via the Higgs mechanism produces non-diagonal mass matrices for all fermions.

On the other hand, in the representation $|0\rangle_{e,\mu}$, the relevant non-linear equations to be solved are

$$\tilde{f} = \tilde{f}(m_e, m_\mu) = \tilde{f}(m_e(\tilde{f},\tilde{h}), m_\mu(\tilde{f},\tilde{h}))\,,$$
$$\tilde{h} = \tilde{h}(m_e, m_\mu) = \tilde{h}(m_e(\tilde{f},\tilde{h}), m_\mu(\tilde{f},\tilde{h}))\,. \quad (69)$$

Again, the explicit forms of $\tilde{f}$ and $\tilde{h}$ are determined from the expectation values $\langle\bar{\psi}_{\rm I}\psi_{\rm I}\rangle_{e,\mu}$ and $\langle\bar{\psi}_{\rm II}\psi_{\rm II}\rangle_{e,\mu}$. Note that the expectation values $\langle\bar{\psi}_{\rm I}\psi_{\rm II}\rangle_{e,\mu} = \langle\bar{\psi}_{\rm II}\psi_{\rm I}\rangle_{e,\mu} = 0$, and so they do not appear in calculations in this representation.

Another important issue to be taken into account is the relative vacuum energy associated with these. The corresponding energetics can be directly read off from the vacuum expectation values (61) and (67). Taking into account that $\sqrt{k^2+x^2}$ is a convex function in x, one can use the Jensen's inequality

$$\sqrt{k^2+(sx+(1-s)y)^2} \le s\sqrt{k^2+x^2} + (1-s)\sqrt{k^2+y^2}\,, \quad (70)$$

valid for any s between 0 and 1. With this the following inequality holds:

$$\sqrt{k^2+m_e^2}+\sqrt{k^2+m_\mu^2} = \sqrt{k^2+(m_1\cos^2\bar{\theta}+m_2\sin^2\bar{\theta})^2}+\sqrt{k^2+(m_1\sin^2\bar{\theta}+m_2\cos^2\bar{\theta})^2}$$
$$\le \sqrt{k^2+m_1^2} + \sqrt{k^2+m_2^2}\,, \quad (71)$$

with equality valid only when $\bar{\theta} = 0$ or $\pi/2$. In deriving (71) we have used the mass relations (59). By inserting the previous inequality into (61) and (67) we obtain that

$$\langle\bar{H}_0\rangle_{e,\mu} \ge \langle\bar{H}_0\rangle_{1,2}\,. \quad (72)$$

The fact that the representation built on $|0\rangle_{1,2}$ has lower energy seems to contradict the above choice of $|0\rangle_{e,\mu}$ as the physical vacuum. In principle, this might be due to the approximation which we have used in the Yang–Feldman equation: one could thus think that by introducing the momentum dependence via higher order corrections one would at low-enough energies observe vacuum-energies level crossing.

Another interesting possibility would be to couple our QFT system with the general relativity (with, e.g., the Robertson–Walker geometry) and to study in each physical phase the interplay between the QFT vacuum energy and the gravitational energy assigned to the curvature change caused by the corresponding QFT condensate. Such an analysis has been performed within the context of the brane-defect-filled Lorentz-Violating vacuum [12], where the extra energy of the vacuum as compared with the defect-free Minkowski vacuum has been interpreted as vacuum energy, of the type observed in the Universe today [26]. Moreover, in this context, due to the Lorentz-violating properties of the microscopic ground state, the physical choice in favour of the flavor vacuum is necessitated on (broken) symmetry grounds, consistent with the findings of [14]. This in fact might be the key to selecting the flavour vacuum over the normal one, and may be realised, for instance, in every finite temperature situation, where Lorentz invariance is broken by the effects of the thermal bath. For early Universe (high-temperature) physics this is probably an accurate description of reality. Even in particle physics context, though, such as neutrino oscillations in the Laboratory, the absolutely zero temperature vacuum is never attained, so it is natural to accept a tiny amount of temperature present which necessitates the use of flavour vacuum for the discussion of mixing.

4. Conclusions and perspectives

In this paper we have analyzed mass and mixing generation for two fermion fields in the context of a dynamical symmetry breaking scenario. By resorting to a formalism in which inequivalent representations of the canonical (anti)-commutation relations are exploited for realizing the dynamical generation of mass [2], we have considered the case of more than one generation, where the mixing terms naturally arise.

By working in the leading order of the Yang–Feldman expansion (corresponding to a mean-field approximation), we found that the representations in which the Hamiltonian is either fully diagonal or it contains a mixing term, are built on unitarily inequivalent Fock spaces, thus describing different physical phases of the system. This is quantitatively reflected in two different sets of gap equations, which however will be discussed in detail elsewhere.

The results obtained in the simple framework here explored seem to confirm the physical relevance of the flavor vacuum, first introduced in Ref.[3]. It is an interesting question, and object of future work, to investigate if such features persist in the context of Higgs mechanism, and how they appear when the analysis is done using a different formalism (e.g. path-integral formulation).

Acknowledgments

We thank G. Vitiello and H. Kleinert for inspiring discussions related to the subject of this paper. P.J. was supported by GAČR Grant No. GA14-07983S. The work of N.E.M. is supported in part by the London Centre for Terauniverse Studies (LCTS), using funding from the European Research Council via the Advanced Investigator Grant 267352 and by STFC (UK) under the research grant ST/J002798/1.

References

[1] Nambu Y and Jona-Lasinio G 1961 *Phys. Rev.* **122** 345; **124** 246.
[2] Umezawa H, Takahashi Y and Kamefuchi S 1964 *Ann. Phys.* **26** 336.
[3] Blasone M and Vitiello G 1995 *Annals Phys.* **244** 283.
[4] Bilenky S M and Pontecorvo B 1978 *Phys. Rept.* **41** 225.
[5] Smy M B (Super-Kamiokande) 2003 *Nucl. Phys. Proc. Suppl.* **118** 25 (2003)
Hallin A L et al. 2003 *Nucl. Phys. Proc. Suppl.* **118** 3
Eguchi K et al. (KamLAND) 2003 *Phys. Rev. Lett.* **90** 021802
Hampel W et al. (GALLEX) 1999 *Phys. Lett.* B **447** 127
Altmann M et al. (GNO) 2000 *Phys. Lett.* B **490** 16
Abdurashitov J N et al. (SAGE) 2002 *J. Exp. Theor. Phys.* **95** 181.
[6] Kienle P *et al.* [Two-Body-Weak-Decays Collaboration] 2013 *Phys. Lett.* B **726** 638; Kleinert H and Kienle P 2009 *Electron. J. Theor. Phys.* **6** 107.
[7] Lambiase G, Papini G and Scarpetta G 2013 *Phys. Lett.* B **718** 998; 2013 *Ann. Phys.* **332** 143.
[8] Abazajian K N et al. 2012 arXiv:1204.5379 [hep-ph].
[9] Alfinito E and Vitiello G 1992 in "Field Theory and Collective Phenomena" (World Scientific); Alfinito E, Blasone M and Vitiello G 1994 Proc. of V-VI Lomonosov Conference on Elementary Particle Physics, Accademia dei Lincei (Roma) p.176.
[10] Kleinert H 1978 in Erice Lectures *Understanding the Fundamental Constituents of Matter*, Plenum Press, New York, A. Zichichi ed., 289. (http://klnrt.de/53)
[11] Blasone M, Henning P A and Vitiello G 1999 *Phys. Lett.* B **451** 140; Blasone M, Jizba P and Vitiello G 2001 *Phys. Lett.* B **517** 471.
[12] Mavromatos N E and Sarkar S 2008 New J. Phys. **10** 073009; Mavromatos N E, Sarkar S and Tarantino W 2009 Phys. Rev. D **80** 084046; 2011 Phys. Rev. D **84** 044050; 2013 Mod. Phys. Lett. A **28** 1350045.
[13] Polchinski J 1998 *String theory. Vol. 2: Superstring theory and beyond*, (Cambridge University Press, Cambridge, UK).
[14] Blasone M, Magueijo J and Pires-Pacheco P 2005 *Europhys. Lett.* **70** 600; Blasone M, Di Mauro M and Lambiase G 2005 Acta Phys. Polon. B **36** 3255.
[15] Blasone M, Jizba P and Vitiello G 2011 *Quantum Field Theory and its Macroscopic Manifestations*

Journal of Physics: Conference Series **538** (2014) 012003 doi:10.1088/1742-6596/538/1/012003

(London: World Scientific & ICP).
[16] Umezawa H, Matsumoto H and Tachiki M 1982 *Thermo Field Dynamics and Condensed States* (North-Holland, Amsterdam).
[17] Miransky V A 1993 *Dynamical Symmetry Breaking in Quantum Field Theories* (World Scientific).
[18] Haag R 1996 *Local Quantum Physics: Fields, Particles, Algebras*, (Springer Verlag, Berlin).
[19] Weyl H 1950 *The Theory of Groups and Quantum Mechanics*, (Dover Publications, London).
[20] Emch G G 1972 *Algebraic Methods in Statistical Mechanics and Quantum Field Theory*, (Wiley & Sons, New York).
[21] Yang C N and Feldman D 1950 *Phys. Rev.* **79** 972.
[22] Bogoliubov N N, Logunov A A, Oksak A I and Todorov I T 1990 *General Principles of Quantum Field Theory* (Kluwer Academic Publishers, Dordrecht).
[23] Blasone M and Jizba P 2002 Ann. Phys. **295** 230.
[24] Bogoliubov N N and Shirkov D V 1979 *Introduction to theory of Quantized Fields*, (John Wiley, New York).
[25] Blasone M, Di Mauro M and Vitiello G 2011 , Phys. Lett. B **697** 238; Blasone M 2011 , J. Phys. Conf. Ser. **306** 012037; Blasone M and Jizba P 2012 J. Phys. Conf. Ser. **380** 012022.
[26] Blasone M, Capolupo A, Capozziello S, Carloni S and Vitiello G 2004 Phys. Lett. A **323** 182; Blasone M, Capolupo A and Vitiello G 2010 Prog. Part. Nucl. Phys. **64** 451.

The Madelung rule, Regge-like sequences and the conformal Lie algebra

Rutwig Campoamor-Stursberg†
† IMI and Dpto. Geometría y Topología
Fac. CC. Matemáticas
Universidad Complutense de Madrid
Plaza de Ciencias, 3
E-28040 Madrid, Spain
E-mail: rutwig@ucm.es

Abstract. The two main group theoretical models related to the Periodic System of Chemical Elements are reviewed, in connection with the Madelung rule and the construction of orthonormal bases of states for the conformal Lie algebra $\mathfrak{so}(2,4)$. Basing on discrepancies between the Madelung rule and chemical experience, a new approach to the conformal symmetry, currently in development, is described.

1. Introduction

Since its first application in Quantum Mechanics [1], group theory has shown to be a powerful tool to understand and interpret various physical phenomena, from the crystalline structure of solids to the classification of particles and nuclei models. In all these applications, the groups are related usually to the symmetries of the system, either as spectrum-generating or dynamical groups, where the Casimir operators of the corresponding Lie algebra and those of distinguished subalgebras play a central role to describe the system. The usual situation to apply group methods is concerned with classification schemes, where irreducible representations of a Lie group have to be decomposed into irreducible representations of a certain subgroup appearing in some relevant reduction chain. In this context, starting from pioneering work by Pauli and Fock on the Hydrogen atom [2], as well as the important contributions by Racah to atomic spectroscopy using Lie groups [3], the theory has been applied systematically in order to obtain a consistent symmetry scheme of the Periodic System of Elements (PSE in short), providing in such manner a formal background to the otherwise empirical "Aufbauprinzip" popularized by Madelung in the 1930s [4].

In this work, we briefly review some aspects of the group theoretical models proposed in connection with the PSE, specifically the role played by the extended conformal group $SO(2,4)\times SU(2)$ obtained by an ascendant procedure from the geometrical symmetry group $SO(3)$ of angular momentum and the obtainment of orthonormal bases of states for $SO(2,4)\times SU(2)$-multiplets, as well as the so-called Madelung rule, indicating some of the serious shortcomings of the latter in order to be an effective and practical tool in Chemistry (beyond the alkali earths). Even if this prescription is not derivable from the theory, it provides an intuitive interpretation[1]

[1] In chemical literature this is usually declared as counterproductive, but from the physicist's point of view,

for the arrangement of elements, as well as a symmetry scheme that has been shown to be of practical use, at least for some special situations. The discussion is finished with a brief overview of a different approach to the conformal symmetry, due to A. Ceulemans and P. Thyssen, based on realizations of Lie algebras in terms of rational functions of enveloping algebras and currently still in progress, for which reason the description is incomplete.

2. Symmetry approaches to the Hydrogen atom and PSE

The symmetry properties of the PSE are directly based on the analysis on geometric and dynamic invariance $SO(3)$ and $SO(4)$ of the Hydrogen atom, from which the symmetry rule is derived as an approximate model. Taking additionally into account the symmetries of the non-relativistic (NR in short) Maxwell equations for classical electrodynamics, this symmetry tower is further enlarged to the conformal group $SO(2,4)$. Finally, the adjunction of spin finally leads to the $SO(2,4) \otimes SU(2)$-models, which constitutes the standard symmetry group considered for the PSE [5, 6]. Actually the conformal group is itself of interest, as it is naturally related to the symplectic group $Sp(8,\mathbb{R})$, corresponding to the dynamical non-invariance group of a four dimensional isotropic harmonic oscillator [7].
We briefly recall the method. For an Hydrogen-like (non-relativistic) atom, we consider the central-field approximation Hamiltonian ($e = \hbar = 1$)

$$H = -\frac{1}{2}\frac{\partial^2}{\partial^2 x_\alpha x^\alpha} - \frac{Z}{r}, \tag{1}$$

where $r = \sqrt{x_\alpha x^\alpha}$. It follows at once that the angular momentum components $L_\alpha = \varepsilon_{\alpha\beta\gamma}L_{\beta\gamma} = \frac{-i}{2}\varepsilon\left(x_\beta\frac{\partial}{\partial x_\gamma} - x_\gamma\frac{\partial}{\partial x_\beta}\right)$, as well as $\mathbf{L}^2$, are constants of motion. Additional ones are given by the Runge-Lenz-Pauli vector

$$M = \frac{1}{2}(\mathbf{p}\times\mathbf{L} - \mathbf{L}\times\mathbf{p}) - Z\frac{\mathbf{r}}{r}. \tag{2}$$

Taken together, the components of these vectors lead to the commutation relations:

$$\begin{aligned} [L_\alpha, L_\beta] &= i\varepsilon_{\alpha\beta\gamma}L_\gamma, \\ [L_\alpha, M_\beta] &= i\varepsilon_{\alpha\beta\gamma}M_\gamma, \\ [M_\alpha, M_\beta] &= -2iH\varepsilon_{\alpha\beta\gamma}L_\gamma, \end{aligned} \tag{3}$$

therefore reproducing the structure of a six dimensional Lie algebra whenever the value of H is fixed. It is immediate that this algebra will be semisimple for any non-zero value of H. In addition, the invariant character of the angular momentum and Runge-Lenz components imply the identities

$$L_\alpha M^\alpha = M_\alpha L^\alpha = 0, \quad H = \frac{\mathbf{M}^2 - \mathbf{Z}^2}{2(1+\mathbf{L}^2)}. \tag{4}$$

As known [2], the analysis divides into the following notable cases:

(i) Discrete spectrum: $H < 0$.
Considering the change of variable $A_\alpha = \sqrt{\frac{-1}{2H}}M_\alpha$, $J_\alpha = \frac{1}{2}(L_\alpha + M_\alpha)$ and $K_\alpha = \frac{1}{2}(L_\alpha - M_\alpha)$, the brackets in (3) reduce to

$$\begin{aligned} [L_\alpha, L_\beta] &= i\varepsilon_{\alpha\beta\gamma}L_\gamma, \\ [L_\alpha, K_\beta] &= 0, \\ [K_\alpha, K_\beta] &= i\varepsilon_{\alpha\beta\gamma}K_\gamma, \end{aligned} \tag{5}$$

symmetry schemes always constitute a guideline.

clearly corresponding to the semisimple Lie algebra $SO(4)$. The corresponding spectrum of the problem is obtained by quantization of the operators $J_3, \mathbf{J}^2, K_3, \mathbf{K}^2$, where $\mathbf{J}^2 = \mathbf{K}^2$ holds. In particular, the eigenvalues (EV in short) $j(1+j)$ of $\mathbf{J}^2$ imply the well known expression

$$E_{2j+1} = -\frac{1}{2}\frac{Z^2}{(2j+1)^2}. \tag{6}$$

It turns out that the energy levels are specified by the representations (j,j) of $SO(4)$.

(ii) Continuous spectrum: $H > 0$.

In this case, the appropriate change of basis is given by $K_\alpha = \sqrt{\frac{1}{2H}} M_\alpha$, from which

$$\begin{aligned} [L_\alpha, L_\beta] &= \ \ i\varepsilon_{\alpha\beta\gamma} L_\gamma, \\ [L_\alpha, M_\beta] &= \ \ i\varepsilon_{\alpha\beta\gamma} M_\gamma, \\ [M_\alpha, M_\beta] &= -i\varepsilon_{\alpha\beta\gamma} L_\gamma, \end{aligned} \tag{7}$$

corresponding to the structure tensor of the Lie algebra of the Lorentz group $SO(1,3)$. In this case, an analogous reasoning to construct an infinite multiplet of the Lorentz group can be developed.

(iii) Zero energy: $H = 0$. Gives rise to the inhomogeneous group $I\mathfrak{so}(3)$. This symmetry is consistent with the fact that the inhomogeneous algebra is obtained as an Inönü-Wigner contraction of both the $\mathfrak{so}(4)$ and $\mathfrak{so}(1,3)$ Lie algebras.

In any case, the kinetic and potential terms are $SO(3)$ invariant, a fact that is immediately inferred from the structure of the Hamiltonian. Back to the case of the discrete spectrum, the Wigner theorem provides the representations associated to the energy level E_{2j+1}

$$\mathcal{D} = \bigotimes_{l=0}^{n-1} (l), \tag{8}$$

Reducibility of the multiplet is observed whenever $n = 2j + 1 \neq 1$. The discrete spectrum therefore exhibits (accidental) degeneracies of $SO(3)$, and the correct group to be considered is $SO(4)$, as was already observed by Fock [8].

The previous situation for the $N = 3$-Coulomb problem can be formally generalized without difficulty to any arbitrary order $N > 3$, due to the properties of the pseudo-orthogonal Lie groups. Therefore, the groups $SO(N+1)$, $SO(1,N)$ and $ISO(N)$ will emerge as the invariance groups for the discrete, continuous and zero value spectra of N-dimensional Coulomb systems, respectively. Again, the zero energy level arises as a contraction of the orthogonal and pseudo-orthogonal groups. A key point to this symmetry approach is to observe that the chain of geometric and dynamical invariance can be naturally enlarged to the non-invariance dynamical group $SO(2, N+1)$, which for the case of $N = 3$ leads to the conformal group $SO(2,4)$.

Schematically, the chain can be presented as:

$$\begin{array}{ccccc} & & SO(1,N) & & \\ & \swarrow_{cont} & & \searrow & \\ I\mathfrak{so}(N) & & & & SO(2,N+1) \\ & \nwarrow_{cont} & & \nearrow & \\ & & SO(1+N) & & \end{array} \tag{9}$$

Equation (9) clearly points out the subgroup chains and symmetry breaking.

3. Analogy between Hydrogen spectrum and PSE

Basing on the central field approximation, various group theoretical models have been developed in order to find a symmetry pattern in the Periodic System of Elements (PSE in short), usually focusing directly on the quantum theoretical approach to the atomic spectra, or some slight variation. Among the different approaches that have been developed in the literature, we briefly comment on two of them, both suggesting a central role of the conformal Lie algebra.

(i) The model based on double-covering group $Spin(4)$, developed in [9], takes the so-called "unitary Coulomb representation" P and defines "chemical states" corresponding to elements.
The main drawbacks of the model are:
  - The representation P is reducible, hence the possibility of accidental degeneracies arises naturally.
  - The disposition of elements with respect to quantum number is not directly related with the underlying symmetry.
  - There is no apparent relation with chemical affinity.

(ii) The Fet-Barut-Konopel'chenko model (short FBK model) proposes a unified treatment of "chemical" and energy spectra based on the extended conformal group $SO(2,4) \times SU(2)$, taking also into account the spin [6, 10]. This approach generalizes naturally the Fock model on $SO(4)$, as well as the geometric invariance $SO(3)$, constituting an "optimal reduction chain". Since $SO(2,4)$ is the largest symmetry group of the (non-relativistic) Maxwell equations [11], the model, in some sense, incorporates the electromagnetic interaction. Although for explicit computation, the group $\widetilde{SO}(2,4) \times SU(2)$ turns out to be a more suitable candidate, the main properties and the quantum numbers, as extracted from the corresponding Lie algebras, remain the same. We will see that this group encompasses the so-called Madelung rule, hence providing and intuitive interpretation of the PSE, even if exactly this fact represents the main drawback of the model, at least when tested for chemical consistency.

In the FBK model, classes of atoms are considered as a multiplet of the group $\widetilde{SO}(2,4) \times SU(2)$, from which an interpretation of the PSE in terms of irreducible representations (short IRREPs) will emerge. The main objective is the construction of an infinite dimensional IRREP R in the Hilbert $\mathcal{H}$ space of two-component Fock functions

$$\psi = \begin{cases} \psi_1(\xi_1, \xi_2, \xi_3, \xi_4) \\ \psi_2(\xi_1, \xi_2, \xi_3, \xi_4) \end{cases} \tag{10}$$

acting on the sphere $\mathbb{S}^3$, thus a Coulomb system. The atoms therefore result from subduced multiplets, and can be identified with properly chosen states in $\mathcal{H}$.
The main features of this symmetry model are :

(i) The basic states are specified by suitable quantum numbers of $\widetilde{SO}(2,4) \times SU(2)$ and the Casimir operators of the reduction chain

$$\widetilde{SO}(2,4) \times SU(2) \supset Spin(4) \supset SU(2)_C \tag{11}$$

(ii) The reduction with respect to the subgroup $SU(2)_C$ provides invariant subspaces R_C^α labelled by $\alpha = (n, \lambda, s_3)$, where n, s_3 are the eigenvalues of group observables and $\lambda(1+\lambda)$ are eigenvalues of the Casimir operator of $SU(2)_C$. An additional quantum number λ is added to separate degeneracies.

(iii) The representation R reproduces, in symmetry terms, those properties derived in the $Spin(4)$-approach.

(iv) A 1-1 correspondence between states $|n, \lambda, \mu, s_3\rangle$ and and eigenvalues for the representation R can be established.

A suitable set of quantum numbers for this model is given by the states

$$|n, \lambda, \mu, s_3\rangle, \quad n = 1 \cdots, \quad \lambda = 0, 1 \cdots n-1; \; \mu = -\lambda \cdots \lambda; \; s_3 = \pm\frac{1}{2}. \tag{12}$$

4. The Madelung rule

The so-called Madelung rule [4] is an empirical method that, in spite of its more than contrasted chemical shortcomings, has been useful for the symmetry models based on group theory. This rule is based on the central approximation and the Bohr atomic model, characterizing electrons by the quantum numbers n, l, m. Here l corresponds to the weight of the $(2l+1)$-dimensional representation of $SO(3)$, while m arises from IRREPS of its Abelian subgroup $SO(2)$. Actually, each value of n provides an n^2-dimensional IRREP of $SO(4)$ according to the dimension formula

$$n^2 = 1 + 3 + 5 + ... + n - 1 = \sum_{l=0}^{\frac{n-2}{2}} \dim(l). \tag{13}$$

Within this empirical interpretation, each pair (n, l) corresponds to an atomic shell. According to the "Aufbauprinzip", the energetic increase is related to the reduction chain $SO(4) \rightarrow SO(3)$, which in particular implies that the energy of shells increase by $n+l$, establishing an "energetic" order

$$[1,1] < [2,2] < [3,2] < [3,3] < [4,3] < [4,4] < [5,3] < [5,4] < ..... \tag{14}$$

or $1s < 2s < 2p < 3s < 3p < 4s < 3d < 4p < 5s < 4d < 5p < 6s < \cdots$ in classical spectroscopic notation [12].

l	0	1	2	3	4	5	..
n							
1	[1,1]						
2	[2,2]	[3,2]					
3	[3,3]	[4,3]	[5,3]				
4	[4,4]	[5,4]	[6,4]	[7,4]			
5	[5,5]	[6,5]	[7,5]	[8,5]	[9,5]		
..	..	..	..	..	..	..	..

Each block $[n+l, l]$ is filled with $2(2l+1)$ "elements" with an increasing atomic number Z (left to right), leading to an arrangement

	s	p	d	f	g
$n=1$	1-2				
$n=2$	3-4	5-10			
$n=3$	11-12	13-18	21-30		
$n=4$	19-20	31-36	39-48	57-70	
$n=5$	37-38	49-54	71-80	89-102	121-138
$n=6$	55-56	81-86	103-112	139-152	..
$n=7$	..	..	..	..	..

The interest of the $SO(2,4) \otimes SU(2)$ model in this context is to provide an interpretation of the table and the rule using symmetries, establishing a correspondence with states in some IRREP. To this extent, for any fixed n, the entries of each block are divided according to the principle

- $l = 0 \longrightarrow m = -\frac{1}{2}, \frac{1}{2}$.
- $l = 1 \longrightarrow m = \{\{-\frac{1}{2}, \frac{1}{2}\}, \quad \{-\frac{3}{2}, -\frac{1}{2}, \frac{1}{2}, \frac{3}{2}\}\}$
- $l = 2 \longrightarrow m = \{\{-\frac{3}{2}, -\frac{1}{2}, \frac{1}{2}, \frac{3}{2}\} , \{-\frac{5}{2}, -\frac{3}{2}, -\frac{1}{2}, \frac{1}{2}, \frac{3}{2}, \frac{5}{2}\}\}$
- $\cdots$

Now, for any $l \neq 0$, the block is divided into two sectors of $2j+1$ entries; those at the left have the value $j = l - \frac{1}{2}$, while those at the right have the value $j = l + \frac{1}{2}$. The corresponding entries are labelled with $m = -j, \cdots, j$. Following this procedure, each entry corresponds to a state noted $|n, l, j, m\rangle$ of the IRREP

$$R = \bigoplus_{n=1}^{\infty} \bigoplus_{l=0}^{n-1} \bigoplus_{j=|l-\frac{1}{2}|}^{l+\frac{1}{2}} (j) = \left[\bigoplus_{n=1}^{\infty} \bigoplus_{l=0}^{n-1} (l) \right] \otimes D_{\frac{1}{2}} \tag{15}$$

of the extended conformal group $SO(2,4) \times SU(2)$.

The previous representation (15) is given over the basis of $SO(2,4)$ obtained from the standard pseudo-rotational basis $L_{\mu\nu}$

$$[L_{\alpha\beta}, L_{\gamma\delta}] = i \left(g_{\alpha\delta} L_{\beta\gamma} + g_{\beta\gamma} L_{\alpha\delta} - g_{\alpha\gamma} L_{\beta\delta} - g_{\beta\delta} L_{\alpha\gamma} \right) \tag{16}$$

through the change of basis [11]:

$$\begin{array}{lll} J_1 = \frac{1}{2}(L_{14} + L_{23}) & J_2 = \frac{1}{2}(L_{24} - L_{13}) & J_3 = \frac{1}{2}(L_{34} + L_{12}) \\ K_1 = \frac{1}{2}(L_{14} - L_{23}) & K_2 = \frac{1}{2}(-L_{24} - L_{13}) & K_3 = \frac{1}{2}(L_{34} - L_{12}) \\ P_1 = \frac{1}{2}(L_{25} - L_{16}) & P_2 = \frac{1}{2}(-L_{15} - L_{26}) & P_0 = \frac{1}{2}(L_{12} + L_{56}) \\ Q_1 = \frac{1}{2}(L_{25} + L_{16}) & Q_2 = \frac{1}{2}(L_{15} - L_{26}) & Q_0 = \frac{1}{2}(L_{56} - L_{12}) \\ S_1 = \frac{1}{2}(L_{36} - L_{45}) & S_2 = \frac{1}{2}(L_{36} + L_{46}) & S_0 = \frac{1}{2}(L_{34} + L_{56}) \\ T_1 = \frac{1}{2}(L_{36} + L_{45}) & T_2 = \frac{1}{2}(L_{35} - L_{46}) & T_0 = \frac{1}{2}(L_{56} - L_{34}) \end{array} \tag{17}$$

subjected to the linear constraints

$$J_3 - K_3 - P_0 + Q_0 = 0, \; J_3 + K_3 - S_0 + T_0 = 0, \; P_0 + Q_0 - T_0 - S_0 = 0. \tag{18}$$

Among the various possibilities given, a suitable choice for a system of creation and annihilation operators is given by

$$\begin{array}{lll} J_\pm = J_1 \pm iJ_2, & K_\pm = K_1 \pm iK_2, & P_\pm = P_1 \pm iP_2, \\ Q_\pm = Q_1 \pm iQ_2, & S_\pm = S_1 \pm iS_2, & T_\pm = T_1 \pm iT_2. \end{array} \tag{19}$$

This choice of generators is adapted to some of the embedded subgroups relevant to the problem, for example

- $SO(3) \otimes SU(2)$: ladder operators indicate movement for fixed (n, l) and variable (j, m).
- $SO(4) \otimes SU(2)$: fixes n, while the remaining parameters can move.
- $SO(1,2)$: l, j, m fixed, n free.

Here, the number Z is given as function of the various Casimir operators

$$\begin{aligned} Z \;=\; & \frac{1}{6}(n+\ell)[(n+\ell)^2 - 1] + \frac{1}{2}(n+\ell+1)^2 - \frac{1}{4}[1 + (-1)^{n+\ell}](n+\ell+1) - 4\ell(\ell+1) \\ & +\ell + j(2\ell+1) + m - 1. \end{aligned} \tag{20}$$

Journal of Physics: Conference Series **538** (2014) 012004 doi:10.1088/1742-6596/538/1/012004

Entries in the table are obtained by means of the ladder operators on the vacuum state following the principal chain

$$SO(2,4)\otimes SU(2)\supset SO(4)\otimes SU(2)\supset SO(3)\otimes SU(2)\supset SO(2)\otimes SU(2). \tag{21}$$

Some structural remarks [2]

- Following this classification, ^{4}He belongs to alkali earth more than to halogens, the latter being specified by triplets $(l,j,m)=\left(1,\frac{3}{2},\frac{3}{2}\right)$.
- The rare earths have a more or less "natural" place within the system, although a displacement in their beginning/end is observed.
- Displacement consistent with the f-shell.[3]
- Predicts a series of "superlanthanides", however positions shifted with respect to the positions suggested by G. T. Seaborg [13].

In all, we should always bear in mind that the conformal group derivation of the PSE is a symmetry scheme whose primary task is to provide an intuitive interpretation of the atomic arrangement. It is not a model developed or derived from experimental data and established to surmount technical difficulties, as is for example the Racah approach to atomic spectra [3]. The conformal group model is developed using a small number of principles, neglecting various fundamental phenomena like orbit-orbit and spin-orbit couplings, hence it cannot be expected to be a faithful description of the periodic system as derived by quantum-chemical methods. The Madelung rule itself is nothing more than an empirical rule, not based on a physical or chemical derivation, but on certain simplified assumptions. The resulting approximation is chemically of interest only for the alkaline earth elements; beyond that point, the anomalies/divergencies observed invalidate the Madelung rule.[4] Some of the deficiencies of the rule are enumerated below [14]:

- The Madelung rule refers to free neutral and unbound atoms. This certainly constitutes a definitive argument against its chemical relevance, as Chemistry is more oriented to the analysis of bonded atoms.
- Implicitly, it is assumed that the electronic configuration of elements is determined by the ground state of the free atom. However, this statement is known to be incorrect for many reasons.
- From the third group onwards, there are divergences of the energetic sequence of orbitals, due to supplementary effects not considered by the rule.
- The Madelung rule is non-relativistic, therefore the relativistic spin-orbit couplings cannot be taken into account. This will inevitably lead to divergencies concerning the ground states.
- The rule does not contemplate or predict the collapse of the PSE from a certain period onwards, as follows from Quantum Mechanics and the stability properties of nuclei [15].

In spite of all these drawbacks, the Madelung rule still has some value, at least as part of a symmetry Ansatz for theoretical Chemistry. Actually it should be mentioned that similar methods have been applied to derive a "periodic system" of molecules, where, at least for the case of diatomic molecules, some useful spectroscopic consequences have been derived [16, 17].

[2] Compare with the IUPAC classification.
[3] For more details on this point, see the book by B. G. Wybourne cited in [3].
[4] Compare to the usual extrapolation of the quantum chemical approach to the Hydrogen atom and alkaline earths. Again, beyond this point the divergences require correction terms.

4.1. Orthonormal $SO(2,4)$-bases of states and the KGR programm

It is well established fact that a semisimple Lie algebra $\mathfrak{g}$ of rank l has exactly l primitive Casimir operators [18], the eigenvalues of which are useful to distinguish irreducible representations of $\mathfrak{g}$. However, in order to characterize the states within an irreducible representation, $f = \frac{1}{2}(\dim \mathfrak{g} - 3l)$ additional operators (sometimes called Racah operators) are necessary to completely classify states and separate degeneracies. This means that any orthonormal basis of states on $\mathfrak{g}$ should consist of

$$i = \frac{1}{2}(\dim \mathfrak{g} + l) \tag{22}$$

normal operators that are simultaneously diagonalizable. The same number of labels is required whenever we use some subalgebra $\mathfrak{h}$ to label the basis states. While the subgroup provides $\frac{1}{2}(\dim \mathfrak{h} + \mathrm{rank}(\mathfrak{h})) + l'$ labels, where l' denotes the number of invariants of $\mathfrak{g}$ that depend only on variables of the subalgebra $\mathfrak{h}$ [18], in addition to the Casimir operators of $\mathfrak{g}$ we need

$$n = \frac{1}{2}\left(\dim \mathfrak{g} - \mathcal{N}(\mathfrak{g}) - \dim \mathfrak{h} - \mathcal{N}(\mathfrak{h})\right) + l' \tag{23}$$

inner operators (missing label operators) for the separation of states. According to these formulae, for the conformal group $SO(2,4)$ an orthonormal basis of states is determined by 9 commuting operators, three of which correspond to the Casimir operators. In the case of $SO(2,4)$, the suitable orthonormal basis should be constructed with respect to the $SO(4)$ sugbgroup. Using the basis (16), we see that the $\mathfrak{so}(4)$ subalgebra is generated by L_{ij} with $1 \leq i < j \leq 4$. The three Casimir operators C_2, C_4 and C_6 of $\mathfrak{so}(2,4)$ provide three labels, while the Casimir operators C'_{21} and C'_{22} of $\mathfrak{so}(4)$ provide two more. The two internal labels for IRREPs of $\mathfrak{so}(4)$ are determined by the eigenvalues of the Casimir C''_2 of the $\mathfrak{so}(3)$ subalgebra generated by the angular momentum operators, as well as the eigenvalues of the projector L_{13}. This makes 7 labels, and according to (23), we need two more labels to separate the degeneracies. In order to determine them, we decompose the Casimir operators as operators of bi-degree $[a,b]$, i.e., as polynomials of degree a in the variables of the subalgebra $\mathfrak{so}(4)$ and degree b in the remaining generators [19]:

$$\begin{aligned} C_2 &= \Theta^{[2,0]} + \Theta^{[0,2]}, \\ C_4 &= \Theta^{[4,0]} + \Theta^{[2,2]} + \Theta^{[1,3]} + \Theta^{[0,4]}, \\ C_6 &= \Theta^{[6,0]} + \Theta^{[4,2]} + \Theta^{[3,3]} + \Theta^{[2,4]} + \Theta^{[1,5]} + \Theta^{[0,6]}. \end{aligned} \tag{24}$$

It can be shown easily that, taking for example the subgroup scalars $\Theta^{[2,2]}$ and $\Theta^{[1,5]}$, we actually have the required commutativity condition $\left[\Theta^{[2,2]}, \Theta^{[1,5]}\right] = 0$. Since these operators commute also with the $\mathfrak{so}(4)$-generators and are independent of the previous seven operators, an orthonormal basis can be given by the eigenvalues of the nine operators $\left\{C_2, C_4, C_6, C'_{21}, C'_{22}, C''_2, L_{13}, \Theta^{[2,2]}, \Theta^{[1,5]}\right\}$.

The KGR programm, first formulated by M. R. Kibler in [20], intends primarily to extend the conformal group approach in quantitative way to describe various properties of atoms, in analogy with the symmetry Ansätze applied in particle physics classification schemes, notably the IBM model. The first step in this direction is to obtain, from $SO(2,4) \otimes SU(2)$ and its more relevant subgroup chains, a set of $9+2$ normal commuting operators that label IRREPs and separate degeneracies. Using the orthonormal basis displayed above for the natural chain $SO(2,4) \supset SO(4) \supset SO(3)$, adding the two labels corresponding to the "spin" subalgebra $\mathfrak{su}(2)$, the implementation of the program reduces to consider linear or functional combinations of these operators and the numerical analysis of their eigenvalues for comparison with various physical properties of atoms, like ionization energies, specific heat or atomic radii. Albeit the problem is formally well established, the main difficulty lies in the numerical part, as the appropriate form of the labelling operators must be determined heuristically from available experimental data, without being based on a theoretical derivation.

5. "Regge" and "Madelung" ladder operators

Another approach to the symmetry scheme of the PSE, based on some observed analogies with the Regge sequences of nuclear physics [21] and currently in progress, has recently been proposed by A. Ceulemans [22]. The main idea of this scheme is the realization of $SO(2,4)$ in terms of creation and annihilation operators that describe diagonal movements in the (n,l) scheme, where the principal diagonal (red) is called the "Regge" sequence, while the counter-diagonal (blue) is called the "Madelung" sequence:[5]

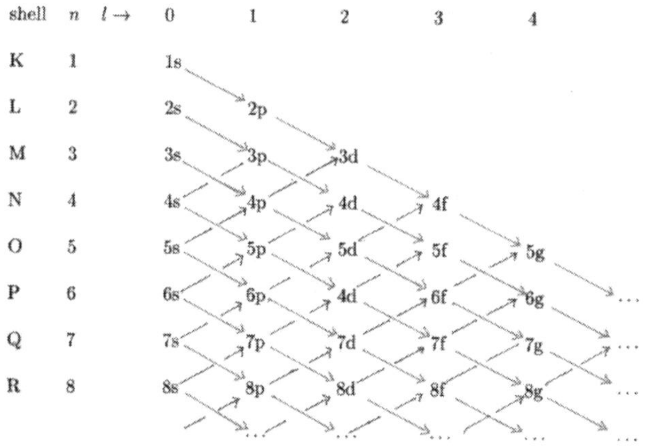

Figure 1. The "Regge" and "Madelung" sequences.

It is clear from this diagram that the classical subgroup chains used in the FBK-model are no longer valid, as both n and l are changed by the action of the creation and annihilation operators. In order to preserve the embedding $SO(3) \subset SO(1,3) \subset SO(2,4)$, it seems plausible to realize the conformal Lie algebra in some enveloping algebra, i.e., considering polynomial generators that reproduce the diagonal movement still preserving angular momentum. Similar constructions have been considered for the Hamilton group in Relativity by means of "virtual" copies of Lie algebras [23].
The first step towards such a realization is a clear identification of the pseudo-rotation generators (16) in terms of their corresponding symmetry subgroups: for $j = 1, 2, 3$, the generators L_{j4} can be seen as the components of a modified Runge-Lenz vector

$$\mathbf{A} = \frac{1}{2}\mathbf{r}p^2 - \mathbf{p}\,(\mathbf{r}\,\mathbf{p}) - \frac{1}{2}\mathbf{r}, \tag{25}$$

while the generators L_{j5} are identified with the components of a conjugated vector

$$\mathbf{B} = \frac{1}{2}\mathbf{r}p^2 - \mathbf{p}\,(\mathbf{r}\,\mathbf{p}) + \frac{1}{2}\mathbf{r}. \tag{26}$$

The $SO(1,2)$ subgroup is given by L_{46}, L_{45}, L_{56}, renamed as Q_1, Q_2, Q_3 respectively, while the remaining elements L_{j6} are identified with the components of a vector $\mathbf{\Gamma}$ proportional to linear momentum.

[5] Observe that the latter sequence corresponds to a reversal of the so-called Pauling diagram.

It can be easily seen that the angular momentum generators L and the components of $\mathbf{A}$ span a copy of $\mathfrak{so}(4)$, recovering the chain $SO(2,4) \supset SO(4)$. Further, the L and Q-operators mutually commute and generate a subgroup $SO(3) \otimes SO(1,2)$, providing the chain $SO(2,4) \supset SO(3) \otimes SO(1,2)$. Finally, the L and B-operators span a copy of $SO(1,3)$ (corresponding to the standard and only embedding of the Lorentz algebra.[6]
With these changes, the bound states of the Hydrogen spectrum ($\hbar = 1$), ignoring the spin quantum number s, are still described primarily by the quantum numbers

$$|n\, l\, m\rangle \tag{27}$$

for $n = 1, 2, \cdots$, $l = 0, 1, \cdots$ and $m = -l, \cdots, l$, as specified by the action of the following operators:

$$\mathbf{L}^2 \,|n\, l\, m\rangle = l\,(l+1)\,|n\, l\, m\rangle\,; \quad Q_3\,|n\, l\, m\rangle = n\,|n\, l\, m\rangle\,; \quad L_{12}\,|n\, l\, m\rangle = m\,|n\, l\, m\rangle\,. \tag{28}$$

The ladder operators are introduced as linear combinations of the components of the $\mathbf{B}$-vector

$$B_\pm = B_x \pm \mathrm{i}\, B_y. \tag{29}$$

As expected, the B_z-operator commutes with L_z and does not change the value of m, while the $B_\pm$ satisfy $[L_z, B_\pm] = \pm B_\pm$ and will raise (lower) the value of m in one unit. Their action on a state shows that we obtain a linear combination of diagonal moves in Figure 1.

$$\begin{aligned} B_z\,|n\, l\, m\rangle &= \alpha_1\,|n\,-1, l-1,\, m\rangle + \alpha_2\,|n\,+1, l-1,\, m\rangle \\ &\quad +\alpha_3\,|n\,-1, l+1,\, m\rangle + \alpha_4\,|n+1, l+1,\, m\rangle \\ B_\pm\,|n\, l\, m\rangle &= \beta_{\pm,1}\,|n\,-1, l-1,\, m\pm 1\rangle + \beta_{\pm,2}\,|n\,+1, l-1,\, m\pm 1\rangle \\ &\quad \beta_{\pm,3}\,|n\,-1, l+1,\, m\pm 1\rangle + \beta_{\pm,4}\,|n\,+1, l+1,\, m\pm 1\rangle\,, \end{aligned}$$

where α_i, $\beta_{\pm,i}$ are functions depending on n, l, m.
However, in order to have proper diagonal moves as depicted by the "Regge" and "Madelung" sequences, the transformed generators of $SO(2,4)$ are not sufficient. In these conditions, it is convenient to introduce a quadratic operator from the enveloping algebra that allows to discard some of the possibilities for the ladder operators [23]. One possibility is given by an operator S subjected to the constraint $S^2 = \left(L^2 + \frac{1}{4}\right)$. This kind of operator, which can be seen as a "counting operator", was first considered by M. Englefield in the context of spherical harmonics [25]. It turns out that S commutes with the angular momentum subalgebra $SO(3)$ as well as with the radial subgroup $SO(1,2)$, but not with the remaining $SO(2,4)$ generators. The interest of S lies in the possibility of constructing a copy of the Lorentz algebra in the enveloping algebra of $SO(2,4)$, by means of a new vector $\mathbf{K}$ defined by the commutators

$$\begin{aligned} K_z &= [B_z, Q_3 + S] \\ K_\pm &= [B_\pm, Q_3 + S] \end{aligned} \tag{30}$$

With this modification, the $[L_i, L_j]$ and $[L_i, K_j]$-commutators preserve the usual rules for a $\mathfrak{so}(1,3)$ subalgebra, while the difference is given in the brackets of the $\mathbf{K}$-components:

$$[K_\pm, K_z] \sim (Q_3 S^{-1} + 1) L_\pm; \quad [K_+, K_-] \sim (Q_3 S^{-1} + 1) L_z. \tag{31}$$

It turns out that $Q_3 - S$ is a constant operator for any Regge-like sequence in Figure 1 and that it appears as a Casimir operator of the Lie algebra generated by the angular momentum and

[6] Actually this is the origin of introducing additional operators and generalized realizations. This follows easily taking into account the branching rules for the defining representation associated to the embedding [24].

K-generators.

This realization of $SO(1,3)$ can be enlarged, using the generators **A**, **Q** and Γ, to form a Lie algebra isomorphic to the conformal algebra. However, while the usual realization of the conformal algebra as pseudo-rotations clearly indicates the reduction chains and the physical significance of the various generators and subgroups, it is not entirely clear, in this realization in the embedding algebra, how to identify the generators external to the Lorentz subalgebra constructed so far with properties of the Regge sequences and the corresponding symmetry breaking patterns.

6. Comments and Open questions

Some of the group theoretic approaches to the Periodic System of Elements have been reviewed in connection with the empirical Madelung rule. The main shortcomings in the chemical (and physical) context have been described, focusing on the value of the rule as a symmetry scheme that allows to extract orthonormal bases of states. Modification of these are suitable candidates to be used in more realistic descriptions of the PSE, or at least for a more close comparison of experimental data and numerical results derived from either the conformal group model or some generalization. One such possible generalization, based on similarities with reduction chains used in the analysis of nuclei, has been briefly commented. While the realization in the enveloping algebra of $SO(2,4)$ works consistently up to the Lorentz subalgebra, the appropriate interpretation of the remaining generators of the conformal algebra is not clear due to the consideration of the external counting operator S, presenting an obstruction not yet solved. Similar non-linear combinations of $SO(2,4)$ and S-operators allow the description of the "Madelung" sequences in Figure 1, albeit leading to the same obstruction in the interpretation of the realization. It is hoped that this problem, properly contextualized within the frame of non-linear extensions and deformations of Lie algebras [26], will be solved in the near future.

Acknowledgements

The author acknowledges R Berger, A Ceulemans, M R Kibler, P G L Leach, V I Man'ko, P Schwerdtfeger and P Thyssen for fruitful discussions, as well as for pointing out some pertinent references.

This work was partially supported by the Ministerio de Ciencia e Innovación (MICINN) of Spain, project MTM2010-18556.

References

[1] Weyl H 1928 *Gruppentheorie und Quantenmechanik* (Hirzel Verlag, Leipzig)
[2] Bohr N 1921 *Nature* **107** 104
Hund F 1925 *Z. Phys.* **33** 855
Lenz W 1924 *Z. Phys.* **34** 197
Pauli W 1926 *Z. Phys.* **36** 336
Condon E U, Shortley G H 1935 *The Theory of Atomic Spectra* (University Press, Cambridge)
Brauer R, Weyl H 1935 *Amer. J. Math.* **57** 425
Fock V A 1936 *Z. Phys.* **98** 145
Bargmann V 1936 *Z. Phys.* **99** 576
[3] Racah G 1949 *Phys. Rev.* **76** 1352
Racah G 1951 *Group Theory and Spectroscopy*, (Princeton Univ. Press, New Jersey)
Racah G 1959 *Nuovo imento Supplemento* **14** 75
Wybourne B G 1965 *Spectroscopic Properties of Rare Earths* (John Wiley, N.Y.)
Bloch F, Cohen S G, de-Shalit A, Sambursky S and Talmi I 1968 *Spectroscopic and Group Theoretic Methods in Physics* (North-Holland, Amsterdam)
[4] Madelung E 1936 *Die mathematischen Hilfsmittel des Physikers* (Springer, Berlin)
Bopp F, Haag R 1950 *Zeit. Naturforsch.* **5a** 644
Pauling L 1965 *Phys. Rev. Lett.* **15** 499
Hara O, Goto T 1968 *Prog. Theor. Phys. Suppl.* **41** 56

Weissmantel C, Lenk R, Forker W, Ludloff R, Hoppe J (Hrsg) 1970 *Atom. Struktur der Materie* (VEB Verlag Enzyklopädie, Leipzig)
Novaro O A, Berrondo M 1972 *J. Phys. B: Atom. Molec. Phys.* **5** 1104
Novaro O A, Berrondo M 1973 *J. Phys. B: Atom. Molec. Phys.* **6** 761
Kibler M R, Négadi T 1983 *J. Phys. A: Math. Gen.* **16** 4265
[5] Малкин И А, Манько В И 1965 *Письма в ЖЭТФ* **2** 230
Barut A O, Kleinert H 1967 *Phys. Rev.* **156** 1541
Малкин И А, Манько В И 1969 *ЯФ* **9** 184
Румер Ю Б 1970 *Докл. Акад. Наук СССР* **191** 64
Фет А И 1975 *Теор. Мат. Физ.* **22** 323
[6] Bédnar M 1973 *Ann. Phys.* **108** 49
Бяков В М, Кулаков Ю И, Румер Ю Б, Фет А И 1976 ИТЕФ-26/1976 ИТЕФ-90/1976 ИТЕФ-07/1977
Kregar M 1985 *Phys. Scripta* **29** 438
Cizek J, Clay M R, Paldus J 1980 *Phys. Rev.* **A22** 793
Adams B G, Cizek J, Paldus J 1982 *Int. J. Quantum Chem.* **21** 153
Komarov L I, Romanova T S 1985 *J. Phys. B: At. Mol. Phys.* **18** 859
Adams B G, Cizek J, Paldus J 1988 *Adv. Quantum Chem.* **19** 1
Kibler M R 1989 *J. Mol. Struct.* **187** 83
Gurskii' L I, Komarov L I, Solodukhin A M 1998 *Int. J. Quantum Chem.* **72** 499
[7] Kibler M R, Négadi T 1983 *Nouvo Cimento* **37** 225
Kibler M R, Négadi T 1984 *Folia Chim. Theor.* **12** 117
Kibler M R, Négadi T 1991 *J. Phys. A: Math. Gen.* **24** 5283
Chen J L, Zhang H B, Wang X B, Jing H, Zhao X G 2000 *Int. J. Theor. Phys.* **39** 2043
Zeng G J, Su K L, Li M 1994 *Phys. Rev.* **A50** 4373
Efstathiou K, Sadovskii' D A 2010 *Rev. Mod. Phys.* **82** 2099
[8] Фок В А 1935 *Изв. Акад. Наук СССР* VII Серия **2** 169
[9] Румер Ю Б, Фет А И 1971 *Теор. Мат. Физ.* **9** 203
Румер Ю Б, Фет А И 1970 *Теория унитарной симетрии* (Наука, М.)
[10] Barut A O, Kleinert H 1967 *Phys. Rev.* **156** 1541
Конопельченко Б Г 1972 *Группа* $SO(2,4)+R$ *и таблица Менделеева*
Фет А И 1975 *Теор. Мат. Физ.* **22** 323
Fet A I, Romanov L N 1996 *Lecture Notes in Computer Science* **1067** 379
[11] Gross L 1964 *J. Math. Phys.* **5** 687
Niederer U 1972 *Helv. Phys. Acta* **45** 802
Niederer U 1974 *Helv. Phys. Acta* **47** 167
Lagu S, Laue H 1974 *Nuovo Cimento* **A20** 217
[12] Lambert D, Kibler R M 1988 *J. Phys. A: Math. Gen.* **21** 308
Barut A O 1972 in *The Structure of Matter* (Proc. Rutherford Centennial Symp., 1971) (Christchurch, New Zealand)
Campoamor-Stursberg R 2008 *J. Phys.: Conf. Ser.* **128** 012052
[13] Keller O L, Seaborg G T 1997 *Ann. Rev. Nucl. Sci.* **25** 139
Indelicato P, Bieroń J and Jönsson P 2001 *Theor. Chim. Acta* **129** 495
Pyykkö P 2011 *Phys. Chem. Chem. Phys.* **13** 161
[14] Goudsmit S A, Richards P L 1964 *Proc. Natl. Acad. Sci. U.S.A* **51** 664
Schwarz W H E, Wang S G 2010 *Int. J. Quantum Chem.* **110** 1455
Schwarz W H E 1990 in *The Concept of the Chemical Bond* Maksic Z B (Ed) (Springer, Berlin)
Wang S G, Schwarz W H E 2009 *Angew. Chem. Int. Ed.* **48** 2
Schwarz W H E, Rich R L 2010 *J. Chem. Ed.* **87** 435
Scerri E R 2007 *The Periodic Table. Its History and its Significance* (Oxford UP, Oxford)
[15] Oganessian Y T 2007 *J. Phys. G: Nucl. Part. Phys.* **34** R165
Hofmann S 2012 *Phys. Unserer Zeit* **43** 30
[16] Judd B R 1963 *Operator Techniques in Atomic Spectroscopy* (McGraw-Hill, N.Y.)
Wybourne B G, Smentek L 1994 *Optical Spectroscopy of Lanthanides*, (CRC Press, Boca Raton)
[17] Hefferlin R A, Campbell R, Gimbel D, Kuhlman H and Cayton T 1979 *J. Quant. Spectrosc. Radiat. Transfer* **21** 315
Hefferlin R A, Kuhlman H, Campbell R and Gimbel D 1979 *J. Quant. Spectrosc. Radiat. Transfer* **21** 337
Kong F A 1982 *J. Mol. Struct.* **90** 17
Hefferlin R A 1989 *Periodic Systems of Molecules and Their Relation to the Systematic Analysis of Molecular*

Data (E Mellin Press, N.Y.)
Hefferlin R A, Sepman A V and Zhuvikin G V 1996 *J. Chem. Comput. Sci.* **36** 399
[18] Sharp R T and Pieper S C 1968 *J. Math. Phys.* **9** 663
Sharp R T 1975 *J. Math. Phys.* **16** 2050
Peccia A and Sharp R T 1976 *J. Math. Phys.* **17** 1313
Louck J D 1976 *Proceedings of the International Symposium on Mathematical Physics* (Mexico City, Mexico)
Campoamor-Stursberg R 2010 *Acta Phys. Pol. B* **42** 53
[19] Abellanas L and Martínez Alonso L 1975 *J. Math. Phys.* **16** 1580
Abellanas L and Martínez Alonso L 1979 *J. Math. Phys.* **20** 437
Couture M, Sharp R T 1989 *J. Phys. A: Math. Gen.* **22** 1525
Campoamor-Stursberg R 2011 *Int. J. Theor. Phys.* **50** 2153
Campoamor-Stursberg R 2011 *J. Phys. A: Math. Theor.* **44** 025204
[20] Kibler M R 2004 in *The Mathematics of the Periodic Table*, eds D H Rouvray and R B King (Baldock: Research Studies Press)
[21] Kleinert H 1968 *Fortschr. Physik* **6** 1
Jaffe R 1971 *J. Math. Phys.* **12** 882
Böhm A 1986 *Phys. Rev.* **D33** 3358
[22] Thyssen P, Binnemans K 2011 in *Handbook on The Physics and Chemistry of Rare Earths* vol 41 (Academic Press, Burlington)
Thyssen P 2013 *Symmetry and Symmetry Breaking in the Periodic Table - Towards a Group-Theoretical Classification of the Chemical Elements* (Ph.D. Thesis, KU Leuven)
[23] Dixmier J 1974 *Algèbres enveloppantes* (Gauthier-Villard, Paris)
Giroux Y and Sharp R T 1987 *J. Math. Phys.* **28** 1671
Giroux Y 1987 *Degenerate Enveloping Algebras of Low-rank Groups* (PhD Thesis, McGill University)
Ruan R, Ruan W 1999 *Phys. Lett. A* **263** 78
Ni Z 1999 *J. Phys. A: Math. Gen.* **32** 2217
Campoamor-Stursberg R and Low S G 2009 *J. Phys. A: Math. Theor.* **42** 065205
[24] Gruber B and Lorente M 1972 *J. Math. Phys.* **13** 1639
[25] Barut A O 1972 *Dynamical Groups and Generalized Symmetries in Quantum Theory* (Univ. Canterbury , Christchurch)
Englefield M J 1972 *Group Theory and the Coulomb problem* (Wiley, N.Y.)
[26] Quesne C 1994 *Phys. Lett. A* **193** 245
Quesne Ch 1995 *J. Phys. A: Math. Gen.* **28** 2847
Ruan D, Ruan W 1999 *Phys. Lett. A* **263** 78
Ruan D, Jia Y and Ruan W 2001 *J. Math. Phys.* **42** 2718

Quantumness of correlations and Maxwell's demons in elementary scattering processes—Energetic consequences

C A Chatzidimitriou-Dreismann[1] and A Dreismann[2]

[1] Institute of Chemistry, Sekr. C2, Technical University of Berlin, D-10623 Berlin, Germany
[2] Department of Physics, Cavendish Laboratory, University of Cambridge, Cambridge, CB3 0HE, UK

E-mail: dreismann@chem.tu-berlin.de

Abstract. The interactions between physical systems generally lead to the formation of correlations. In this paper we consider the phenomena of entanglement and "quantumness of correlations", such as quantum discord, with particular emphasis on their energetic consequences for the participating systems. We describe a number of theoretical models that are commonly employed in this context, highlighting the general character of one of their most intriguing results: In contradiction to conventional expectations, erasure (decay, consumption) of quantum correlations may be a source of work, i.e. may have "negative energetic costs". We report experimental evidence of this surprising effect obtained within the framework of an elementary scattering experiment, namely ultrafast neutron Compton scattering from normal-state liquid ^{4}He. The general theory of quantumness of correlations provides a natural way of interpreting the reported results, which stand in blatant contrast to the conventional theory of scattering, where neutron-atom-environment quantum correlations and decoherence play no role. Moreover, they provide a new operational meaning of discord and related measures of quantumness.

1. Introduction: "Information is Physical"

Correlations between physical (chemical, biological etc.) systems are created by interactions and thus are ubiquitous in Nature. They play a significant role in classical and quantum physics. The emergent behaviour of many complex systems cannot simply be derived from their individual components, since their properties fundamentally depend on the delicate correlations between their subsystems; cf. [1]. Such a situation might be expressed by the old saying: "the whole is more than the sum of its parts", which expresses a holistic in contrast to reductionist view of nature.

To prevent possible confusion, it should be emphasized that the well-known quantum mechanical correlations due to the indistinguishability of identical particles (as treated in most textbooks of quantum mechanics) play no role in the following investigations. Namely we will deal with quantum correlations between *distinguishable quantum systems* only. For instance, below we will consider specific quantum correlations between a neutron (which is a fermion) and a ^{4}He atom (a boson). Additionally, classical correlations (e.g. as appearing in classical statistical mechanics and condensed matter theory) are also excluded from our considerations.

In quantum mechanics, correlations can be essential for the description of quantum systems, in ways that, in some cases, are in blatant contrast to every expectation based on classical physics. A particular kind of quantum correlations which have been shown to exhibit distinctive non-classical features are known as quantum entanglement [2, 3, 4, 5, 6, 7], and are related to the phenomenon of non-locality [8]. Nowadays entanglement is recognized to be an indispensable resource for quantum information, computing and communication, and thus plays a central role in the associated field of quantum information processing (QIP); see e.g. [9].

The correlations between two quantum systems (say, A and B) are associated with information, as for instance in the case of DNA, where the correlations between base pairs encode genetic information. In statistical physics and thermodynamics, the crucial role of correlations became obvious when Maxwell [10], in an attempt to clarify the limitations of thermodynamics, introduced his famous demon, which also plays an important role in information theory [11]. Generally, thermodynamics and information theory possess several links, as e.g. indicated by the formal similarity of the formulas of the Shannon and the von Neumann entropies. Finally Landauer [12] succeeded to "exorcize" Maxwell's demon through his information erasure theorem; see also the related work by Bennett [13, 14]. There is no doubt that the intricate information-thermodynamics relation also implies a thermodynamic cost for information acquisition. Likewise, any information acquisition process is expected to be ultimately limited by the Second Law of thermodynamics. In short, as Landauer puts it: "Information is physical". For recent comprehensive works about the physics of (classical and quantum) Maxwell's demons and their relation to correlation, information and thermodynamics, see Refs. [15, 16, 11].

In 2001, the latest chapter in the story of the relation between correlations and information theory began when the pioneering works by Ollivier and Zurek [17] and Henderson and Vedral [18] revealed that, beyond entanglement, there exists another, more general kind of correlation in quantum physics. These correlations are popularly known as "quantum discord" [19], or more generally "quantumness of correlations". Discord and related concepts (like e.g. quantum (work) deficit, measurement-induced disturbance [20] etc.) have recently entered the field of QIP and are subject to numerous investigations regarding their operational meaning; see e.g. the review articles [21, 22]. As they concern the physics of Maxwell's demon, they also have entered the foundations of (quantum) thermodynamics.

In the present paper, we consider the possibility of applying these phenomena (i.e., entanglement, discord, decoherence, etc.) and their dynamics in a concrete experimental context, namely elementary scattering processes. We will particularly point out the "energetic costs" of quantumness of correlations and their experimental measurability, which then establishes additional physical insight into the phenomenon. We then proceed by employing the presented concepts to interpret experimental results obtained from neutron scattering (qualitatively, for the time being), which contradict conventional expectations. We believe that these results illustrate a new operational meaning (or interpretation) of quantum correlations in the context of a real experiment.

In this regard it should be mentioned that the importance of the phenomena described above has thus far not been recognized in the context of scattering experiments, e.g. neutron scattering [24, 25], electron scattering [26], or inelastic x-ray scattering from condensed matter [27]. Although it is, for instance, well known that neutron scattering from H_2 molecules may reveal quantum correlations between the two protons of the H_2 (such as the different scattering properties of ortho-H_2 and para-H_2 [23]), possible quantum correlations (or even entanglement) between the neutron and the struck nucleus, which may be created by the scattering interaction, are absent in conventional theory; see section 4.

The paper is organized as follows.

(A) Section 2 is devoted to certain aspects of the aforementioned quantum correlations. First, in sect. 2.1 we quickly consider the well established concepts of entanglement and non-locality. Second, in sect. 2.2, we present in more detail the new concept or quantumness of correlations beyond entanglement like discord. Needless to say, this section is not meant to provide a complete account of the topic under consideration.

(B) Section 3 considers some thermodynamic and energetic consequences of quantum correlations, starting with two examples about "quantum thermal engines" (sect. 3.1), and then presenting two specific theoretical examples of direct energetic consequences of decaying quantum correlations, i.e. decoherence. The unexpected common feature of both these theoretical results is that, surprisingly, decoherence and erasure of quantum correlations can lead to work production (instead to dissipate work).

(C) An experimental observation of, or related with, this phenomenon is described in section 4, which also presents a short description of a conceptually simple scattering experiment and related experimental results obtained with neutron Compton scattering from liquid ^{4}He. The findings illustrate the significance of quantum correlations and their dynamics in the framework of neutron scattering. This extends the physical insights provided in the context of QIP by offering an additional operational meaning of quantumness of correlations.

2. Quantumness of Correlations

2.1. On Non-Locality and Quantum Entanglement

Quantum entanglement (QE) has been recognized as the most emblematic feature, or even the essence, of standard (non-relativistic) quantum theory [2, 4, 5], raising widespread interest in various branches of natural science as well as quantum information processing (QIP) and the related emerging quantum technologies; cf. [9].

It was Einstein, Podolsky, Rosen [2] and Schrödinger [3] who first recognized a "spooky action at a distance" feature of the formalism of quantum mechanics which has absolutely no classical analogue or any intuitive interpretation. This novel phenomenon, known as quantum entanglement (or simply entanglement, originally called by Schrödinger "Verschränkung"), lies at the center of interest of physics and quantum technology of the 21st century. Entanglement implies the existence of global states of a composite system which cannot be written as a product of the states of individual subsystems. As a consequence, there exist quantum statistical relations between subsystems of a compound quantum system which are by far much stronger than any conceivable classical correlations between the subsystems. Thus it may be said that entanglement represents the most nonclassical manifestation of the quantum formalism. Entanglement also contradicts the so-called Einstein's locality principle. Based on this, Einstein et al. concluded that the quantum description of physical reality is not complete [2].

About 30 years later, this conclusion, and related suggestions for introducing so-called "hidden variables" in order to make quantum mechanics a "complete theory", led Bell to the discovery of the famous Bell inequalities [4] which are experimentally testable. With his work, Bell proved that quantum mechanics is an inherently nonlocal theory that is incompatible with any physical theory in which the principle of locality holds. For instance, two spatially well separated observers measuring states of two entangled systems observe strong correlations between the results of their measurements—correlations that can be so strong that they violate Bell's inequalities. This phenomenon, called quantum non-locality, has been repeatedly observed and confirmed experimentally—which also invalidates the conclusions of [2]. For insightful presentations, one may consult the textbook by Peres [6] or the recent review by Horodecki et al. [7].

In this context, it may be noted that entanglement and non-locality are often considered as two facets of the same physical phenomenon. But now they are recognized as two different physical resources, as Gisin and collaborators have demonstrated [8]. However, the relation between them is yet to be fully understood. While entanglement is necessary for quantum non-locality, it is not always sufficient. Indeed, entangled states exist which are local, i.e. which do not violate any Bell's inequality.

Despite its fundamental physical significance, however, quantum entanglement was considered by many as a rather "philosophical issue" having no concrete impact in "real physical fields" (like e.g. solid state physics) and/or concrete technological applications (like electronic devices and telecommunication). So it has had to wait about 70 years to enter laboratories as a novel resource as real as e.g. energy. For compound quantum systems, entanglement involves nonclassical correlations between subsystems and has been recognized to have potential for novel quantum processes even of technological relevance, like e.g. quantum cryptography, quantum dense coding and quantum teleportation [7, 9].

2.2. Quantumness of Correlations Beyond Entanglement—Discord

In this subsection, we mainly follow the presentation in Ref. [21] by Modi et al.

Coherent interactions that generate negligible entanglement can still exhibit unique quantum behaviour. This observation has motivated a search beyond entanglement for a complete description of all quantum correlations [17, 18]. Classicality and quantumness of correlations belong to the realm of information theory.

Quantum correlations, seminally quantified by the quantum discord and related measures of quantumness [21], are general manifestations of non-classicality in composite systems. They can be revealed in the process of locally measuring a subsystem, even in states where entanglement or non-locality are absent. Despite a massive surge in recent studies investigating interpretation, quantification, and applications of discord and related quantifiers of quantum correlations, cf. [21], it is a fact that these newly discovered quantities remain far less understood than entanglement. Therefore, going beyond the beautiful mathematical results achieved in this fast developing field, every result about the concrete physical meaning and/or "operational meaning", of discord (and the other measures) is of highest importance.

In simple terms one can say that two systems are correlated if together they contain more information than taken separately. If we measure the lack of information by entropy, this definition of correlations is captured by the mutual information

$$I(A:B) \equiv S(A) + S(B) - S(AB), \tag{1}$$

where $S(X)$ is the Shannon entropy

$$S(X) = -\sum_x p_x \log p_x \tag{2}$$

if X is a classical variable with values x occurring with probability p_x, or $S(X)$ is the von Neumann entropy

$$S(X) = -\mathrm{Tr}(\rho_X \log \rho_X) \tag{3}$$

if ρ_X is a quantum state of system X (all logarithms are base two). For classical variables, Bayes' rule defines a conditional probability as $p_{x|y} = p_{xy}/p_y$. This implies an equivalent form for the classical mutual information

$$J_{\mathrm{cl}}(B|A) = S(B) - S(B|A), \tag{4}$$

where the conditional entropy

$$S(B|A) = \sum_a p_a S(B|a) \tag{5}$$

is the average of entropies

$$S(B|a) = -\sum_b p_{b|a} \log p_{b|a}. \tag{6}$$

The classical correlations can therefore be interpreted as information gain about one subsystem resulting from a measurement on the other.

In contradistinction to the classical case, in the quantum analog there are many different measurements that can be performed on a system, and measurements generally disturb the quantum state. A measurement on subsystem A is described by a positive-operator-valued measure (POVM) [9] with elements $E_a = M_a^\dagger M_a$, where M_a is the measurement operator and a is the classical outcome of the associated measurement. The initial state ρ_{AB} is transformed under the measurement (with unknown result) to

$$\rho_{AB} \to \rho'_{AB} = \sum_a M_a \rho_{AB} M_a^\dagger, \tag{7}$$

where the subsystem A yields outcome a with probability $p_a = \mathrm{Tr}(E_a \rho_{AB})$ and B has the conditional state $\rho_{B|a} = \mathrm{Tr}_A(E_a \rho_{AB})/p_a$. This allows us to define a so-called classical-quantum version of the conditional entropy,

$$S(B|\{E_a\}) \equiv \sum_a p_a S(\rho_{B|a}), \tag{8}$$

and introduce classical correlations of the state ρ_{AB} in analogy with Eq. 4, [18]:

$$J(B|\{E_a\}) \equiv S(B) - S(B|\{E_a\}). \tag{9}$$

To quantify the classical correlations of the state *independently* of a measurement, $J(B|\{E_a\})$ is maximized over all measurements (on subsystem A),

$$J(B|A) \equiv \max_{\{E_a\}} J(B|\{E_a\}). \tag{10}$$

When the measurement is carried out by a set of rank-one orthogonal projections $\{\Pi_a\}$, the state on the right hand side of Eq. 7 has the form

$$\chi_{aB} = \sum_a p_a \Pi_a \otimes \rho_{B|a}, \tag{11}$$

which involves only fully-distinguishable states for A and some indistinguishable states for B. Such states are often called classical-quantum states. It is important to note that for a classical-quantum state there exists a von Neumann measurement of A which does not perturb the state. By exchanging the roles of A and B, one defines quantum-classical states (QC) by

$$\chi_{Ab} = \sum_b p_b \rho_{A|b} \otimes \Pi_b, \tag{12}$$

The *quantum discord* of a state ρ_{AB} under a measurement $\{E_a\}$ is defined as a difference between total correlations, as given by the quantum mutual information in Eq. (1), and the classical correlations Eq. (9), [17]:

$$\begin{aligned} D(B|A) &\equiv I(A:B) - J(B|A) \\ &= \min_{\{E_a\}} \sum_a p_a S(\rho_{B|a}) + S(A) - S(AB). \end{aligned} \tag{13}$$

Note that the minimization here is equivalent to maximization in Eq. 10. This is just a difference between two classically-equivalent versions of conditional entropy

$$D(B|A) = \min_{\{E_a\}} S(B|\{E_a\}) - S(B|A), \tag{14}$$

where $S(B|A) = S(AB) - S(A)$ is the usual conditional entropy, see e.g. [9]. This equivalence holds for rank-one POVM measurements [9] which, in classical theory, correspond to questions about a value of a classical random variable.

Here some properties of quantum discord may be noted [21, 22, 23]:

(*a*) It is not symmetric, i.e. in general $D(B|A) = D(A|B)$, which may be expected because conditional entropy is not symmetric. Some theorists consider this as a weakness. However, in the context of concrete experiments this asymmetry makes sense, as the following example shows: Knowledge about system A (say, a neutron in the n-He scattering experiment presented below) is *not* equivalent to knowledge about system B (say, the struck He). Or in other terms, it is natural to expect that measurements on A may provide another information content about the composite system AB than measurements on B; cf. Eq. (13).

(*b*) Discord is nonnegative, $D \geq 0$, which is a direct consequence of the concavity of conditional entropy.

(*c*) Discord is invariant under local unitary transformations, i.e. it is the same for state ρ_{AB} and state $(U_A \otimes U_B)\rho_{AB}(U_A \otimes U_B)^\dagger$. This follows from the fact that discord is defined via entropies, and the value obtained for measurement $\{E_a\}$ on the state ρ_{AB} can also be achieved with measurement $\{U_A E_a U_A^\dagger\}$ on the transformed state. Note that discord is not contractive under general local operations, and therefore should not be regarded as a strict measure of correlations. However, $J(B|A)$ is contractive under general local operations.

(*d*) Quantum discord $D(B|A)$ vanishes if and only if the state is classical-quantum [28].

In the last part of this section we point at three particular specializations of the above topics which may be appropriate in context with the experimental scattering measurements presented below (see section 4). These are focusing on the connection of quantum discord (and related measures of quantumness, including entanglement) with the process of quantum measurement. At last, an additional subsection about the recently discovered thermodynamic meaning of negative conditional entropy is presented.

2.2.1. Using Measurement-Induced Disturbance to Characterize Correlations. The above mentioned quantifiers of quantumness of correlations are based on an extremalization procedure involving the set of POVM's, which is a very difficult task to do. Additionally, while in the classical description of nature measurements can be carried out without disturbance, in the quantum description, generic measurements usually disturb the system. Thus classical states are characterized in terms of non-disturbance under a quantum measurement. Formalizing these observations, Luo has proposed a correlation quantifier based on the disturbance that the measurement processes causes in a system [20], called *measurement-induced disturbance* (MID).

This quantity, is defined as the difference between the quantum mutual information of the state, ρ_{AB}, and that one of the completely dephased state, χ_{AB}

$$MID\left(\rho_{AB}\right) = I\left(\rho_{AB}\right) - I\left(\chi_{AB}\right). \tag{15}$$

The dephasing takes place in the marginal basis, leaving the marginal states unchanged. The suitability of MID and its relation to quantum discord and various other correlation measures are discussed in Refs. [22, 21].

It should be mentioned that all quantum correlation quantifiers discussed in this article vanish for the so-called classical-classical (CC) states

$$\chi_{AB} = \sum_{a,b} p_{ab} \Pi_a \otimes \Pi_b. \tag{16}$$

Since all projectors in this decomposition correspond to fully distinguishable states, the probability p_{ab} can be regarded as a classical joint probability of random variables a and b.

It was then proved that that certain separable states, e.g. quantum-quantum states

$$\rho^{AB} = \sum_i p_i \rho_i^A \otimes \rho_i^B, \tag{17}$$

(where ρ_i^A and ρ_i^B are local states pertinent to subsystems A and B) still possess correlations of a quantum nature. He also showed that the specific quantum correlations contained in these mixed quantum-quantum states correspond to a "generalization" of the concept of entanglement [20].

2.2.2. Linking Quantum Discord to Entanglement in a Measurement. The focus of this work, as mentioned earlier, lies on the experimental context of elementary scattering processes in condensed systems and molecules. In this context, the interacting two systems A and B are "disturbed" (or "measured", or "observed") by another system C, e.g. by adjacent particles of the "environment". Thus it is natural to mention here some related works in which the measurement process may have effect on the quantumness of correlations of the composite system $A + B$.

Recently Streltsov et al. [29] have introduced an alternative approach to quantum correlations via an interpretation of a quantum measurement, especially on one of the subsystems of a composite system.

In carrying out a von Neumann measurement on a system S in the quantum state ρ^S, correlations between the system and the measurement apparatus M must be created. E.g., consider a von Neumann measurement in the eigenbasis $|i\rangle$ of the mixed state

$$\rho^S = \sum_i p_i |i^S\rangle\langle i^S| \tag{18}$$

with the eigenvalues p_i. One then expects the final state of the total system to be

$$\rho_{\text{final}} = \sum_i p_i |i^M\rangle\langle i^M| \otimes |i^S\rangle\langle i^S|, \tag{19}$$

where $|i^M\rangle$ are orthogonal states of the measurement apparatus M. This ρ_{final} is a classical-classical state (see above), i.e. the correlations between M and S are purely classical.

The situation changes completely if we consider *partial* von Neumann measurements, i.e. von Neumann measurements which are restricted to a subsystem A of the system AB. (Indeed, these correspond to the usual case in real experiments; cf. section 4.) Streltsov et al. showed that in this case creation of entanglement is usually unavoidable. They also showed the close connection of this approach to the one-way information deficit [30, 22, 21] and the quantum discord. In short, they showed that the one-way information deficit is equal to the minimal distillable entanglement [15, 7] between the measurement apparatus M and the system AB which has to be created in a von Neumann measurement on subsystem A. The quantum discord is then equal to the corresponding minimal partial distillable entanglement. (The minimization is done over all unitaries U which realize a von Neumann measurement on A; see Theorem 2 of [29].) Additionally, it was shown that this approach can be generalized to multipartite von Neumann measurements [29].

2.2.3. On the Thermodynamic Meaning of Negative Conditional Entropy. In the preceding derivations of quantumness of correlations, the concept of conditional entropy (or conditional infomation) plays a considerable role. Very recently, del Rio et al. [31] showed that the standard formulation and implications of Landauers principle [12, 13] (cf. Introduction) are no longer valid in the presence of quantum information. Their main result is that the work cost of erasure of information is determined by the entropy of the system, conditioned on the quantum information an observer has about it: the more an observer knows about the system carrying the information, the less it costs to erase the information. Obviously this result gives a direct thermodynamic significance to *conditional entropy*, which plays a dominant role in the theory of quantumness of correlations, see above.

The main result characterizes the work, $W(A|Q)$, that an observer with access to a quantum memory Q correlated with the system A, needs to perform to erase system A. In the "thermodynamic limit", where the observer erases many identical copies of A jointly, the authors have found a specific erasure process [31] whose work cost does not exceed

$$W(A|Q) = S(A|Q)\,kTln(2) \tag{20}$$

per copy of A; (k: Boltzmann constant, T: temperature). Here $S(A|Q)$ is the conditional von Neumann entropy,

$$S(A|Q) = S(AQ) - S(Q). \tag{21}$$

This work cost is shown to be optimal, under the assumption that Landauers principle holds for a classical observer.

In the quantum case, however, novel features may arise. In particular, the last equation implies that the work required for erasure may be negative for an observer with a quantum memory Q: the process may result in a net *gain of work*. For instance, the combined system AQ may be closed and in a pure state, and thus its von Neumann entropy will be zero, $S(AQ) = 0$, whereas the reduced state of the memory Q is mixed and has positive entropy $S(Q) > 0$, which finally yields $S(A|Q) < 0$. This provides a thermodynamic operational meaning for negative conditional entropies, which earlier only had information-theoretical interpretations [31, 21]. Moreover, it was noted that the obtained results suggest that quantum discord can quantify the difference between the respective work costs of erasure using quantum and classical memories.

Furthermore, it was pointed out that the above surprising result does not violate the Second Law of thermodynamics, because the proposed process is not cyclic. That is, the negative work cost is associated with the consumption of entanglement between system and quantum memory, which can only be restored by doing work (i.e., with positive work costs). The overall work costs are then expected to be positive or zero; see addendum of [31].

3. Quantum Correlations and Decoherence—Thermodynamic and Energetic Consequences

Decoherence is the ubiquitous phenomenon that destroys quantum correlations (state superpositions, quantum interference, quantum phases) thus leading to the "appearance of a classical world in quantum theory"; cf. [32]. Decoherence, similarly to many other irreversible mechanisms, is a time-oriented process, i.e. it breaks the time-inversion invariance of the Schrödinger equation. Moreover, it is known to be much faster than energy dissipation (say, due to friction or T_1 relaxation). A considerable number of models of decoherence, based on a variety of physical motivations, have been proposed and investigated in the scientific literature, cf. [32].

3.1. On the Quantum Carnot Engines of Lloyd and of Scully et al.—Thermodynamic Consequences

The quantum heat engine concept has attracted considerable interest since it provides novel insights into the fundamental physics of heat–information–energy conversion. Hence this also concerns our purposes, see section 4 below.

Lloyd [16] analyzed the effects of the quantum Maxwell demon. He found that a quantum device acquiring information in a measurement process and the associated decoherence disturb the system and act as a source of thermodynamic inefficiency. The quantum demon has been realized in the concrete context of nuclear magnetic resonance (NMR) [16, 9]. Of particular interest appears the result that this cost of quantum measurement is realized when decoherence occurs, which destroys the non-diagonal matrix elements of the system's density matrix ρ. (Studying special processes with ρ remaining always diagonal, the quantum demon reaches the well known Carnot efficiency.) In other words, the extra information introduced by quantum measurement and decoherence has been identified as the cause of *decrease* of the efficiency of the Carnot engine [16].

This interesting result pointing out the role of decoherence should be contrasted with the associated result of the well known Lindblad equation presented below in subsection 3.2.1.

Here let us proceed to a second example of a quantum heat engine. Scully et al. [33] proposed and analyzed a new kind of quantum Carnot engine powered by a special quantum heat bath. It was shown to allow extraction of work even from a single thermal reservoir. For this type of heat engine the piston is driven by radiation pressure. The working fluid (say, analogous to steam of a conventional machine) finds its counterpart in the radiation, which is generated by a beam of hot atoms (constituting the thermal reservoir).

In the "regular" case of two-level thermal atoms, the engine's efficiency cannot exceed the well-known Carnot limit. In the case of hot specific three-level atoms constituting a heat bath things look quite different, i.e. when the nearly degenerate lower levels feature a small amount of quantum coherence. The corresponding quantum phase ϕ can be varied as a control parameter to effectively increase the temperature of the radiation field. In this scenario work is obtained even when only one single heat bath is present. Consumption of quantum coherence is shown to produce work.

The deep physics behind the Second Law of thermodynamics, however, is not violated. The assumed atomic coherence itself causes energetic costs, as e.g. it must be generated by the passage of the atoms through a suitable microwave field. An explicit estimation of the necessary microwave field's energy to produce the assumed appreciable coherence between the atomic levels was shown to be greater than the extracted work in the above process. Consequently and in line with the Second Law of thermodynamics, the total entropy of the whole system is constantly increasing [33].

3.2. Pure Decoherence—Energetic Consequences

Pure decoherence (i.e. without dissipation) is usually described with the well-known master equations of Lindblad form, which ensure positivity of the systems reduced density operator.

Quite unexpectedly, it turns out that "pure" decoherence may have a perplexing, peculiar consequence. Notably, as shown below (subsections 3.2.1-2) in the frame of two independent theoretical models, pure decoherence turns out to be intrinsically connected with an increase (!) of mean energy of the system, i.e. with negative energy costs — in the absence of any direct interaction with other systems. This is certainly unexpected, since erasure of quantum phases and/or correlations is not widely acknowledged as a source of energy. The presentations below will show that this theoretical result can be of rather general character, under the restrictive condition that the characteristic time of the process is sufficiently short [50].

In the following two subsections we mainly follow the presentation of Ref. [34].

3.2.1. Lindblad Equation and Spontaneous Energy Increase. To describe the dynamics of open quantum systems and decoherence, various generalizations of the Schrödinger equation have been proposed; see the textbooks [32]. Among these theoretical approaches, the Born-Markov master equation plays an enormously important role. Master equations of the so-called Lindblad form refer to a particular (albeit quite general) class of Markovian master equations, which ensure positivity of the reduced density operator $\rho(t)$ describing the system, i.e. $\langle\rho(t)\rangle \geq 0$, for any pure state $|\psi\rangle$ of the system and for all t. The most general mathematical form of such equations was derived by Gorini, Kossakowski and Sudarshan [35] and Lindblad [36].

Consider the simplest Lindblad-type ansatz for the master equation for the statistical operator ρ of an open quantum system, which includes only a single Lindblad operator L; in a real system we would have a multitude of such dynamical variables. We set

$$\frac{\partial\rho}{\partial t} = -\frac{i}{\hbar}[H,\rho] - \Lambda\,[L,[L,\rho]] \tag{22}$$

where $\Lambda > 0$ is a positive constant and H is the Hamiltonian. The first term on the right-hand side (rhs) describes the usual unitary time evolution of the state. The double commutator term describes decoherence (and/or dephasing). It may be noted that this equation, by suitable choice of L, may describe "pure" decoherence only. Namely, it does not contain any term describing friction explicitly, as is done in Caldeira-Leggett-type equations, cf. [32]. In the following we consider only the case of pure decoherence (i.e. destruction of quantum coherence).

Considering Eq. (22) in the position representation, the double commutator term takes the form

$$-\Lambda(r-r')^2\langle r|\rho|r'\rangle$$

which, since $\Lambda > 0$, leads to an exponential decay of the non-diagonal elements of the density matrix:

$$\langle r|\rho(t)|r'\rangle = \langle r|\rho(0)|r'\rangle \exp(-\Lambda(r-r')^2 t) \qquad \text{for } 0 \leq t \tag{23}$$

where $t = 0$ is the initial time [32]. Obviously, this result does not hold for $t \leq 0$. In other words, coherent superpositions of different states $|r\rangle$ and $|r'\rangle$ will be suppressed over time. The most well-known specialization of Eq. (22) is the case of an one-particle system that interacts with an external (thermal) environment, in which case L is taken to be the position operator, q. This corresponds to the Joos-Zeh master equation [37].

It is well known that Eq. (22) preserves the normalization $Tr\rho = 1$ [36], which is satisfactory. However, Ballentine [38] realized that the expectation value of the energy, $\langle H\rangle = Tr(H\rho)$ is in general not conserved. Namely,

$$\frac{d\langle H\rangle}{dt} = \frac{d}{dt}Tr(H\rho) = Tr\left(H\frac{\partial\rho}{\partial t}\right) = -\Lambda\,Tr([H,L][L,\rho]) \tag{24}$$

(Recall the operator identity $Tr(A[B,C]) = Tr([A,B]C)$.) Obviously, if the Hamiltonian and the Lindblad operator L do not commute, $\langle H\rangle$ is not conserved for every state ρ.

The significance of this result becomes immediately obvious in the special case of a free particle —e.g. represented by a wave packet, *not* by a plane wave with constant wavevector— moving in one dimension with Hamiltonian $H = p^2/2m$. In various well-known theoretical models, the decoherence in position owing to the "environment" is described by taking L to be the particle's position operator q; for details see [37, 38, 32]. Then, since $[p,q] = \hbar/i$ and

$$[p^2,q] = p^2q - qp^2 = (p^2q - pqp) + (pqp - qp^2) = p[p,q] + [p,q]p = \frac{2\hbar}{i}p$$

one obtains

$$\begin{aligned}
\frac{d\langle H\rangle}{dt} &= -\Lambda\,Tr([p^2/2m,q][q,\rho]) \\
&= -\frac{\Lambda}{2m}\frac{2\hbar}{i}Tr(p[q,\rho]) = -\frac{\Lambda}{2m}\frac{2\hbar}{i}Tr([p,q]\rho) \\
&= +\frac{\Lambda\hbar^2}{m} > 0
\end{aligned} \tag{25}$$

Thus the system appears to steadily gain energy at a constant rate [38].

This surprising result holds also in three dimensions, and for inclusion of scalar and/or vector potentials. It also remains valid for more general cases, e.g. even in cases with the decoherence factor Λ becoming a function of the "distance" $|r - r'|$; for details see [38, 39].

Undoubtedly, this result seems quite disturbing as it contradicts every conventional expectation about what the consequences of decoherence should be (besides the expected destruction of coherent superpositions of quantum states). Namely, theorem (25) seems to represent a serious weakness of the theory, because it implies that the mean energy of a free paricle should always increase; or, in other terms, the environment causing decoherence seems to act as an inexhaustible source of kinetic energy. Moreover, this continuous gain of kinetic energy is clearly incompatible with the system's attainment of equilibrium [38]. For a thorough discussion of these paradoxical findings, see [38, 39, 40].

Parenthetically, another surprising consequence of the Lindblad equation and pure decoherence may be shortly noted here. Recently, a "first principles" description of scattering from open quantum systems subject to a Lindblad-type dynamics was provided [41]. It was shown that this time evolution may cause a reduction of the system's transition rate being effectuated by scattering. This is tantamount to a shortfall of scattering intensity [41, 42], or "intensity deficit", which represents a witness of quantumness of correlations. This effect has been observed experimentally [43, 44, 45, 46, 47, 48, 49].

3.2.2. The Theoretical Model by Schulman and Gaveau. Surprisingly, a recent independent and more general theoretical analysis by Schulman and Gaveau [50, 51] appears to lead essentially to the same result. A short description of the general model is as follows.

A quantum system A, e.g. a quantum oscillator, with free Hamiltonian H_A makes an elastic collision with a second system B with free Hamiltonian H_B. Let the interaction potential be V_{AB}. The total Hamiltonian is $H = H_A + H_B + V_{AB}$. Before the collision, the two systems are assumed to be not entangled and so the complete density matrix $\rho(0)$ should be $\rho(0) = \rho_A(0) \otimes \rho_B(0)$. In general, subsequent to their collision they become entangled and the exact density operator

$$\rho(t) = U(t)\,\rho(0)\,U^\dagger(t)$$

($U(t)$: time evolution operator) is not a product state of individual density operators $\rho_A(t) = \mathrm{Tr}_B\rho(t)$ and $\rho_B(t) = \mathrm{Tr}_A\rho(t)$.

However, it is widely believed that once the particles are separated the quantum correlations can be dropped (provided one does not perform an experiment of Einstein-Podolsky-Rosen type), simply because measurements of physical quantities of each of the two particles cannot depend on their correlations. Thus the replacement

$$\rho(t) \to \rho_A(t) \otimes \rho_B(t)\ , \tag{26}$$

i.e. the erasure of quantum correlations, is usually assumed to be "innocuous" and e.g., not affect the energies of the systems.

The striking result by Schulman and Gaveau [50] contradicts this intuitive expectation. Putting

$$\Delta\rho(t) = \rho_A(t) \otimes \rho_B(t) - \rho(t)$$

and for a particular form of the interaction Hamiltonian, they show that for sufficiently *short* times the following relation holds:

$$\Delta E \equiv Tr(\Delta\rho(t)H) = Tr(\Delta\rho(t)\,V_{AB}) > 0\ . \tag{27}$$

See Ref. [51] for a detailed derivation of this inequality. In simple words, the replacement of the entangled $\rho(t)$ by the non-entangled state $\rho_A(t) \otimes \rho_B(t)$ necessarily increases the system's energy [50, 51]. This appears highly paradoxical since, as Schulman and Gaveau put it: "...losing quantum correlations should not heat the gas. You do not burn your finger because of a partial trace over a density matrix" [50].

Moreover, this result was shown to be valid for a large class of potentials, e.g. for two-body interactions, although it does not hold universally [51]. This is an interesting detail, and therefore the following details should be mentioned. The spin-boson model Hamiltonian

$$H = H_A + H_B + V_{SB} = \omega_a a^\dagger a + \omega_b b^\dagger b + g(a^\dagger + a)(b^\dagger + b)$$

was shown to exhibit the considered effect, in clear contrast to the related Jaynes-Cummings model

$$H = H_A + H_B + V_{JC} = \omega_a a^\dagger a + \omega_b b^\dagger b + g(a^\dagger b + b^\dagger a) \ ,$$

in which the effect is absent [50].

In view of the counter-intuitive character of the result (27), one may object that it is unphysical since seems to violate energy conservation. However this is not the case, as the detailed discussions of Ref. [50] explained. It was stressed that, in the situation contemplated, the coupling Hamiltonian must be considered as time dependent, because the physical approach and separation of the particles leads to a time-dependent coupling coefficient. Thus, energy conservation need not apply. Moreover, it was discussed that this "additional" energy $\Delta E > 0$ is supplied by the *translational* degrees of freedom of A and B [50], which do not appear explicitly in the Hamiltonian and thus may be considered to represent an effective "environment".

4. Experimental Consequences

4.1. Conventional Neutron Scattering Formalism

Let us briefly consider some basic formulas of conventional neutron scattering theory [24, 52, 53]. The partial differential cross section $d^2\sigma/d\omega d\Omega$ and the associated dynamic structure factor $S(\mathbf{q},\omega)$ are determined from the experiment. For scattering by a system of N identical atoms they are related according to:

$$\frac{d^2\sigma}{d\omega d\Omega} = Nb^2 \frac{k_1}{k_0} S(\mathbf{q},\omega) \tag{28}$$

(b: bound scattering length of atom; k_0, k_1: absolute values of wavevectors of incident and scattered neutrons; $d\Omega$: small solid angle subtended by the neutron detector in direction of $\mathbf{k}_1$ at the target). $\hbar\mathbf{q}$ and $\hbar\omega$ are the momentum and energy transfers from the neutron to a scattering nucleus, respectively, i.e, $\hbar\mathbf{q} = \hbar\mathbf{k}_0 - \hbar\mathbf{k}_1$ and $\hbar\omega = E_0 - E_1$. The subscripts "0" and "1" refer to quantities before and after the collision.

It is assumed that the neutron-nucleus interaction is well represented by the Fermi pseudo-potential [52, 24]

$$V(\mathbf{r}) = \frac{2\pi\hbar^2}{m}\, b\, \delta(\mathbf{r}) \tag{29}$$

(m: neutron mass; $\delta(\mathbf{r})$: delta function) which phenomenologically describes the short-range strong interaction. According to the basic van Hove theory [52] $S(\mathbf{q},\omega)$ is given by

$$S(\mathbf{q},\omega) = \frac{1}{2\pi}\int_{-\infty}^{\infty} \exp(-i\omega t)\, F(\mathbf{q},t)\, dt \tag{30}$$

where

$$F(\mathbf{q},t) = \frac{1}{N}\sum_{j,k}^{N} \langle \exp(-i\mathbf{q}\cdot\mathbf{r}_j(0)) \exp(-i\mathbf{q}\cdot\mathbf{r}_k(t))\rangle . \tag{31}$$

is the so-called intermediate correlation function. $\langle ... \rangle$ stands for the thermodynamic average of the quantum expectation values of the enclosed operators, and the $\mathbf{r}_j(t)$ denotes the position operator of atom j at time t in the *Heisenberg* representation [52, 24].

The derivations of the above formulas are based on Fermi's golden rule, which, for scattering processes, is equivalent to the first Born approximation [52, 24]. Recall that both these approximations are special cases of general first-order perturbation theory, which, however, does not hold for a singular potential like Eq. (29). The justification for the use of Fermi's golden rule in neutron scattering is that, in combination with Eq. (29), it gives the required result of isotropic (s-wave) scattering for a single fixed nucleus; see [24] for further discussions.

In the limit of sufficiently high momentum and energy transfers, $q = |\mathbf{q}| \gg 2\pi/d$ with d being the nearest-neighbor distance of scattering particles, terms with $j \neq k$ are negligible (i.e. the incoherent approximation holds [24, 25]) and one obtains

$$F(\mathbf{q}, t) = \langle \exp(-i\mathbf{q} \cdot \mathbf{r}(0)) \exp(+i\mathbf{q} \cdot \mathbf{r}(t)), \rangle \tag{32}$$

see [24, 25]) for the derivation. Note that $\mathbf{r}(t)$ is in general a N-body operator, despite the (rather misleading) notation of "position operator of a scattering atom". Furthermore, for sufficiently fast scattering, the impulse approximation (IA) [25, 54] applies, and the corresponding expression for the dynamic structure factor simplifies to:

$$S_{IA}(\mathbf{q}, \omega) = \int n(\mathbf{p})\ \delta(\hbar\omega - \hbar\omega_r - \hbar\mathbf{q} \cdot \mathbf{p}/M)\, d\mathbf{p}\ , \tag{33}$$

Here, $n(\mathbf{p})$ is the (one-body) momentum distribution of the scattering nucleus before the collision,

$$\hbar\omega_r = \hbar^2 q^2/2M \tag{34}$$

is the so-called recoil energy of the nucleus, and M is the mass of the scattering nucleus. The delta-function represents energy conservation in the binary neutron-nucleus scattering process.

Eq. (33) is the basic formula describing neutron Compton scattering (NCS) [25, 54], also known as deep inelastic neutron scattering (DINS). It follows that the measured recoil peak is centred at an energy transfer $\hbar\omega_r$, i.e. at the recoil energy. The width of the peak is given by the term $\hbar\mathbf{q} \cdot \mathbf{p}/M$, i.e. the projection of the nuclear momentum before the collision on the measured momentum transfer. This term represents the well-know Doppler effect [25, 54]. Generally, the recoil peak obtained from liquid ^{4}He is broadened due to the momentum of the classical atomic motion and the atomic confinement in the effective potential caused by the atom's environment [54].

From the viewpoint of conventional scattering theory [52, 24, 25], NCS is expected to measure "single particle" quantities, e.g. the mean kinetic energy of an atom. In various cases, an effective Born-Oppenheimer potential for the atom is assumed to exist, which then can be extracted from the measured Compton profile [25].

With respect to electron–atom Compton scattering [47, 48], Bonham et al. provided a thorough analysis in the frame of conventional theory, which (as also NCS) is based on the first Born and Born-Oppenheimer approximations [55].

In this context, it is important to stress that the characteristic scattering time of the processes under consideration is very short; e.g. it is about one femtosecond in the real experiment considered below. This is of the order of the characteristic electronic rearrangement processes in molecules and condensed matter.

The above considerations show that neutron-nucleus entanglement and/or more general quantum correlations do not play any role in conventional neutron scattering theory.

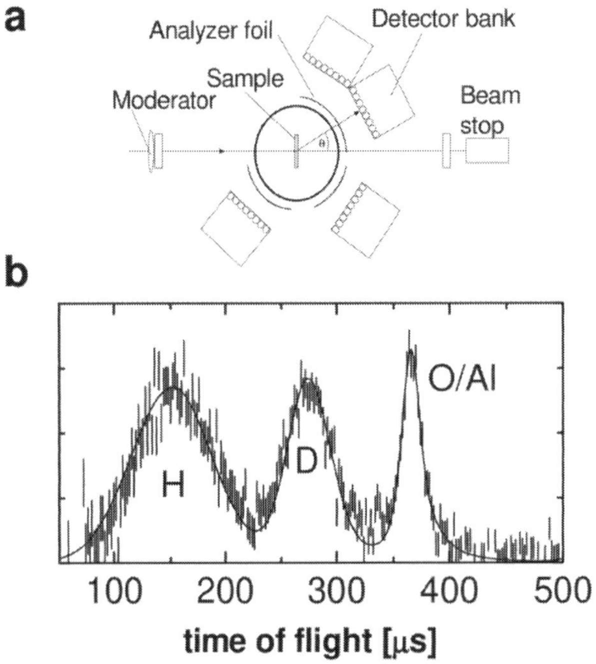

Figure 1. (a) Schematic representation of time-of-flight (TOF) spectrometer eVS/Vesuvio of ISIS. (b) A TOF-spectrum of a 20:80 H_2O-D_2O mixture in an Al cell. The H and D recoil peaks are well separated from each other and from the joint O/Al peak; adapted from [43].

4.2. On Time-of-Flight (TOF) Technique

The time-of-flight (TOF) scattering technique (cf. Fig. 1) allows the measurement of the partial differential cross section $d^2\sigma/d\omega d\Omega$ and the associated dynamic structure factor $S(\mathbf{q}, \omega)$ [56]. The TOF t of each detected neutron is determined by

$$t = \frac{L_0}{v_0} + \frac{L_1}{v_1} + t_0 \ . \tag{35}$$

Here L_0 is the source–sample distance, L_1 is the sample–detector distance. The detector is positioned at the scattering angle θ. v_0 and v_1 are the velocities of the incident and scattered neutron, respectively. t_0 is a small time offset due largely to electronic delays. We used values for L_0, L_1, θ and t_0 for the individual detectors as provided by the instrument parameter file IP0002 of the ISIS experimental report [57]; see Table 1.

The used spectrometer [57] is a so-called "inverse geometry" instrument meaning that the final velocity v_1 (and thus the wavevector k_1) of the neutrons is fixed, and v_0 varies. The final neutron energy is fixed at $E_1 = 4906$ meV (the resonance energy of ^{197}Au used as analyzer foil [58]), corresponding to a velocity $v_1 = 3.063 \times 10^4$ m/s and wavevector $k_1 = 48.663\,\text{Å}^{-1}$.

Note that the geometric quantities L_0, L_1 and θ can be determined by appropriate methods; cf. [57]. It follows from Eq. (35) that each t corresponds to an initial velocity v_0 and thus to an energy transfer

$$\hbar\omega = \frac{1}{2}mv_0^2 - \frac{1}{2}mv_1^2 = \frac{(\hbar k_0)^2}{2m} - \frac{(\hbar k_1)^2}{2m} \tag{36}$$

For the corresponding momentum transfer $\hbar q$ from the neutron to the struck atom, $\hbar\mathbf{q} = \hbar\mathbf{k_0} - \hbar\mathbf{k_1}$, where

$$q = \sqrt{k_0^2 + k_1^2 - 2k_0k_1\cos\theta}\ . \tag{37}$$

It should be emphasized that, for each value of t, the associated momentum ($\hbar\mathbf{q}$) and energy ($\hbar\omega$) transfers from the neutron to the struck particle are uniquely determined. (Note that both q and ω vary over the TOF-range of the recoil peak, which is measured by a detector positioned at a fixed scattering angle θ.)

For scattering from free atoms with zero initial momentum, $p = 0$, conservation of kinetic energy and momentum in an elastic neutron-atom collision yield the kinematic relation

$$\frac{v_1}{v_0} = \frac{k_1}{k_0} = \frac{\cos\theta + \sqrt{(M/m)^2 - \sin^2\theta}}{M/m + 1}. \tag{38}$$

m and M are the masses of the neutron and struck nucleus, respectively. This equation corresponds to neutrons detected at the center of the measured recoil peak; cf. Fig. 1. For a concise account of the above kinematic formulas, see [44].

4.3. Conventional Results in View of Quantumness of Correlations

Let us first "reformulate" the preceding basic formula (33) of NCS by including the neutron state explicitly into the considerations.

First, recall that momentum conservation

$$\mathbf{p}/\hbar + \mathbf{k}_0 = (\mathbf{p}/\hbar + \mathbf{q}) + (\mathbf{k}_0 - \mathbf{q}) = (\mathbf{p}/\hbar + \mathbf{q}) + \mathbf{k}_1^\theta \tag{39}$$

($\mathbf{k}_0 - \mathbf{q} = \mathbf{k}_1^\theta$ by definition; superscript θ indicates the scattering angle at which a specific detector is positioned) and energy conservation

$$\frac{(\hbar k_0)^2}{2m} - \frac{(\hbar k_1)^2}{2m} = \frac{(\mathbf{p} + \hbar\mathbf{q})^2}{2M} - \frac{(\mathbf{p})^2}{2M} = \hbar\omega_r + \hbar\mathbf{q}\cdot\mathbf{p}/M \tag{40}$$

hold in the binary collision. The assumed quantum state of the neutron-nucleus system before collision is factorizable,

$$\rho_{\text{init}} = \left(\int d\mathbf{p}\, n(\mathbf{p})\ |\mathbf{p}\rangle\langle\mathbf{p}|\right) \otimes |\mathbf{k}_0\rangle\langle\mathbf{k}_0| \tag{41}$$

Since conventional theory takes the neutron's final state to be pure (a plane wave with well defined and fixed $\mathbf{k}_1^\theta$), the state of the neutron-scatterer system after scattering is still factorizable (i.e. not entangled),

$$\rho_{\text{final}}^\theta = \left(\iint^\star d\mathbf{q}\, d\mathbf{p}\, n(\mathbf{p})\ |\mathbf{p}/\hbar + \mathbf{q}\rangle\langle\mathbf{p}/\hbar + \mathbf{q}|\right) \otimes |\mathbf{k}_1^\theta\rangle\langle\mathbf{k}_1^\theta| \tag{42}$$

where the $\star$ on the integral denotes restriction to momentum and energy conservation. Recall that $\mathbf{k}_0$ is not fixed, and thus the conservation conditions can be fulfilled by various pairs of values of $\mathbf{p}$ and $\mathbf{q}$. (We use both notations $|\mathbf{p}/\hbar + \mathbf{q}\rangle$ and $|\mathbf{p} + \mathbf{q}\hbar\rangle$ to represent the same physical state.) Clearly, these two-body initial and final states of conventional theory are not entangled and, moreover, classical-classical; see section 2.2. Thus they have zero quantum discord.

4.3.1. Relation to Quantumness of Correlations. We now proceed in reconsidering the derivations and the basic formula (33) of NCS in the light of the modern understanding of the quantumness of correlations; see section 2. Against this background, the basic formulas of the conventional theory of NCS appear to have a serious weakness. Notably, in conventional theory the neutron represents a classical particle, its quantum degrees of freedom are not taken into account. ($\mathbf{k}_0$, $\mathbf{k}_1$ and also $\mathbf{q}$ are not operator quantities but only c-numbers.) Moreover the quantum states of the neutron "disappear", since the average operation $\langle ... \rangle$ in the formulas of section 4.1 contains quantum states of the scattering particles only.

According to basic quantum mechanical principles, on the other hand, one generally would expect that a collision of two particles A and B results in entanglement of these particles. E.g., for s-wave scattering, the final state is given by a spherical outgoing wavefunction (in the center-of-mass system) of their relative coordinate $r = |\mathbf{r}_A - \mathbf{r}_B|$,

$$\Psi_{AB} \propto \frac{\exp(ikr)}{r} \tag{43}$$

(k: suitable constant) and, due to the strict validity of momentum conservation, A and B will generally become entangled in momentum space after scattering. [This entangling process should not be assumed to be equivalent to the well-known separation of centre-of-mass and relative coordinate motions, which appears also in the two-body problem of classical mechanics.]

Then why the neutron-scatterer entanglement (or more general quantum correlations) does not play a role in the formulas of conventional theory? The final state of the neutron is here *assumed* to be a well defined plane wave $|\mathbf{k}_1\rangle$, which is a pure one-particle state. This assumption is necessary in order to apply the first Born approximation (and the Fermi golden rule) and to derive the scattering probability in the $\mathbf{k}_1$-direction in which the neutron detector stands. Obviously, this is tantamount to tacitly presume a collaps of the entangled (or quantum correlated) neutron-nucleus state and assume a factorizable final state $|\mathbf{k}_1\rangle \otimes |\mathbf{p}/\hbar + \mathbf{q}\rangle$, in which the neutron and the scatterer are in pure final states $|\mathbf{k}_1\rangle$ and $|\mathbf{p}/\hbar + \mathbf{q}\rangle$, with well defined momenta. In other words, the existing entanglement between neutron and scatterer has been "removed by hand". As shown below, this "removal" appears to have measurable energetic consequences which contradict conventional theory.

In the context of the following considerations it is important to observe that the scattering atom (nucleus) is not free in a real experiment but interacts with its environment. Moreover, the characteristic scattering time τ_{scatt} of NCS is very short but finite (about 1 femtosecond and less [43, 44]), and thus this interaction is expected to cause a partial degradation of the considered entanglement, which may still lead to states containing quantum discord.

The same conclusion is achieved in the general frame of the theoretical work by Streltsov et al. [29] discussed above: Namely, M of [29] corresponds to the neutron, the subsystem A to the single scattering nucleus X, B to the environment $\mathcal{E}$ of the stuck nucleus or atom, and AB to the complete N-body system constituting the sample, $X + \mathcal{E}$. The latter, being an interacting system, may be naturally expected to be quantum correlated with nonzero discord. Then, the neutron scattering process creates entanglement between the neutron and the scattering sample [29].

Now let us consider the concrete physical context of NCS from normal liquid ^{4}He, which we also study in the context of a real experiment (see below). Recalling the fact that NCS transfers momentum $\hbar q$ to one He atom (with, say $q \sim 100$ Å^{-1}), we can reasonably assume the struck atom to strongly interact with adjacent He atoms (i.e. the environment $\mathcal{E}$) a few Ångströms apart, after an ultrashort time (of the order of one femtosecond) following the impulsive neutron-He collision. Note that the scattered neutron is still very near, also a few Ångströms apart. The scattered neutron reaches the detector without further disturbance, at a distance of about 0.6 m away and after some tens of microseconds (see below)—about 9 orders of magnitude later than the previous process.

These estimates suggest the following important insights:

At and shortly after the moment of collision, the neutron and the struck He should become entangled, as explained above, see Eq. (43). Due to the "monogamy" of entanglement [21], one may expect that any entanglement of the struck He atom with adjacent atoms existing previously (i.e. before collision) will be removed.

The first object that interacts with (or "measures") the neutron-He binary system and disturbs and/or reduces its state (say, by a POVM) is not the neutron detector, but the environment $\mathcal{E}$ of the struck He. (Recall that we are considering liquid He.) This happens on timescales of the order of 1 femtosecond after the initial collision, which is a time difference similar to the so-called characteristic scattering time τ_{scatt} of NCS [44, 25]. Due to its many-body characteristics, this He-$\mathcal{E}$ interaction can not reasonably be assumed to lead to a pure "final" state of the He atom. So let ρ_λ^{He} (index λ indicating the states) instead be one of the typical mixed reduced states of He, and correspondingly let ρ_λ^n be the associated mixed reduced states of the scattered neutron. These states may still be partially entangled. Nevertheless one might make the ansatz of a separable state

$$\rho^{n,He}(t \sim \tau_{scatt}) \approx \int d\lambda\, p(\lambda)\, \rho^n(\lambda) \otimes \rho^{He}(\lambda), \tag{44}$$

of the type of Eq. (17), section 2.2.1, which is a quantum-quantum state and thus in general exhibits discord. In other words, the environment $\mathcal{E}$ is continuously "measuring" the "neutron+He" system, thus continuously changing its quantum correlations.

Later, after a characteristic relaxation or equilibration time τ_{relax}, the struck He should be well equilibrated with its environment and again constitute an N-body state $\rho^{He,\mathcal{E}}$ of the same physical nature as that before the collision. The neutron assumes a mixed state ρ^n, which due to the process of detection is further selected and can approximately represented as a pure state $|\mathbf{k}_1^\theta\rangle\langle\mathbf{k}_1^\theta|$ of conventional theory. The final measurement of the neutron by the detector can, due to the principle of causality, not affect the dynamics and/or the correlations of the neutron-He collisional process.

The following presents a *schematic* summary of the physical processes that, according to our point of view, describe the dynamics of a NCS experiment:

$$t = -\infty \quad : \quad \rho^n \otimes \rho^{He,\mathcal{E}} \tag{45}$$

$$\text{collision}: \quad t \approx 0 \quad : \quad \rho^{n,He} \otimes \rho^{\mathcal{E}} \tag{46}$$

$$t \leq \tau_{scatt} \sim 10^{-15}\text{s} \quad : \quad \left(\int d\lambda\, p(\lambda)\, \rho^n(\lambda) \otimes \rho^{He}(\lambda)\right) \otimes \rho^{\mathcal{E}} \tag{47}$$

$$t > \tau_{relax} \quad : \quad \rho^{He,\mathcal{E}} \otimes \rho^n \tag{48}$$

$$t = +\infty \quad : \quad \rho^{He,\mathcal{E}} \otimes |\mathbf{k}_1^\theta\rangle\langle\mathbf{k}_1^\theta| \tag{49}$$

(Note: A symbol appearing in more than one line may refer to varying physical quantities; e.g. $\rho^{He,\mathcal{E}}$ in first line should be significantly different from that in line 4.)

4.4. Experiment

We investigated NCS from normal-state liquid ^{4}He at $T \approx 2$ K and saturated vapor pressure. The original data [59] were kindly provided by ISIS under the "ISIS Data Policy", cf. [60].

Here we present results obtained from TOF-spectra measured with nine detectors (see Table 1) in backscattering, at mean scattering angle $\overline{\theta} = 134.594° \pm 0.2324°$. The mean sample-detector distance is $\overline{L_1} = (0.6392 \pm 0.0038)$ m; the mean time offset is $\overline{t_0} = (-0.3246 \pm 0.0943) \cdot 10^{-6}$ s. The L_0 length is given as $L_0 = 11.0050$ m. These specific detectors were selected because their parameters, and especially their angles, are very similar and thus the TOF spectra can be

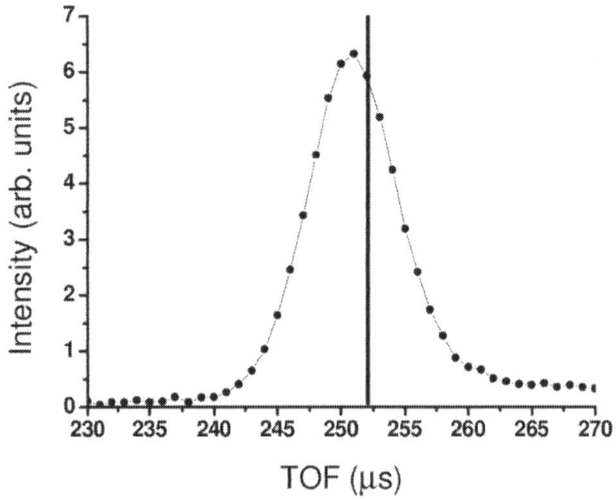

Figure 2. The accumulated TOF spectrum (9 detectors, see Table 1; as measured) of ^{4}He at about 2 K and saturated vapor pressure; see the text. According to the instrument parameters, Table 1, the conventionally expected position of the peak is given by the vertical bar. The observed peak displacement corresponds to a transfer energy increased by ca. 110 meV. This effect has no conventional interpretation; see the text.

Table 1. The instrument parameters of the detectors used in the data analysis; reproduced from file IP0002.DAT [57]

Det. No	θ (degrees)	t_0 (μsec)	L_0 (m)	L_1 (m)
10	134.3881	-0.2896000	11.0050	0.6410817
11	134.8883	-0.3632000	11.0050	0.6379817
12	134.7490	-0.1396000	11.0050	0.6374661
14	134.5436	-0.2448000	11.0050	0.6318235
18	134.9788	-0.3489000	11.0050	0.6433660
54	134.3012	-0.4116000	11.0050	0.6416131
56	134.4089	-0.3321000	11.0050	0.6421713
58	134.5938	-0.4649000	11.0050	0.6356288
62	134.4992	-0.3263000	11.0050	0.6414817

directly accumulated in order to improve signal to noise ratio. The resulting accumulated TOF spectrum (as measured) is shown in Figure 2.

Table 1 contains the data of the used detectors, and is reproduced from the file containing instrument parameters, IP0002.DAT, in Ref. [57], pages 39-41.

Using the aforementioned final neutron energy $E_1 = 4906$ meV, one obtains $k_1 = 48.663\,\text{Å}^{-1}$. For scattering from ^{4}He, and using the formulas of subsection 4.2, the corresponding initial wavevector at $\overline{\theta}$ is $k_0 = 75.517\,\text{Å}^{-1}$, and the related momentum transfer is $\hbar q$ with $q = 115.027\,\text{Å}^{-1}$. The associated recoil energy (or energy transfer) is $\hbar\omega_r = 6908.7$ meV.

These numerical values, which are obtained from conventional theory, and the basic formula Eq. (35) predict the TOF position of the ^{4}He recoil (Compton) peak to be at $t_{conv} = 252.067$ μs. This value is shown with a vertical bar in Figure 2. One observes a significant displacement of the measured recoil peak to shorter TOF's, and equivalently to larger energy transfers (since the instrument is of "inverse geometry"; see above). The observed displacement $\Delta_Q t$ of the peak maximum is about one microsecond, $\Delta_Q t \approx 1.0\ \mu s$. This corresponds to an increased initial velocity (and energy) of the neutron, and thus to an energy transfer increased by

$$\Delta_Q E \approx 110\ \text{meV}. \tag{50}$$

Due to the uncertainty of the TOF-position of the peak, we estimate the error of $\Delta_Q E$ to be about $\pm 10\%$.

It may be noted that the intrinsic width of the recoil peak of ^{4}He is very narrow, and thus the width shown in Figure 2 is mostly due to the instrumental resolution. Hence $\Delta_Q E$ should not be considered as being "too small".

4.4.1. Quantum Maxwell demon and $\Delta_Q E > 0$. Note that the epithermal neutron, the struck ^{4}He atom, and the adjacent ^{4}He atoms (being strongly disturbed by the former) do not represent thermalized systems or baths, due to the smallness of the characteristic time window τ_{scatt}, see above. Thus the experimental result (50) cannot be directly compared with the well know theoretical result

$$\Delta W = W_Q - W_C = kTD(A|B) \tag{51}$$

(k: Boltzmann constant; T: temperature) connecting quantum discord [61], Eq. (13), with the difference between the work-extraction efficiency of the quantum and classical demons, which may extract works W_Q and W_C, respectively. The quantity ΔW is also called work deficit, cf. [21].

However, using the traditional demon's scenario, one may obtain an associated result as follows. Let the neutron correspond to the subsystem A, and the struck ^{4}He atom together with certain adjacent ^{4}He atoms interacting with it to subsystem B. The demon continuously "reads off", or erases, quantum correlations (i.e. information) encoded in $A+B$ during the time window τ_{scatt} of the collision. For a classical demon, this erasure should have an inherent positive work cost [12], e.g. it may dissipate some kinetic energy of A, resulting to a *smaller* energy transfer $\hbar\omega_r - W_C$ for the binary neutron-^{4}He collision than the conventional recoil energy $\hbar\omega_r$. In the case of a quantum demon, however, this process may have "negative costs of erasure" [31], e.g. it may "extract" some additional work W_Q, resulting to a *larger* energy transfer $\hbar\omega_r + W_Q$ for the binary collision than the conventionally expected value $\hbar\omega_r$. Thus we can define the "efficiency" of the quantum demon by the ratio:

$$\frac{W_Q - W_C}{\hbar\omega_r} \geq \frac{\text{measured E-transfer} - \hbar\omega_r}{\hbar\omega_r} = \frac{110\ \text{meV}}{6908.7\ \text{meV}} \approx 1.6\ \% \tag{52}$$

In other words, the Maxwell demon observing the actual NCS process from ^{4}He is a quantum one—and about 1.6% more efficient than any classical demon.

5. Discussion and Conclusions

At first glance, one obvious objection to the theoretical arguments presented is that the observed shift of the NCS peak might have a more simple explanation. For instance, the following possibility might be proposed: The observed mean energy transfer is slightly higher than the one predicted by standard theory. Could it not be interpreted simply as being caused by a slightly higher effective mass of the scatterer? The assumption of a free scatterer is clearly a crude (although very common) approximation. If the scatterer is partially bound with adjacent particles, one scatters on a seemingly little heavier object, as the environment to which it is bound has a certain mass. This effective increase of mass might be due to some simply classical binding (e.g., the extreme case would be analogous to the well known Mössbauer effect, where the crystal as a whole acts as a scatterer. Similarly in solid state physics it is common to talk about heavy electrons, which is nothing more than simply a breakdown of the independent particle scatterer approximation due to correlation which causes an apparent heavier mass). However, the proposed consideration demonstrates quite the opposite of what is intended. Namely, according to standard theory of binary collisions, the mentioned higher effective mass of the scattering ^{4}He must necessarily cause a lower (not higher!) energy transfer (e.g. collision with a scatterer of infinite mass causes no energy transfer at all). In other terms, and in the light of the above remarks, it is clear that no classical partial bonding to the environment can lead to a smaller effective mass of the scatterer—which would then provide a conventional interpretation of the experimentally measured increased energy transfer.

The experimental results discussed in the previous section can be understood by noting that erasure of quantum correlations in the system "neutron+^{4}He", which is interacting with the environment, may have negative work costs [31] that affect the quantum dynamics of neutron scattering on ultrafast timescales. First indications of this effect have been recently observed in scattering from molecular H_2 [62] and D_2 [63].

The aforementioned theoretical considerations for the interpretation of the experimental finding (50) in the frame of quantum information theory are qualitative; a quantitative theoretical model for calculating the positive energy-transfer shift of the NCS-recoil peak is presently not available.

The phenomena of entanglement, discord and decoherence are not considered in conventional neutron scattering theory [25, 24, 52, 53]. As mentioned above, the observed positive energy-transfer deviation, $\Delta_Q E > 0$, contradicts all conventional expectations. This becomes even more evident by considering the opposite (and conventionally expected) case of a negative $\Delta_Q E$, which is attributed to so-called final-state-effects [25]. Such effects arise if the energy transfer is not sufficient to validate the impulse approximation [25, 54], in which case the struck particle is not "fully free" and a part of the neutron's kinetic energy must be used to overcome the particle-environment binding forces, resulting in a reduction of the energy transferred to the struck particle.

In contrast to conventional theory, and based on the aforementioned theoretical understanding of quantum correlations and their dynamics accompanying elementary scattering processes, we attribute the result $\Delta_Q E > 0$ to "negative energetic costs" of correlations-erasure discussed in sections 2 and 3. In particular we refer to the qualitatively similar results following from the Lindblad equation [38] (sect. 3.2.1), the result of the Schulman-Gaveau analysis [50] (sect. 3.2.2), and the "negative work costs" of negative conditional entropy by del Rio et al. [31] (sect. 2.2.3). Note that in all these cases the quantumness of correlations leads to higher work-values as compared to the associated processes in the absence of quantum correlations. The same holds for quantum Maxwell demons, which can extract more work from quantum correlations than classical demons; cf. [21, 22, 61].

In view of these remarks and considerations, our scattering effect offers new physical insights into entanglement, discord and other measures of quantumness of correlations, as well as on their operational meaning. Moreover, the general character of its causes suggests that this effect may be observed in other experimental areas involving scattering (e.g. inelastic x-ray scattering, electron-atom Compton scattering, etc.), and also in a lower energy-transfer range (e.g. in vibrational spectroscopy with neutrons [64, 65, 66, 67]). Related work is in progress [68].

Acknowledgments

We thank ISIS, Rutherford Appleton Laboratory, for providing the ^{4}He experimental data used in this paper under the ISIS Data Policy.

References

[1] Prigogine I 1980 *From Being to Becoming* (San Francisco: Freeman)
[2] Einstein A, Podolsky B and Rosen N 1935 *Phys. Rev.* **47** 777
[3] Schrödinger E 1935 *Naturwissenschaften* **23** 807, 823 and 844
[4] Bell J S 1964 *Physics* **1** 195
[5] Gisin N 1991 *Phys. Lett. A* **154** 201
[6] Peres A 1995 *Quantum theory: Concepts and methods* (Dordrecht: Kluwer Academic Publishes)
[7] Horodecki R, Horodecki P, Horodecki M and Horodecki K 2009 *Rev. Mod. Phys.* **81** 865
[8] Brunner N, Gisin N and Scarani V 2005 *New J. Phys.* **7** 88
[9] Nielsen M A and Chuang I L 2000 *Quantum Computation and Quantum Information* (Cambridge: Cambridge University Press)
[10] Maxwell J C 1871 *Theory of Heat* (London: Appleton)
[11] Maruyama K, Nori F and Vedral V 2009 *Rev. Mod. Phys.* **81** 1
[12] Landauer R 1961 *IBM J. Res. Develop.* **5** 148
[13] Bennett C H 1982 *Int. J. Theor. Phys.* **21** 905
[14] Bennett C H 2003 *Stud. Hist. Phil. Mod. Phys.* **34** 501
[15] Plenio M B and Vitelli V 2001 *Contemp. Phys.* **42** 25
[16] Lloyd S 1997 *Phys. Rev.* A **56** 3374
[17] Ollivier H and Zurek W H 2001 *Phys. Rev. Lett.* **88** 017901
[18] Henderson L and Vedral V 2001 *J. Phys. A: Math. Gen.* **34** 6899
[19] Merali Z 2011 *Nature* **474** 24
[20] Luo S 2008 *Phys. Rev.* A **77** 022301
[21] Modi K, Brodutch A, Cable H, Paterek T and Vedral V 2012 *Rev. Mod. Phys.* **84** 1655
[22] Lang M D, Caves C M and Shaji A 2011 *Int. J. Quantum Inform.* **9** 1553
[23] (a) Luo S 2008 *Phys. Rev.* A **77** 042303; (b) Luo S and Sun W 2010 *Phys. Rev.* A **82** 012338
[24] Squires G L 1996 *Introduction to the Theory of Thermal Neutron Scattering* (Mineola: Dover)
[25] Watson G I 1996 *J. Phys.: Condens. Matter* **8** 5955
[26] Weigold E and McCarthy I E 1999 *Electron Momentum Spectroscopy* (New York: Kluwer Academic/Plenum Publishers)
[27] Sinha S K 2001 *J. Phys.: Condens. Matter* **13** 7511
[28] Datta A 2008 PhD thesis; available at arXiv:0807.4490[quant-ph].
[29] Streltsov A, Kampermann H and Bruß D 2011 *Phys. Rev. Lett.* **106** 160401
[30] Oppenheim J, Horodecki M, Horodecki P and Horodecki R 2002 *Phys. Rev. Lett.* **89** 180402
[31] del Rio L, Åberg J, Renner R, Dahlsten O and Vedral V 2011 *Nature* **474** 61; and Addendum at doi:10.1038/nature10395
[32] (a) Breuer H P and Petruccione F 2002 *The Theory of Open Quantum Systems* (Oxford: Oxford University Press); (b) Mensky M B 2000 *Quantum Measurements and Decoherence* (Dordrecht: Kluwer); (c) Schlosshauer M 2007 *Decoherence and the Quantum-to-Classical Transition* (Berlin: Springer Verlag); (d) Giulini D, Joos E, Kiefer C, Kupsch J, Stamatescu I O and Zeh H D 1996 *Decoherence and the Apperance of a Classical World in Quantum Theory* (Berlin: Springer Verlag)
[33] Scully M O, Zubairy M S, Agarwal G S and Walther H 2003 *Science* **299** 862
[34] Chatzidimitriou-Dreismann C A, Gray E MacA and Blach T P 2012 *J. Phys.: Conf. Ser.* **380** 012013
[35] Gorini V, Kossakowski A and Sudarshan E C G 1976 *J. Math. Phys.* **17** 821
[36] Lindblad G 1976 *Commun. Math. Phys.* **48** 119
[37] Joos E and Zeh H D 1985 *Z. Phys.* B **59** 223
[38] Ballentine L E 1991 *Phys. Rev.* A **43** 9
[39] Gallis M R and Fleming G N 1991 *Phys. Rev.* A **43** 5778
[40] Stenholm S 1993 *Phys. Scr.* **47** 724
[41] Chatzidimitriou-Dreismann C A and Stenholm S 2005, in: *Decoherence, Entanglement and Information Protection in Complex Quantum Systems*, NATO Science Series II – Vol 189, ed V M Akulin, A Sarfati, G Kurizki and S Pellegrini (Dordrecht: Springer) pp 555-562; also available at arXiv:quant-ph/0702038v1
[42] Chatzidimitriou-Dreismann C A 2005 Laser Physics **15** 780
[43] Chatzidimitriou-Dreismann C A, Abdul-Redah T, Streffer R M F and Mayers J 1997 *Phys. Rev. Lett.* **79** 2839
[44] Karlsson E B, Chatzidimitriou-Dreismann C A, Abdul-Redah T, Streffer R M F, Hjörvarsson B, Öhrmalm J and Mayers J 1999 *Europhys. Lett.* **46** 617
[45] Karlsson E B Abdul-Redah T, Streffer R M F, Hjörvarsson B, Mayers J and Chatzidimitriou-Dreismann C A 2003 *Phys. Rev.* B **67** 184108
[46] Chatzidimitriou-Dreismann C A, Abdul-Redah T and Kolarić B 2001 *J. Am. Chem. Soc.* **123** 11945

[47] Chatzidimitriou-Dreismann C A, Vos M, Kleiner C and Abdul-Redah T 2003 *Phys. Rev. Lett.* **91** 057403
[48] Cooper G, Hitchcock A P and Chatzidimitriou-Dreismann C A 2008 *Phys. Rev. Lett.* **100** 043204
[49] Karlsson E B 2012 *Int. J. Quant. Chem.* **112** 587
[50] Schulman L S and Gaveau B 2006 *Phys. Rev. Lett.* **97** 240405
[51] Gaveau B and Schulman L S 2010 *J. Phys. A: Math. Theor.* **43** 055308
[52] van Hove L 1954 *Phys. Rev.* **95** 249
[53] Tietje I C 2012 *J. Phys.: Conf. Ser.* **380** 012014
[54] Hohenberg P C and Platzman P M 1966 *Phys. Rev.* **152** 198
[55] Bonham R A, Cooper G and Hitchcock A P 2009 *J. Chem. Phys.* **130** 144303
[56] Windsor C 1981 *Pulsed Neutron Scattering* (London: Taylor and Francis)
[57] Mayers J and Adams M A 2009 *Calibration of the Electron Volt Spectrometer VESUVIO at ISIS*; Technical Report RAL-TR-2009-022
[58] Mughabghab S F 1984 *Neutron Cross Sections* (Orlando: Academic Press)
[59] Data files 14777-14812 from March 2010.
[60] See http://www.isis.stfc.ac.uk/groups/computing/data/icat11680.html
[61] Zurek W H 2003 *Phys. Rev.* A **67** 012320
[62] Chatzidimitriou-Dreismann C A, Gray E MacA and Blach T P 2011 *AIP Advances* **1** 022118
[63] Chatzidimitriou-Dreismann C A, Gray E MacA and Blach T P 2012 *Nucl. Instr. Meth.* A **676** 120
[64] Kearley G J, Fillaux F, Baron M H, Bennington S and Tomkinson J 1994 *Science* **264** 1285
[65] Anovitz L M, Mamontov E, Ben Ishai P and Kolesnikov A I 2013 *Phys. Rev.* E **88** 052306
[66] Spencer E C, Huang B, Parker S F, Kolesnikov A I, Ross N L and Woodfield B F 2013 *J. Chem. Phys.* **139** 244705
[67] Mitchell P C H, Parker S F, Ramirez-Cuesta A J and Tomkinson J 2005 *Vibrational Spectroscopy with Neutrons* (New Jersey: World Scientific)
[68] Chatzidimitriou-Dreismann C A 2014 *Z. Naturforsch. A*, in press

Bounded motion for classical systems with position-dependent mass

S. Cruz y Cruz, C. Santiago-Cruz

Unidad Profesional Interdisciplinaria en Ingeniería y Tecnologías Avanzadas del Instituto Politécnico Nacional
Av. Instituto Politécnico Nacional 2580, Col. La Laguna Ticomán, C. P. 07340, México D. F. Mexico

E-mail: sgcruzc@ipn.mx,csantiagoc0500@ipn.mx

Abstract. In this work the dynamical equations for a system with position-dependent mass are considered. The phase space trajectories are constructed by means of the factorization method for classical systems. To illustrate how this formalism works the phase space trajectories for position-dependent mass oscillator, Scarf and Pöschl-Teller potentials with Gaussian and singular masses are presented.

1. Introduction

The problem of describing the dynamics of systems with position-dependent mass has gain much interest in the last decades. Yet, it appears in many physical areas such as semiconductor theory [1–3], molecular dynamics [4], geometric optics [5] and astrophysics [6] among others. Of particular importance are its applications in the study of the properties of semiconductors. In this context one can mention, *e. g.*, the evaluation of linear and nonlinear optical properties and the binding energies of square, parabolic, V-shaped and Pöschl-Teller-type quantum dots and quantum wells, in the effective mass approximation. A wide range of mass functions have been used including constant, exponential and inverse of quadratic polynomials depending on the structure and chemical composition of each system [7–12]. Some other important applications include the description of the dynamics of systems in curved spaces [13–15] as well as nonlinear oscillators [16, 17]. In these cases some masses varying as the inverse of quadratic functions evolving under the influence of quadratic potentials have been considered. Further, in the generation of inversion potentials of molecules in density theory, some masses with singularities are used as the reduced mass of the molecules as functions of the inversion coordinate [4], and some power law masses model the galactic mass loss [6] and the deploying of a cable on a reel [18].

The position-dependent mass concept is, by itself, a fundamental problem which is far from being understood. On this matter, many contributions have been developed all over the years from different approaches [19–29]. In particular, the factorization method [30–32] has been applied in constructing exactly solvable position-dependent mass potentials [21–28]. From the classical point of view, the factorization of the Hamiltonian in terms of two functions that together with the Hamiltonian induce a particular Poisson algebra has been discussed in [27,33].

In this regime, the position-dependent mass functions lead to dynamical equations including terms quadratic in the velocity, a problem which have been continuously discussed [18, 34–39]. In this work we consider the dynamical equations for classical systems with position-dependent mass in the Newton, Lagrange and Hamilton approaches. The factorization method applied to an *energy invariant* is used to construct, in an algebraic form, the phase space trajectories for systems under the effect of harmonic oscillator, Scarf and Pöschl-Teller potentials. The formalism works equally well for regular and singular masses. To illustrate this fact, the cases of a Gaussian mass and a mass with a quadratic singularity are considered in order to connect our approach to some of the applications mentioned above.

The paper is organized as follows. In section 2 we derive the Lagrangian and Hamiltonian equations of motion for a position-dependent mass system. These equations include a non-inertial force quadratic in the velocity. Since the corresponding Hamiltonian is not conserved, an integral of motion in energy units is constructed that allow to express the dynamical equations in the standard form. In section 3, the factorization method is applied to this *energy invariant* in order to construct the phase space trajectories. In section 4 the underlying Poisson algebras for oscillator, Scarf and Pöschl-Teller position-dependent mass systems are established and the corresponding phase space trajectories for Gaussian and singular masses are presented. The paper close with some concluding remarks.

2. The position-dependent mass problem in the classical framework

Consider a classical system with position-dependent mass $m(x) > 0$. Suppose that a force $F(x, \dot{x})$ is acting on the system. The Newton second law can be then stated as (assuming null velocity for the accreted or ablated mass)

$$F(x, \dot{x}) = \frac{dp}{dt} = m'(x)\dot{x}^2 + m(x)\ddot{x}, \tag{1}$$

where $p = m(x)\dot{x}$ is the linear momentum of the system and $'$ stand for the derivative with respect to the position. A trajectory in the (x, v) plane can be then constructed by solving the nonlinear set of equations

$$\dot{x} = v, \quad \dot{v} = \frac{1}{m(x)}\left[F(x, \dot{x}) - m'(x)\dot{x}^2\right]. \tag{2}$$

Since $\dot{x}^2 > 0$, it is clear that for an increasing mass, *i.e.*, $m'(x) > 0$ the system is decelerated, while it is accelerated for a decreasing mass $m'(x) < 0$.

2.1. The Lagrangian and Hamiltonian approaches

For external conservative forces $F(x) = -V'(x)$ and applying de D'Alembert principle we obtain that

$$\frac{d}{dt}\left(\frac{\partial T}{\partial \dot{x}}\right) - \frac{\partial T}{\partial x} = F(x) + \tilde{R}(x, \dot{x}), \tag{3}$$

where the kinetic energy T and the reacting thrust $\tilde{R}$ [33] are given by

$$T(x, \dot{x}) = \frac{1}{2}m(x)\dot{x}^2, \quad \tilde{R}(x, \dot{x}) = -\frac{1}{2}m'(x)\dot{x}^2. \tag{4}$$

Since the force is velocity independent the Lagrangian form of the Newton second law (1) reads

$$\frac{d}{dt}\left(\frac{\partial \mathcal{L}}{\partial \dot{x}}\right) - \frac{\partial \mathcal{L}}{\partial x} = \tilde{R}(x,\dot{x}), \quad \mathcal{L}(x,\dot{x}) = T(x,\dot{x}) - V(x). \tag{5}$$

The canonical momentum obtained from the Lagrangian $\mathcal{L}$ turn out to be the linear momentum

$$p = \frac{\partial \mathcal{L}}{\partial \dot{x}} = m(x)\dot{x}. \tag{6}$$

Now the position-dependent mass Hamiltonian $\mathcal{H}$ is obtained from the Legendre transformation

$$\mathcal{H}(x,p) = p\dot{x} - \mathcal{L}(x,\dot{x}) = \frac{p^2}{2m(x)} + V(x). \tag{7}$$

Note that this Hamiltonian is not time invariant since

$$\frac{d}{dt}\mathcal{H} = -\frac{1}{2}m'(x)\dot{x}^3 \neq 0. \tag{8}$$

The dynamical equation (1) can be expressed in canonical form as

$$\dot{x} = \{x, \mathcal{H}\} = \frac{\partial \mathcal{H}}{\partial p}, \quad \dot{p} = \{p, \mathcal{H}\} = -\frac{\partial \mathcal{H}}{\partial x} + R, \tag{9}$$

with

$$\{f, g\} = \frac{\partial f}{\partial x}\frac{\partial g}{\partial p} - \frac{\partial f}{\partial p}\frac{\partial g}{\partial x} \tag{10}$$

the Poisson bracket and

$$R(x,p) = \tilde{R}\left(x, \dot{x}(x,p)\right) = -\frac{m'(x)}{m(x)}\left(\frac{p^2}{2m(x)}\right) \tag{11}$$

the thrust in the phase space.

So far, we have constructed the Lagrangian and Hamiltonian description of the dynamics of a position-dependent mass system. We have shown that, even when the Lagrangian and Hamiltonian functions are of the standard form, the Lagrange and Hamilton equations are not. Additionally, we have found that the Hamiltonian $\mathcal{H}$ is not time independent, its time rate of change is cubic in the velocity leading to non standard canonical equations and to the problem of dissipative systems [36, 40]. It turns out, however, that the dynamical equations can be expressed in the conventional form by means of an *energy constant of motion*.

2.2. Energy invariant and phase space trajectories

In this Section we construct an *energy invariant* and show that the corresponding dynamical equations have the form of the standard Hamilton ones. Let the function $I(x,v)$ be such an invariant, then, it must satisfy

$$\frac{d}{dt}I(x,v) = \dot{x}\frac{\partial}{\partial x}I(x,v) + \dot{v}\frac{\partial}{\partial v}I(x,v) = 0. \tag{12}$$

Substituting $\dot{x}$ and $\dot{v}$ from (2), this equation can be written as

$$v\frac{\partial I}{\partial x} - \frac{1}{m(x)}\left[V'(x) + m'(x)\dot{x}^2\right]\frac{\partial I}{\partial v} = 0. \tag{13}$$

A simple calculation leads to the function [33]

$$I = \frac{m^2(x)v^2}{2m_0} + \int_{x_0}^{x} \frac{m(s)}{m_0} V'(s)ds = \frac{p^2}{2m_0} + \int_{x_0}^{x} \frac{m(s)}{m_0} V'(s)ds, \tag{14}$$

with m_0 and x_0 two constants with units of mass and position respectively. This invariant can be considered as the conservative Hamiltonian of the system. It is important to stress that it is reduced to the constant mass Hamiltonian $H = \frac{p^2}{2m_0} + V$ in the case that $m = m_0$. Now defining new variables

$$P = \frac{m(x)}{m_0} p = \frac{m^2(x)}{m_0} v = M(x)v, \quad M(x) = \frac{m^2(x)}{m_0}, \tag{15}$$

the energy invariant in the (x, P) phase space is

$$H = \frac{P^2}{2M} + V_{\text{eff}}(x), \tag{16}$$

where

$$V_{\text{eff}}(x) = \int_{x_0}^{x} \frac{m(s)}{m_0} V'(s)ds \tag{17}$$

is the effective potential experienced by the position-dependent mass system.

It is not difficult to show that the dynamical equation (1) can be written in the form

$$\frac{\partial H}{\partial P} = \{x, H\}_{x,P} = \dot{x}, \quad \frac{\partial H}{\partial x} = \{P, H\}_{x,P} = -\dot{P}. \tag{18}$$

Additionally, the time evolution of an observable $O(x, P; t)$ is given by

$$\frac{d}{dt} O(x, P; t) = \{O, H\}_{x,P} + \frac{\partial O}{\partial t}. \tag{19}$$

This approach shows that, in order to preserve the canonical equations, the mass m, the linear momentum p and the potential V must be transformed into M, P and V_{eff} by (15) and (17). The phase space trajectories of a particular system, for an arbitrary mass function $m(x)$, can be now constructed by solving the set of equations (18).

3. Factorization method and phase space trajectories for position-dependent mass systems

Once we have the Hamiltonian description for the position-dependent mass systems, it is worthwhile to mention that one can reach to the solution in two different approaches. In the first one it is assumed that the potential $V(x)$ and the mass function $m(x)$ are known and the phase space trajectories must be determined by integrating the dynamical equations (18). In the second one it is considered that the underlying algebra associated to the dynamics of the system and its mass are given and the potentials and phase space trajectories must be determined. In this work we will adopt the second approach. Thus, we will suppose that the algebraic structure of the system is known and the corresponding position-dependent mass potential and phase space trajectories must be determined. The problem will be solved by means of the factorization method in the (x, P) plane. A complete discussion of the factorization method for one dimensional classical systems can be found in [41].

Suppose that the Hamiltonian

$$H = \frac{P^2}{2M} + V_{\text{eff}}(x) \tag{20}$$

can be factorized as

$$H = A^+A^- + \epsilon, \tag{21}$$

with

$$A^\pm = \mp i f(x)\frac{P}{\sqrt{2M(x)}} + \sqrt{\gamma H}g(x) + \varphi(x), \tag{22}$$

where $f(x)$, $g(x)$, $\varphi(x)$ are functions to be determined and ϵ is a constant. Here we consider the possibility of bound states for negative energies, then γ take the values $+1$ and -1 for $H > 0$ and $H < 0$ respectively. The factorization approach (20-22) leads to the relation

$$(f^2 + \gamma g^2 - 1)H - f^2 V_{\text{eff}} + 2\varphi g\sqrt{\gamma H} + \varphi^2 + \epsilon = 0, \tag{23}$$

from which it is clear that $g(x)$ and $\varphi(x)$ can not be both different from zero.

Next, we demand that the functions H and $A^\pm$ define the Poisson algebra (for the sake of simplicity, from now on, we drop the subindex x, P from the Poisson brackets)

$$i\left\{H, A^\pm\right\} = \pm\alpha(H)A^\pm \tag{24}$$

$$i\left\{A^-, A^+\right\} = \beta(H) \tag{25}$$

with $\alpha(H)$ and $\beta(H)$ functions of $\sqrt{\gamma H}$ to be fixed. Note that (24) leads to

$$i\left\{H, A^+A^-\right\} = 0,$$

meaning that A^+A^- is a function of H, consistently with (21). Additionally, the Jacobi identity

$$\left\{H, i\left\{A^-, A^+\right\}\right\} + \left\{A^-, i\left\{A^+, H\right\}\right\} + \left\{A^+, i\left\{H, A^-\right\}\right\} = 0$$

implies that

$$\left\{H, i\left\{A^-, A^+\right\}\right\} = 0,$$

meaning that the Poisson bracket $i\left\{A^-, A^+\right\}$ can be written as (25).

Furthermore,

$$i\left\{H, A^\pm\right\} = \pm\frac{2}{\sqrt{2M}}f\left(\sqrt{H}g' + \varphi'\right)A^\pm, \tag{26}$$

which, together with (23) lead to the set of equations

$$f^2 V_{\text{eff}} = \left(f^2 + \gamma g^2 - 1\right)H + \varphi^2 + \epsilon, \tag{27}$$

$$\alpha(H) = \frac{2}{\sqrt{2M}}f\left(\sqrt{\gamma H}g' + \varphi'\right) \tag{28}$$

to determine the functions f, g, φ and α.

On the other hand, the algebraic relations (24) allow us to write two non autonomous integrals of motion

$$Q^\pm(x, P; t) = A^\pm e^{\mp i\alpha(H)t} \tag{29}$$

satisfying

$$\left|Q^\pm\right|^2 = Q^+Q^- = A^+A^- = H - \epsilon. \tag{30}$$

If $q^\pm$ denote the values of $Q^\pm$, then

$$q^\pm = \sqrt{E-\epsilon}e^{\pm i\theta_0}, \quad \theta_0 \in \mathbb{R}, \quad E - \epsilon > 0.$$

Now the set of equations

$$q^{\pm} = Q^{\pm}\left(x\left(t\right), P\left(t\right); t\right) = \sqrt{E-\epsilon}\, e^{\pm i\theta_0} \tag{31}$$

can be used to determine algebraically the phase trajectories in terms of two constants of motion (E, θ_0) fixed by the initial conditions

$$\sqrt{\gamma E} g\left(x(t)\right) + \varphi\left(x(t)\right) = \sqrt{E-\epsilon}\,\cos\left(\theta_0 + \alpha(E)t\right), \tag{32}$$

$$P(t) = -\sqrt{\frac{2(E-\epsilon)M\left(x(t)\right)}{f^2\left(x(t)\right)}}\,\sin\left(\theta_0 + \alpha(E)t\right). \tag{33}$$

4. Position-dependent mass oscillator, Scarf and Pöschl-Teller potentials

4.1. The harmonic oscillator

Consider again the set of equations (27-28). If we assume that $g(x) = 0$ and $\gamma = 1$, then $\alpha(H) = \beta(H) = \alpha = \text{constant}$ and

$$\varphi(x) = \sqrt{\frac{m_0\alpha^2}{2}}\int_c^x J(s)ds, \quad J(x) = \sqrt{\frac{M(x)}{m_0}}. \tag{34}$$

Without loss of generality we may set $f(x) = f_0 = 1$. The potential is then defined by (27) as

$$V_{\text{eff}} = \frac{m_0\alpha^2}{2}\left(\int_c^x J(s)ds\right)^2 + \epsilon. \tag{35}$$

Hence, given the mass $M(x)$, a potential of the harmonic oscillator form (35) is such that the algebraic structure becomes the constant mass oscillator algebra [41]

$$i\left\{A^-, A^+\right\} = \alpha, \quad i\left\{H, A^{\pm}\right\} = \pm\alpha A^{\pm}. \tag{36}$$

4.1.1. Gaussian mass-function Consider the mass function

$$M(x) = m_0 e^{-2\lambda^2 x^2}, \tag{37}$$

with $m_0 > 0$ and $\lambda > 0$. This function is defined on the whole real line, as well as the corresponding harmonic oscillator ($c = 0$)

$$V_{\text{eff}}(x) = \frac{\pi m_0\alpha^2}{8\lambda^2}\left(\text{Erf}\left(\lambda x\right)\right)^2. \tag{38}$$

The phase space trajectories read

$$x(t) = \frac{1}{\lambda}\text{Erf}^{-1}\left[\sqrt{\frac{8\lambda^2}{\pi m_0\alpha^2}}(E-\epsilon)\cos\left(\theta_0 + \alpha t\right)\right], \tag{39}$$

$$P(t) = -\sqrt{2m_0}e^{-\lambda^2 x^2(t)}\sqrt{E}\,\sin\left(\theta_0 + \alpha t\right). \tag{40}$$

The potential, phase space trajectories $(x(t), P(t))$ and the 3D phase space surface $(x(t), P(t), E)$ are shown in Figure 1 for different values of the parameters. The mass takes its maximum value at $x = 0$, as well as the canonical momentum P. As $|x| \to \infty$, the system losses mass and the momentum decrease until the particle reaches the turning points, where P changes sign.

Journal of Physics: Conference Series **538** (2014) 012006 doi:10.1088/1742-6596/538/1/012006

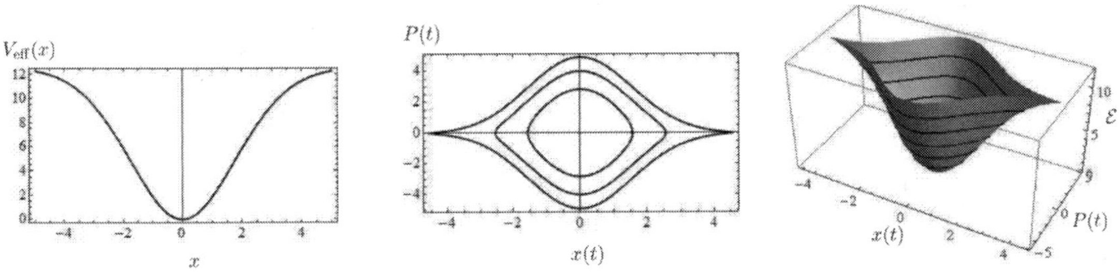

Figure 1. The potential (left), the phase space trajectories (middle) and the 3-D phase space (right) for the position-dependent mass harmonic oscillator with a Gaussian mass. In this plots $\alpha = 2$, $m_0 = 1$, $\lambda = 0.5$ and $\epsilon = 0$.

4.1.2. Singular mass Consider now the mass function

$$M(x) = \frac{m_0}{\lambda^2 x^2}, \tag{41}$$

with $m_0 > 0$ and $\lambda > 0$. This mass is defined in the half real line. The corresponding harmonic oscillator potential ($c = 1$)

$$V_{\text{eff}} = \frac{m_0 \alpha^2}{2\lambda^2} (\ln \lambda x)^2, \tag{42}$$

is defined in the same domain. The corresponding phase space trajectories have the form

$$x(t) = \frac{1}{\lambda} \exp\left[\sqrt{\frac{2\lambda^2(E-\epsilon)}{m_0\alpha^2}} \cos(\theta_0 + \alpha t)\right] \tag{43}$$

$$P(t) = -\sqrt{2m_0(E-\epsilon)} \exp\left[-\sqrt{\frac{2\lambda^2(E-\epsilon)}{m_0\alpha^2}} \cos(\theta_0 + \alpha t)\right] \sin(\theta_0 + \alpha t). \tag{44}$$

Figure 2 shows the potential, the phase space trajectories $(x(t), P(t))$ and the 3D phase space surface $(x(t), P(t), E)$ for different values of the parameters. In this case the behavior of the system is different from the previous one. The mass function has a singularity at $x = 0$. As the particle approaches this point the mass increases as well as the momentum P. When the particle reaches the turning point $x = 0$ the momentum change sign and the motion is reverted.

4.2. Scarf and Pöschl-Teller potentials

Concerning equations (27-28), now consider $g(x) \neq 0$ but$\varphi(x) = 0$. We can see that one simple solution to this set of equations can be set on the form

$$f^2 + \gamma g^2 = 1, \quad \alpha(H) = \beta(H) = \alpha\sqrt{\gamma H}, \tag{45}$$

$$g(x) = \begin{cases} \sin\left[\sqrt{\frac{m_0\alpha^2}{2}} \int_c^x J(s)ds\right] & \gamma = 1 \quad H > 0, \\ \sinh\left[\sqrt{\frac{m_0\alpha^2}{2}} \int_c^x J(s)ds\right] & \gamma = -1 \quad H < 0. \end{cases} \tag{46}$$

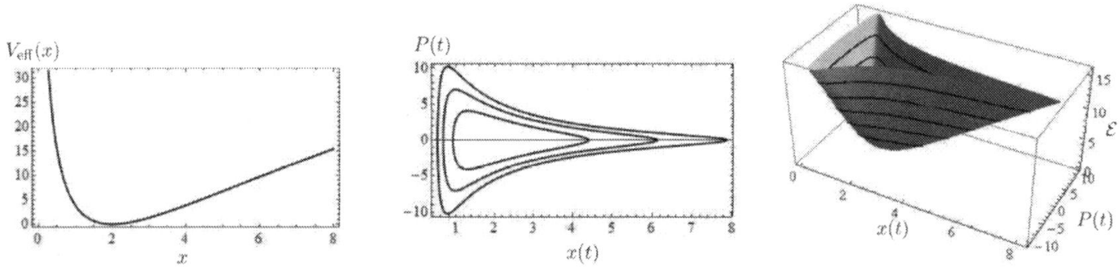

Figure 2. The potential (left), the phase space trajectories (middle) and the 3-D phase space (right) for the position-dependent mass harmonic oscillator with singular mass. In this plots $\alpha = 2$, $m_0 = 1$, $\lambda = 0.5$ and $\epsilon = 0$.

The Scarf and Pöschl-Teller potentials in this case have the form

$$V_{\rm eff}(x) = \frac{\epsilon}{1-\gamma g^2(x)} = \begin{cases} \dfrac{\epsilon}{\cos^2\sqrt{\frac{m_0\alpha^2}{2}}\int_c^x J(s)ds} & \gamma = 1 \quad H > 0, \\ \dfrac{\epsilon}{\cosh^2\sqrt{\frac{m_0\alpha^2}{2}}\int_c^x J(s)ds} & \gamma = -1 \quad H < 0 \end{cases} \tag{47}$$

respectively.

Now the functions H, $A^{\pm}$ obey the algebraic structure

$$i\left\{A^-, A^+\right\} = \alpha\sqrt{\gamma H}, \quad i\left\{H, A^{\pm}\right\} = \pm\alpha\sqrt{\gamma H}. \tag{48}$$

By defining new functions

$$a^{\pm} = \frac{1}{\alpha}A^{\pm}, \quad a^0 = \gamma\frac{1}{\alpha}\sqrt{\gamma H}, \tag{49}$$

it is possible to see that a^0, $a^{\pm}$ close the $su(11)$ algebra

$$i\left\{a^-, a^+\right\} = 2\alpha^0, \quad i\left\{a^0, a^{\pm}\right\} = \pm a^{\pm}, \tag{50}$$

for $H > 0$ $(\gamma = 1)$ and the $su(2)$ algebra

$$i\left\{a^-, a^+\right\} = -2\alpha^0, \quad i\left\{a^0, a^{\pm}\right\} = \pm a^{\pm} \tag{51}$$

for $H < 0$,$(\gamma = 1)$.

4.2.1. Gaussian mass Consider again the mass function (37). The Scarf and Pöschl-Teller position-dependent mass potentials in this case are

$$V_{\rm eff} = \begin{cases} \dfrac{\epsilon}{\cos^2\sqrt{\frac{\pi m_0\alpha^2}{8\lambda^2}}{\rm Erf}(\lambda x)} & \gamma = 1 \quad H > 0, \\ \dfrac{\epsilon}{\cosh^2\sqrt{\frac{\pi m_0\alpha^2}{8\lambda^2}}{\rm Erf}(\lambda x)} & \gamma = -1 \quad H < 0, \end{cases} \tag{52}$$

defined in the domains

$$\mathcal{D} = \left\{x \in \mathbb{R} \,\middle|\, -\frac{\pi}{2} < \sqrt{\frac{\pi m_0\alpha^2}{8\lambda^2}}{\rm Erf}(\lambda x) < \frac{\pi}{2}\right\}, \tag{53}$$

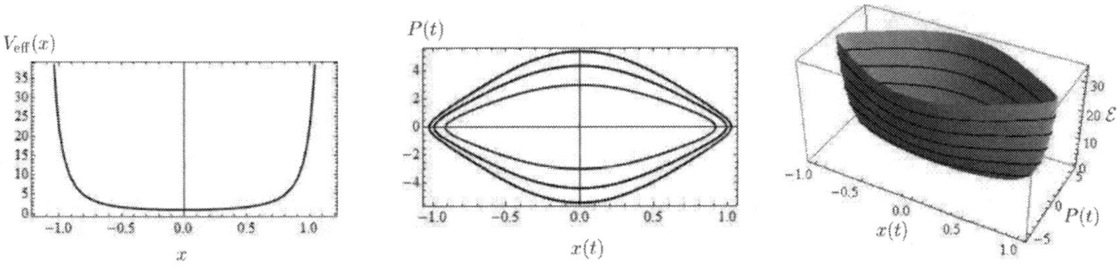

Figure 3. The potential (left), the phase space trajectories (middle) and the 3-D phase space (right) for the position-dependent mass Scarf potential with Gaussian mass. In this plots $\alpha = 2$, $m_0 = 1$, $\lambda = 0.5$ and $\epsilon = 1$.

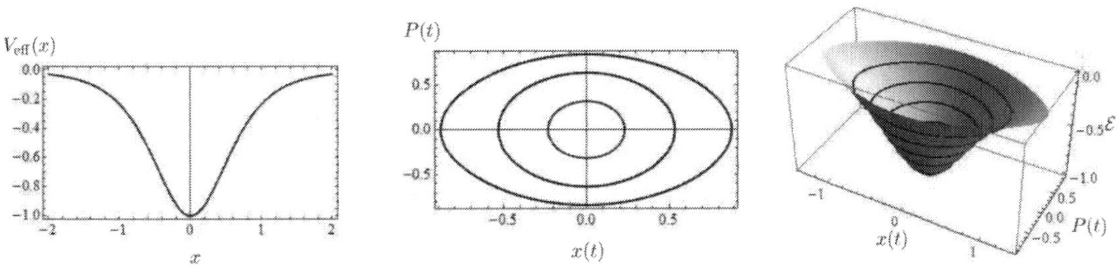

Figure 4. The potential (left), the phase space trajectories (middle) and the 3-D phase space (right) for the position-dependent mass Pöschl-Teller potential with Gaussian mass. In this plots $\alpha = 2$, $m_0 = 1$, $\lambda = 0.5$ and $\epsilon = -1$.

and $\mathbb{R}$ respectively. The phase space trajectories are given by

$$x(t) = \frac{1}{\lambda}\mathrm{Erf}^{-1}\left[\sqrt{\frac{8\lambda^2}{\pi m_0 \alpha^2}}\arcsin\left(\sqrt{\frac{E-\epsilon}{E}}\cos\left(\theta_0 + \alpha\sqrt{E}t\right)\right)\right], \tag{54}$$

$$P(t) = -\sqrt{\frac{2Em_0}{E-(E-\epsilon)\cos^2\left(\theta_0+\alpha\sqrt{E}t\right)}}e^{-\lambda^2 x^2}\sin\left(\theta_0+\alpha\sqrt{E}t\right), \tag{55}$$

for the position-dependent mass Scarf potential and by

$$x(t) = \frac{1}{\lambda}\mathrm{Erf}^{-1}\left[\sqrt{\frac{8\lambda^2}{\pi m_0 \alpha^2}}\mathrm{arcsinh}\left(\sqrt{\frac{E-\epsilon}{-E}}\cos\left(\theta_0 + \alpha\sqrt{-E}t\right)\right)\right], \tag{56}$$

$$P(t) = -\sqrt{\frac{2Em_0}{E+(E-\epsilon)\cos^2\left(\theta_0+\alpha\sqrt{-E}t\right)}}e^{-\lambda^2 x^2}\sin\left(\theta_0+\alpha\sqrt{-E}t\right). \tag{57}$$

for the position-dependent mass Pöschl-Teller potential.

Figures 3 and 4 shows respectively the potential $V_{\rm eff}$, the phase space trajectories $(x(x), P(t))$ and the 3D phase space surface $(x(t), P(t), E)$ for the Scarf and Pöschl-Teller potentials. The description of the behavior of the system is similar to that given in Section *4.1.1*.

Journal of Physics: Conference Series **538** (2014) 012006 doi:10.1088/1742-6596/538/1/012006

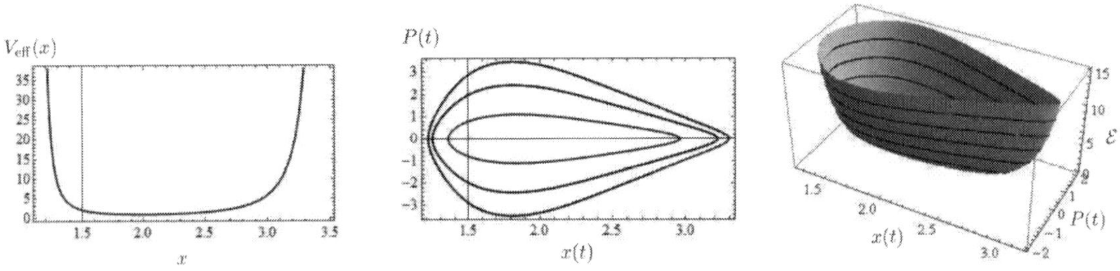

Figure 5. The potential (left), the phase space trajectories (middle) and the 3-D phase space (right) for the position-dependent mass Scarf potential with singular mass. In this plots $\alpha = 2$, $m_0 = 1$, $\lambda = 0.5$ and $\epsilon = 1$.

4.2.2. Singular mass Consider again the mass function (41). The Scarf and Pöschl-Teller position-dependent mass potentials in this case are

$$V_{\rm eff} = \begin{cases} \dfrac{\epsilon}{\cos^2\sqrt{\frac{m_0\alpha^2}{2\lambda^2}}\ln(\lambda x)} & \gamma = 1 \quad H > 0, \\ \dfrac{\epsilon}{\cosh^2\sqrt{\frac{m_0\alpha^2}{2\lambda^2}}\ln(\lambda x)} & \gamma = -1 \quad H < 0, \end{cases} \tag{58}$$

defined in the domains

$$\mathcal{D} = \left\{ x \in \mathbb{R} \middle| -\frac{\pi}{2} < \sqrt{\frac{m_0\alpha^2}{2\lambda^2}}\ln(\lambda x) < \frac{\pi}{2} \right\}, \tag{59}$$

and $(0, \infty)$ respectively. The phase space trajectories are given by

$$x(t) = \frac{1}{\lambda}\exp\left[\sqrt{\frac{2\lambda^2}{m_0\alpha^2}}\arcsin\left(\sqrt{\frac{E-\epsilon}{E}}\cos\left(\theta_0 + \alpha\sqrt{E}t\right)\right)\right], \tag{60}$$

$$P(t) = -\sqrt{\frac{2Em_0}{E-(E-\epsilon)\cos^2\left(\theta_0+\alpha\sqrt{E}t\right)}}\frac{1}{\lambda x(t)}\sin\left(\theta_0 + \alpha\sqrt{E}t\right) \tag{61}$$

for the position-dependent mass Scarf potential and by

$$x(t) = \frac{1}{\lambda}\exp\left[\sqrt{\frac{2\lambda^2}{m_0\alpha^2}}\operatorname{arcsinh}\left(\sqrt{\frac{E-\epsilon}{-E}}\cos\left(\theta_0 + \alpha\sqrt{-E}t\right)\right)\right], \tag{62}$$

$$P(t) - \sqrt{\frac{2Em_0}{E+(E-\epsilon)\cos^2\left(\theta_0+\alpha\sqrt{-E}t\right)}}\frac{1}{\lambda x(t)}\sin\left(\theta_0 + \alpha\sqrt{-E}t\right). \tag{63}$$

for the position-dependent mass Pöschl-Teller potential.

Figures 5 and 6 shows respectively the potential $V_{\rm eff}$, the phase space trajectories $(x(t), P(t))$ and the 3D phase space surface $(x(t), P(t), E)$ for the Scarf and Pöschl-Teller potentials. The description of the behavior of the system is similar to that given in Section *4.1.2*.

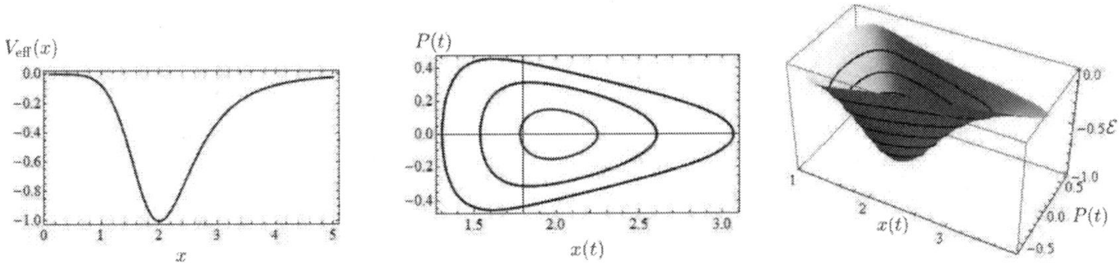

Figure 6. The potential (left), the phase space trajectories (middle) and the 3-D phase space (right) for the position-dependent mass Pöschl-Teller potential. In this plots $\alpha = 2$, $m_0 = 1$, $\lambda = 0.5$ and $\epsilon = -1$.

5. Concluding remarks

The equations of motion for position-dependent mass systems have been considered from the Lagrangian and Hamiltonian approaches. While the Lagrangian and Hamiltonian functions have standard forms, namely

$$\mathcal{L} = \frac{m(x)v^2}{2} + V, \quad \mathcal{H} = \frac{p^2}{2m(x)} + V,$$

a term quadratic in the velocity appears in the dynamical equations as a consequence of the position-dependence of the mass. Since the Hamiltonian is not time invariant, an *energy constant of motion* H was explicitly constructed leading to dynamical equations having the Hamilton form in the phase space $(x(t), P(t))$. From the factorization of this invariant in terms of two functions $A^{\pm}$, and demanding that these functions, together with H close some specific Poisson algebras we have found two integrals of motion $Q^{\pm}$ which allow the construction of the phase space trajectories for bounded motion in different cases. Three kind of potentials were discussed: the harmonic oscillator potentials, connected with the oscillator algebra in terms of Poisson brackets, and the Scarf and Pöschl-Teller potentials related to the $su(1,1)$ and $su(2)$ Poisson algebras respectively. Finally, we have to mention that some other potentials, as singular oscillators, generalized Pöschl-Teller and Morse potentials can be included in this approach for different factorization functions $A^{\pm}$.

Acknowledgments

This work is supported by the IPN Project SIP-20131783. The authors thank the invaluable comments of O Rosas-Ortiz. SCyC thanks the organizers of the Symposium Symmetries in Science XVI for the invitation to participate in the Conference and for the kind hospitality at Bregenz.

References

[1] Wannier G H 1937 *Phys. Rev.* **52** 191
[2] Bastard G 1998 *Wave mechanics applied to semiconductor heterostructures* (Paris: Les Ulis Editions de Physique)
[3] von Roos M 1983 *Phys. Rev.* B **35** 5493
[4] Aquino N, Campoy G and Yee-Madeira H 1998 *Chem. Phys. Lett.* **296** 111
[5] Wolf K B 2004 *Geometric optics on phse space* (Berlin: Springer-Verlag)
[6] Richstone D O and Potter M D 1982 *Astrophys. J* **254** 451
[7] Herling G H and Rustgi M L 1992 *J. Appl. Phys.* **71** 796
[8] Zhao F Q, Liang X X and Ban L 2003 *Eur. Phys. J.* B **33** 3
[9] Yildirim H and Tomak M 2006 *J. Appl. Phys.* **99** 093103
[10] Khordad R 2010 *Physica* E **42** 1503
[11] Mora-Ramos M E, Barseghyan M G and Duque C A 2011 *Phys. Satus Solidi* B **248** 1412

[12] Khordad R 2011 *Physica* B **406** 3911
[13] Ballesteros A, Enciso A, Herranz J F, Ragnisco O and Riglioni D 2011 *Phys. Lett.* A **375** 1431
[14] Cariñena J F, Rañada M and Santander M 2012 *J. Phys. A: Math Theor.* **45** 265303
[15] Cariñena J F, Rañada M and Santander M 2012 *J. Math. Phys.* **53** 102109
[16] Cariñena J F, Perelomov A M, Rañada M and Santander M 2008 *J. Phys. A: Math. Theor.* **41** 085301
[17] Ballesteros A, Enciso A, Herranz J F, Ragnisco O and Riglioni D 2011 *Int. J. Theor. Phys.* **50** 2268
[18] Pesce C P 2003 *J. Appl. Mech.* **70** 751
[19] Lévy-Leblond J M 1995 *Phys. Rev.* A **52** 1845
[20] Alhaidari A D 2002 *Phys. Rev.* A **66** 042116
[21] Milanović V and Ikonić Z 1999 *J. Phys. A: Math. Gen.* **32** 7001
[22] Plastino A R, Rigo A, Casas M, Gracias F and Plastino A 1999 *Phys. Rev.* A **60** 4318
[23] Gönül B, Gönül B, Tutcu D and Özer O 2002 *Mod. Phys. Lett.* A **17** 2057
[24] Bagchi B and Tanaka T 2008 *Phys. Lett.* A **372** 5390
[25] Quesne C 2006 *Ann. Phys.* **321** 1221
[26] Roy B and Roy P 2005 *Phys. Lett.* A **340** 70
[27] Cruz y Cruz S, Negro J and Nieto L M 2007 *Phys. Lett.* A **369** 400
[28] Cruz y Cruz S and Rosas-Ortiz O 2009 *J. Phys. A: Math. Theor.* **42** 185205
[29] Ganguly A, Kuru Ş, Negro J and Nieto L M 2006 *Phys. Lett.* A **360** 228
[30] Mielnik B 1984 *J. Math. Phys.* **25** 3387
[31] Andrianov A A, Borisov N V and Ioffe M V 1984 *Theor. Math. Phys.* **61** 1078
[32] Mielnik B and Rosas-Ortiz O 2004 *J. Phys. A: Math. Gen.* **37** 10007
[33] Cruz y Cruz S and Rosas-Ortiz O 2013 *SIGMA* **9** 004
[34] Cveticanin L 1993 *Trans. ASME J. Appl. Mech.* **60** 954
[35] Plastino A R and Muzzio J C 1992 *Celestial Mech. Dynam. Astronom.* **53** 227
[36] Musielak Z E 2008 *J. Phys. A: Math. Theor.* **41** 055205
[37] Stuckens C and Kobe D H 1986 *Phys. Rev.* A **34** 3565
[38] Bagchi B, Das S, Ghosh S and Poria S 20013 *J. Phys. A: Math. Theor.* **46** 032001
[39] Bagchi B, Das S, Ghosh S and Poria S 20013 *J. Phys. A: Math. Theor.* **46** 368002
[40] Castaños O, Schuch D and Rosas-Ortiz O 2013 *J. Phys. A: Math. Theor.* **46** 075304
[41] Kuru Ş and Negro J 2008 *Ann. Phys.* **323** 413

Symmetries in Science XVI IOP Publishing
Journal of Physics: Conference Series **538** (2014) 012007 doi:10.1088/1742-6596/538/1/012007

Some applications of the Kronecker product in Hubbard representation

Marco Enríquez and Oscar Rosas-Ortiz
Physics Department, Cinvestav, A.P. 14-740, 07000 México D.F., Mexico
E-mail: menriquezf@fis.cinvestav.mx, orosas@fis.cinvestav.mx

Abstract. The properties of the Kronecker product are revisited in terms of Hubbard operators. The simplest representation of a Hubbard operator $X_n^{i,j}$ is a square matrix of size n with an entry equal to 1 and zero elsewhere. This framework simplifies the calculation of the Kronecker product of arbitrary matrices no matter the size or the number of the involved factors. Some applications are presented, these include the algebra of permutation matrices, the Hadamard matrix, the XXX Heisenberg model and the interaction of an atom with radiation fields.

1. Introduction

The Kronecker (tensor or direct) product '$\otimes$' is widely used in several areas of mathematics and theoretical physics [1–9]. This is particularly useful in the description of multipartite quantum systems $\mathcal{S} = \mathcal{S}_1 + \mathcal{S}_2 + \cdots + \mathcal{S}_N$, the pure states of which are represented by vectors in the Hilbert space $\mathcal{H} = \mathcal{H}_1 \otimes \mathcal{H}_2 \otimes \cdots \otimes \mathcal{H}_N$, with $\mathcal{H}_k$ the pure state space of the kth subsystem $\mathcal{S}_k$ [10–12]. For finite-dimensional Hilbert spaces ($\dim \mathcal{H}_k = n_k < \infty$) one has $\dim \mathcal{H} = n_1 \cdots n_N$, so that any operator A_k defined to act on $\mathcal{H}_k$ can be represented by a square matrix of order n_k^2, this can be also promoted to act on the entire space $\mathcal{H}$ as follows

$$\mathbf{A}_k \leftrightarrow \mathbb{I}_1 \otimes \mathbb{I}_2 \otimes \cdots \otimes \mathbb{I}_{k-1} \otimes A_k \otimes \mathbb{I}_{k+1} \otimes \cdots \mathbb{I}_N,$$

where $\mathbb{I}_k$ is the identity operator in $\mathcal{H}_k$. Thus, $\mathbf{A}_k$ must be represented by a square matrix of order $(n_1 \cdots n_N)^2$. As the Kronecker product of two matrices of any order is the block matrix $A \otimes B = [a_{i,j}B]$, we realize that the promoted operator $\mathbf{A}_k$ has at most $n_1 n_2 \cdots n_{k-1} n_k^2 n_{k+1} \cdots n_N < (n_1 \cdots n_N)^2$ entries different from zero. This last fact suggests the looking for simple algorithms to calculate the Kronecker product of operators.

In a recent work we have applied the Hubbard operators as the building-blocks of the Kronecker product $A \otimes B \otimes C \otimes \cdots$ [13]. These are two-indexed operators $X^{i,j}$ fulfilling the properties [14–16]:

i) $X^{i,j} X^{k,m} = \delta_{jk} X^{i,m}$ (multiplication rule)

ii) $\sum_k X^{k,k} = \mathbb{I}$ (completeness)

iii) $(X^{i,j})^\dagger = X^{j,i}$ (non-hermiticity)

iv) $[X^{i,j}, X^{k,m}]_\pm = \delta_{jk} X^{i,m} \pm \delta_{mi} X^{k,j}$ (commutation rules)

Let $\{|\psi_k\rangle, k \in \mathcal{I} \subseteq \mathbb{Z}^+\}$ be an orthonormal basis of the pure state space $\mathcal{H}$ with $|\psi_k\rangle$ the column vector having 1 in the kth row and zero elsewhere. The Hubbard operators $X^{i,j}$ cause the transition from the state $|\psi_j\rangle$ to the state $|\psi_i\rangle$ and can be expressed as the outer products

$$X^{i,j} = |\psi_i\rangle\langle\psi_j|, \quad i,j \in \mathcal{I}.$$

That is, in such a representation the Hubbard operator $X^{i,j}$ corresponds to an square matrix that has all the entries equal to zero except the one at the ith row and jth column, where it takes the value 1. In this context any linear operator $O : \mathcal{H} \to \mathcal{H}$ admits the X-representation

$$O = \sum_{i,j \in \mathcal{I}} o_{i,j} X_n^{i,j}, \quad o_{i,j} = \langle\psi_i|O|\psi_j\rangle.$$

Using the X-representation the algebra of square matrices can be operated in compact form, no matter the size or the number of the factors.

In this contribution we follow the approach introduced in [13] to summarize some of the Kronecker product properties in terms of Hubbard operators. In X-representation the algebraic properties associated to multipartite systems can be analyzed in simple form since even complicated calculations involving large matrices are reduced to simple relations of subscripts. Our interest is to address some applications of this representation in the study of permutation matrices, quantum logic gates, and some interaction models as the Heisenberg or the Jaynes-Cummings ones. With this aim we introduce the notation and basic definitions in Section 2. The properties of the Kronecker product in X-representation are revisited in Section 3 and some of the applications are reported in Section 4. Some final remarks are presented in Section 5.

2. Notation and basic definitions

Let $\mathbb{K}^n = \mathsf{sp}\{|e_i^n\rangle\}_{i=1}^n$ be a vector space of dimension n defined on the field $\mathbb{K}$. The basis vectors $|e_k^n\rangle$ are orthonormal n-tuples with 1 in the position k and zero elsewhere. Any element $|x\rangle \in \mathbb{K}^n$ is represented by a column vector

$$|x\rangle = \sum_{k=1}^n x_k |e_k^n\rangle \equiv (x_1, x_2, \ldots, x_n)^T, \quad x_\ell = \langle e_\ell^n|x\rangle \in \mathbb{K}.$$

The Hermitian transpose of the ket-vector $|x\rangle$ is represented by the bra-vector

$$|x\rangle^\dagger := \langle x| = (x_1^\dagger, x_2^\dagger, \ldots, x_n^\dagger) = \sum_{k=1}^n x_k^\dagger \langle e_k^n|, \quad x_\ell^\dagger = \langle x|e_\ell^n\rangle \in \mathbb{K},$$

with $x_k^\dagger = x_k$ if $\mathbb{K} = \mathbb{R}$ and $x_k^\dagger = \overline{x}_k$ for $\mathbb{K} = \mathbb{C}$. Here $\overline{z}$ stands for the complex conjugate of $z \in \mathbb{C}$. The inner product of two vectors $|x\rangle$ and $|y\rangle$ yields

$$\langle x|y\rangle = \sum_{k,\ell=1}^n x_k^\dagger y_\ell \langle e_k^n|e_\ell^n\rangle = \sum_{k=1}^n x_k^\dagger y_k.$$

Therefore $(\langle x|y\rangle)^\dagger = \langle y|x\rangle$, so that the vector space $\mathbb{K}^n$ is Euclidean (Hermitian) with linear (sesquilinear) metric if $\mathbb{K} = \mathbb{R}$ ($\mathbb{K} = \mathbb{C}$) [17].

The outer product of $|x\rangle$ and $|y\rangle$ yields

$$|x\rangle\langle y| = \sum_{i,j=1}^n x_i y_j^\dagger |e_i^n\rangle\langle e_j^n| \equiv \sum_{i,j=1}^n x_i y_j^\dagger X_n^{i,j},$$

where the "dyadic" matrices

$$X_n^{i,j} = |e_i^n\rangle\langle e_j^n|, \quad i,j = 1,2,\ldots,n \tag{1}$$

are Hubbard operators. Indeed, they satisfy (i) the multiplication rule

$$X_n^{i,j} X_n^{k,\ell} = |e_i^n\rangle\langle e_j^n|e_k^n\rangle\langle e_\ell^n| = \delta_{jk} X_n^{i,\ell}, \tag{2}$$

are (ii) complete

$$\mathbb{I}_n = \sum_{i=1}^n X_n^{i,i}, \tag{3}$$

and (iii) non-hermitian since

$$\left(X_n^{i,j}\right)^T = (|e_i^n\rangle\langle e_j^n|)^T = ((\langle e_j^n|)^T(|e_i^n\rangle))^T = |e_j^n\rangle\langle e_i^n| = X_n^{j,i} \tag{4}$$

implies $\left(X_n^{i,j}\right)^\dagger = \left(X_n^{i,j}\right)^T = X_n^{j,i}$. Finally, they satisfy the (iv) commutation relationships

$$[X_n^{i,j}, X_n^{k,m}]_\pm = X_n^{i,j} X_n^{k,m} \pm X_n^{k,m} X_n^{i,j} = \delta_{jk} X_n^{i,m} \pm \delta_{mi} X_n^{k,j}. \tag{5}$$

The action of these X operators on the vector space $\mathbb{K}^n$ is defined by the rule

$$X_n^{i,j}|e_k^n\rangle = \delta_{jk}|e_i^n\rangle \quad \Rightarrow \quad X_n^{i,j}|x\rangle = x_j|e_i^n\rangle, \tag{6}$$

so that their matrix elements are easily computed

$$\langle e_i^n|X_n^{k,\ell}|e_j^n\rangle = \delta_{\ell j}\langle e_i^n|e_k^n\rangle = \delta_{\ell j}\delta_{ik}. \tag{7}$$

Using the completeness (3) we can write any n-square matrix A as the appropriate linear combination of Hubbard operators

$$A = \sum_{i,j=1}^n a_{i,j} X_n^{i,j}, \quad a_{i,j} \in \mathbb{K}. \tag{8}$$

Therefore the complex conjugate $\overline{A}$, the transpose A^T, and the adjoint $A^\dagger$ of A are computed as follows

$$\overline{A} = \sum_{i,j=1}^n \overline{a}_{i,j} X_n^{i,j}, \qquad A^T = \sum_{i,j=1}^n a_{i,j} X_n^{j,i}, \qquad A^\dagger = \sum_{i,j=1}^n a_{i,j}^\dagger X_n^{j,i}. \tag{9}$$

Considering (8) and (6) one gets

$$A|e_j^n\rangle = \sum_{k=1}^n a_{k,j}|e_k^n\rangle, \tag{10}$$

so that

$$\langle e_i^n|A|e_j^n\rangle = \sum_{k,\ell=1}^n a_{k,\ell}\delta_{i,k}\delta_{\ell j} = a_{i,j} \quad \Rightarrow \quad \mathrm{Tr}A = \sum_{i=1}^n \langle e_i^n|A|e_i^n\rangle = \sum_{i=1}^n a_{i,i}, \tag{11}$$

and

$$A|x\rangle = \sum_{k,j,\ell=1}^n a_{k,j} x_\ell X_n^{k,j}|e_\ell^n\rangle = \sum_{k,\ell=1}^n a_{k,\ell} x_\ell |e_k^n\rangle. \tag{12}$$

Finally, the usual matrix product is closed

$$AB = \left(\sum_{i,j=1}^{n} a_{i,j} X_n^{i,j}\right)\left(\sum_{k,\ell=1}^{n} b_{k,\ell} X_n^{k,\ell}\right) = \sum_{i,\ell=1}^{n}\left(\sum_{k}^{n} a_{i,k} b_{k,\ell}\right) X_n^{i,\ell} = C, \tag{13}$$

with C the n-square matrix

$$C = \sum_{i,\ell=1}^{n} c_{i,\ell} X_n^{i,\ell}, \quad c_{i,\ell} = \sum_{k=1}^{n} a_{i,k} b_{k,\ell}, \tag{14}$$

as expected.

3. The X-representation of the Kronecker algebra

The Kronecker product follows from the next definition

Definition 1. Let $A = [a_{i,j}]$ and $B = [b_{r,s}]$ be respectively matrices of order $m \times n$ and $k \times \ell$ over the field $\mathbb{K}$. The Kronecker product $A \otimes B$ is the matrix of order $mk \times n\ell$ over the field $\mathbb{K}$ defined as $A \otimes B = [a_{i,j} B]$.

Let us consider the vector space $\mathbb{K}^{n_1 n_2}$ on which the matrix $A \otimes B$ acts on. Here A and B are square matrices defined to act on $\mathbb{K}^{n_1}$ and $\mathbb{K}^{n_2}$ respectively. We have a second definition

Definition 2. Let $|e_{i_1}^{n_1}\rangle$ and $|e_{i_2}^{n_2}\rangle$ be basis vectors of $\mathbb{K}^{n_1}$ and $\mathbb{K}^{n_2}$ respectively. The Kronecker product $|e_{i_1}^{n_1}\rangle \otimes |e_{i_2}^{n_2}\rangle$ is the $n_1 n_2$-tuple having 1 at $(i_1 - 1)n_2 + i_2$, and zero in all other entries.

The vector space $\mathbb{K}^{n_1 n_2}$ is built according to the following proposition which is introduced without a proof.

Proposition 1. Let $\mathbb{K}^{n_1} = \mathrm{Sp}\left\{|e_{i_1}^{n_1}\rangle\right\}_{i_1=1}^{n_1}$ and $\mathbb{K}^{n_2} = \mathrm{Sp}\left\{|e_{i_2}^{n_2}\rangle\right\}_{i_2=1}^{n_2}$ be vector spaces defined on the field $\mathbb{K}$. The set of all the Kronecker products $|e_{i_1}^{n_1}\rangle \otimes |e_{i_2}^{n_2}\rangle$ is orthonormal and spans a vector space of dimension $n_1 n_2$, written $\mathbb{K}^{n_1 n_2} = \mathrm{Sp}\{|e_{i_1}^{n_1}\rangle \otimes |e_{i_2}^{n_2}\rangle, i_1 = 1, \ldots, n_1; i_2 = 1, \ldots, n_2\}$, with the following axioms ($\alpha, \beta, \gamma, \eta$ are elements of $\mathbb{K}$):

(i) $(\alpha|e_{i_1}^{n_1}\rangle) \otimes |e_{i_2}^{n_2}\rangle = \alpha(|e_{i_1}^{n_1}\rangle \otimes |e_{i_2}^{n_2}\rangle) = |e_{i_1}^{n_1}\rangle \otimes (\alpha|e_{i_2}^{n_2}\rangle)$

(ii) $(\alpha|e_{i_1}^{n_1}\rangle + \beta|e_{i_1}^{n_1}\rangle) \otimes |e_{i_2}^{n_2}\rangle = \alpha(|e_{i_1}^{n_1}\rangle \otimes |e_{i_2}^{n_2}\rangle) + \beta(|e_{i_1}^{n_2}\rangle \otimes |e_{i_2}^{n_2}\rangle)$

(iii) $|e_{i_1}^{n_1}\rangle \otimes (\gamma|e_{i_2}^{n_2}\rangle + \eta|e_{i_2}^{n_2}\rangle) = \gamma(|e_{i_1}^{n_1}\rangle \otimes |e_{i_2}^{n_2}\rangle) + \eta(|e_{i_1}^{n_1}\rangle \otimes |e_{i_2}^{n_2}\rangle)$

In general one has two different forms to write the vector $|x\rangle \in \mathbb{K}^{n_1 n_2}$; this can be done by using either two independent indices i_1, i_2,

$$|x\rangle = \sum_{i_1=1}^{n_1}\sum_{i_2=1}^{n_2} x_{i_1,i_2} |e_{i_1}^{n_1}\rangle \otimes |e_{i_2}^{n_2}\rangle,$$

or a single index k,

$$|x\rangle = \sum_{k=1}^{n_1 n_2} \widetilde{x}_k |e_k^{n_1 n_2}\rangle, \quad k = (i_1 - 1)n_2 + i_2.$$

Notice that the latter expression considers the construction of the Kronecker products $|e_{k=(i_1-1)n_2+i_2}^{n_1 n_2}\rangle = |e_{i_1}^{n_1}\rangle \otimes |e_{i_2}^{n_2}\rangle$ as this is indicated in Definition 2. In correspondence, there are $n_1 n_2$ different coefficients x_{i_1,i_2} that can be mapped into the $\widetilde{x}_k$ ones and vice versa. The generalization of Definition 1 and Proposition 1 to the case of N factors is straightforward. In that case the state space $\mathbb{K}^{n_1 n_2 \ldots n_N}$ is spanned by the basis vectors $|e_{k'}^{n_1 n_2 \ldots n_N}\rangle = |e_{i_1}^{n_1}\rangle \otimes |e_{i_2}^{n_2}\rangle \otimes \cdots \otimes |e_{i_N}^{n_N}\rangle$, with k' properly defined.

Proposition 2. Let $X_m^{i,j}$ and $X_n^{k,\ell}$ be two Hubbard operators of order n and m respectively. Then the $\otimes$-product

$$X_m^{i,j} \otimes X_n^{k,\ell} = X_{mn}^{n(i-1)+k,n(j-1)+\ell}, \tag{15}$$

is a Hubbard operator of order mn.

Proof. The proof follows from Definition 1, explicitly

$$\begin{aligned}
&X_m^{i,j} \otimes X_n^{k,\ell} \\
&= \begin{pmatrix} & & & 0 & & & \\ & 0_{(i-1)\times(j-1)} & & \vdots & & 0_{(i-1)\times(n-j)} & \\ & & & 0 & & & \\ 0 & \cdots & 0 & 1 & 0 & \cdots & 0 \\ & & & 0 & & & \\ & 0_{(n-i)\times(j-1)} & & \vdots & & 0_{(n-i)\times(n-j)} & \\ & & & 0 & & & \end{pmatrix} \otimes X_n^{k,\ell} \\
&= \begin{pmatrix} & & & 0 & & & \\ & 0_{[n(i-1)+k-1]\times[n(j-1)+\ell-1]} & & \vdots & & 0_{[n(i-1)+k-1]\times[mn-n(j-1)-\ell]} & \\ & & & 0 & & & \\ 0 & \cdots & 0 & 1 & 0 & \cdots & 0 \\ & & & 0 & & & \\ & 0_{[mn-n(i-1)-k]\times[n(j-1)+\ell-1]} & & \vdots & & 0_{[mn-n(i-1)-k]\times[n(j-1)+\ell-1]} & \\ & & & 0 & & & \end{pmatrix} \\
&= X_{mn}^{n(i-1)+k,n(j-1)+\ell}. \quad \square
\end{aligned}$$

This last result shows that the tensor product is closed on the set of Hubbard operators. The following properties are also fulfilled

Proposition 3. Let $X_\alpha^{\beta,\gamma}$ be Hubbard operators of order α and take $\lambda \in \mathbb{K}$. Then

i) $X_m^{i,j} \otimes X_n^{k,\ell} \neq X_n^{k,\ell} \otimes X_m^{i,j}$ in general.

ii) $\left(X_m^{i,j} \otimes X_n^{k,\ell}\right)^T = \left(X_m^{i,j}\right)^T \otimes \left(X_n^{k,\ell}\right)^T$.

iii) $\left(\lambda X_m^{i,j}\right) \otimes X_n^{k,\ell} = \lambda\left(X_m^{i,j} \otimes X_n^{k,\ell}\right) = X_m^{i,j} \otimes \left(\lambda X_n^{k,\ell}\right)$.

iv) $\left(X_m^{i,j} + X_m^{r,s}\right) \otimes X_n^{k,\ell} = X_m^{i,j} \otimes X_n^{k,\ell} + X_m^{r,s} \otimes X_n^{k,\ell}$.

v) $X_n^{k,\ell} \otimes \left(X_m^{i,j} + X_m^{r,s}\right) = X_n^{k,\ell} \otimes X_m^{i,j} + X_n^{k,\ell} \otimes X_m^{r,s}$

vi) $\left(X_m^{i,j} \otimes X_n^{k,\ell}\right) \otimes X_p^{r,s} = X_m^{i,j} \otimes \left(X_n^{k,\ell} \otimes X_p^{r,s}\right)$.

Proof. See Ref. [13], pp. 230.

Thus, the product $\otimes$ is distributive over ordinary matrix addition (iv), (v), associative (vi), compatible with ordinary matrix transposition (ii) as well as with matrix multiplication by an scalar (iii) and, in general, non-abelian (i). We can use now the X-representation of n-square matrices (8), and the properties included in the Propositions 1–3, to define the $\otimes$-product of an arbitrary number of operators A_k in terms of simple relations of subscripts. We have the next proposition.

Proposition 4. Let $A_r = \left[a_{i,j}^{(r)}\right]$ be a square matrix of order n_r. The Kronecker product $A_1 \otimes A_2 \otimes \cdots \otimes A_{k+1}$ is the square matrix of order $n^{(k)} = n_1 n_2 \cdots n_{k+1}$, expressed as the following linear combination of Hubbard operators

$$A = A_1 \otimes A_2 \otimes \cdots \otimes A_{k+1} = \sum_{p,q=1}^{n^{(k)}} \widetilde{a}_{p,q}^{\,(k)}\ X_{n^{(k)}}^{p,q}, \quad k \geq 1, \tag{16}$$

where

$$\widetilde{a}_{p,q}^{\,(k)} = a_{p_k,q_k}^{(1)} \prod_{s=0}^{k-1} a_{p_s+n_{k-s+1}-n_{k-s+1}p_{s+1},\ q_s+n_{k-s+1}-n_{k-s+1}q_{s+1}}^{(k-s+1)}, \tag{17}$$

and

$$p_s = \left\lceil \frac{p}{\prod_{\ell=1}^{s} n_{k-\ell+2}} \right\rceil, \qquad q_s = \left\lceil \frac{q}{\prod_{\ell=1}^{s} n_{k-\ell+2}} \right\rceil. \tag{18}$$

Here

$$\lceil x \rceil = \min\{z \in \mathbb{Z} : x \leq z\} \tag{19}$$

is the ceiling function of $x \in \mathbb{R}$ (this yields the smallest integer greater than or equal to x).

Proof. See Ref. [13], pp. 238-39.

The results reported in the previous propositions allow to recover all the known properties of the Kronecker product of square matrices. Most of them are straightforwardly extended to matrices of arbitrary order though some caution is necessary (see the discussion on the matter in Section 2.2 of [13]). In this context we present the next theorems and propositions without a proof (the proofs in X-representation can be found in our review [13], for conventional approaches see e.g. [7–9]).

Theorem 1. Let A, B and C be n-square matrices and $\lambda \in \mathbb{K}$. Then

i) In general, $A \otimes B \neq B \otimes A$.
ii) $(A \otimes B)^\dagger = A^\dagger \otimes B^\dagger$ and $(A \otimes B)^T = A^T \otimes B^T$.
iii) $(\lambda A) \otimes B = \lambda(A \otimes B) = A \otimes (\lambda B)$.
iv) $(A + B) \otimes C = A \otimes C + B \otimes C$.
v) $A \otimes (B + C) = A \otimes B + A \otimes C$.
vi) $(A \otimes B) \otimes C = A \otimes (B \otimes C)$.

Theorem 2. Let $A = [a_{i,j}], C = [c_{p,q}]$, and $B = [b_{k,\ell}], D = [d_{r,s}]$, be pairs of n and m-square matrices respectively. The usual matrix product of the nm-square matrices $A \otimes B$ and $C \otimes D$ fulfills

$$(A \otimes B)(C \otimes D) = AC \otimes BD. \tag{20}$$

That is, the Kronecker product of square matrices is compatible with ordinary matrix multiplication.

Proposition 5. Let $A = [a_{i,j}]$ and $B = [b_{k,\ell}]$ be two square matrices of order n and m respectively. Then

$$\mathrm{Tr}(A \otimes B) = \mathrm{Tr}(A)\mathrm{Tr}(B) = \mathrm{Tr}(B \otimes A).$$

Proposition 6. Let $A = [a_{i,j}]$ and $B = [b_{k,\ell}]$ be two n-square matrices. Then

$$\mathrm{Det}(A \otimes B) = (\mathrm{Det}\, A)^n (\mathrm{Det}\, B)^n. \tag{21}$$

Additional properties (and proofs) of the Kronecker product of matrices can be found in the books [5–7]. The most recent summary of the properties of the $\otimes$ operation has been reported in [9] (see also [8]).

4. Applications

4.1. Permutation matrices

As a first example consider a permutation defined by the bijection π of the set of natural numbers $S = \{1, \ldots, n\}$ onto itself. In the Cauchy's two-line notation this map reads as

$$\pi = \begin{pmatrix} 1 & 2 & \cdots & n \\ \pi(1) & \pi(2) & \cdots & \pi(n) \end{pmatrix}.$$

The set of all $n!$ permutations of S forms the symmetric (or permutation) group S_n. The identity element π_e is defined as $\pi_e(k) = k$ for all k in S. A linear representation of S_n is obtained by assigning a matrix P_π per each permutation π; this is a square matrix of order n that has only one entry 1 per row and column, and is zero elsewhere. In X-representation we have

$$P_\pi = \sum_{j=1}^{n} X_n^{j,\pi(j)}. \tag{22}$$

Then the product of P_σ and P_π, two permutation matrices of order n, yields

$$P_\sigma P_\pi = \sum_{k,\ell=1}^{n} X_n^{k,\sigma(k)} X_n^{\ell,\pi(\ell)} = \sum_{k,\ell=1}^{n} \delta_{\sigma(k),\ell} X_n^{k,\pi(\ell)} = \sum_{k=1}^{n} X_n^{k,\pi(\sigma(k))} = P_{\pi\circ\sigma}. \tag{23}$$

This shows that the composition $\pi \circ \sigma$ of two permutations π and σ is obtained from the product of the corresponding matrices. It also follows that in general the product of permutation matrices is non-commutative as $P_\pi P_\sigma = P_{\sigma\circ\pi}$ and $\sigma \circ \pi \neq \pi \circ \sigma$. The inverse of P_π is the matrix P_π^T, indeed

$$P_\pi P_\pi^T = \sum_{k,\ell=1}^{n} \delta_{\pi(k),\pi(\ell)} X_n^{k,\ell} = \sum_{k=1}^{n} X_n^{k,k} = \mathbb{I}_n. \tag{24}$$

In similar form, $P_\pi^T P_\pi = \mathbb{I}_n$, so that $(P_\sigma P_\pi)^{-1} = P_\pi^{-1} P_\sigma^{-1}$. On the other hand, the action of a permutation matrix P_π on a vector $|x\rangle \in \mathbb{K}^n$ reads

$$P_\pi|x\rangle = \sum_{j,k=1}^{n} x_k X_n^{j,\pi(j)} |e_k^n\rangle = \sum_{j,k=1}^{n} x_k \delta_{\pi(j),k} |e_j^n\rangle = \sum_{j=1}^{n} x_{\pi(j)} |e_j^n\rangle. \tag{25}$$

Now we consider the application of the Kronecker product to construct permutation matrices.

Theorem 3. The square matrix

$$\Pi = \sum_{i,j=1}^{n} X_n^{i,j} \otimes X_n^{j,i} \tag{26}$$

is a permutation matrix of order n^2, defined by the rule

$$\pi(p) = n(p + n - 1) - (n^2 - 1)p', \qquad p = 1, 2, \ldots, n^2, \tag{27}$$

with

$$p' = \left\lceil \frac{p}{n} \right\rceil \tag{28}$$

the ceiling function applied on $\frac{p}{n}$.

Proof. See Ref. [13], pp 231.

To get some insight on the meaning of the permutation matrix (26) let us consider an arbitrarycontravariant tensor of rank 2:

$$|x_1\rangle \otimes |x_2\rangle = \sum_{i_1,i_2=1}^{n} x_{i_1} x_{i_2} |e^n_{i_1}\rangle \otimes |e^n_{i_2}\rangle = \sum_{i_1,i_2=1}^{n} x_{i_1} x_{i_2} |e^{n^2}_{(i_1-1)n+i_2}\rangle. \tag{29}$$

The action of Π on this last vector reads $\Pi(|x_1\rangle \otimes |x_2\rangle) = |x_2\rangle \otimes |x_1\rangle$. Thus, relative to the indices labeling the contravariant tensor (29), the operator Π corresponds to the bijection $\pi_2 : (1,2) \to (2,1)$. Hence $\Pi \equiv P_{\pi_2} \in S_2$. Indeed, there are only $2! = 2$ different permutations on the set $\{1,2\}$, these are the identity $\pi_e \equiv \pi_1$ and π_2. In Hubbard representation we have

$$P_{\pi_1} = \sum_{i=1}^{n} X_n^{i,i} \otimes X_n^{i,i} \equiv \sum_{i,j=1}^{n} \delta_{ij} X_n^{i,j} \otimes X_n^{j,i}, \quad P_{\pi_2} = \Pi. \tag{30}$$

Therefore one arrives at the symmetrization operator

$$\mathcal{S}_{(2)} = \tfrac{1}{2}(P_{\pi_1} + P_{\pi_2}) = \frac{1}{2} \sum_{i,j=1}^{n} (1 + \delta_{ij}) X_n^{i,j} \otimes X_n^{j,i}, \tag{31}$$

together with the anti-symmetrization operator

$$\mathcal{A}_{(2)} = \tfrac{1}{2}\left[\chi(\pi_1) P_{\pi_1} + \chi(\pi_2) P_{\pi_2}\right] = \frac{1}{2} \sum_{i,j=1}^{n} \left[\chi(\pi_1) + \chi(\pi_2)\delta_{ij}\right] X_n^{i,j} \otimes X_n^{j,i}, \tag{32}$$

where $\chi(\pi)$ is the parity of the bijection π [17]. The generalization of the above results to tensors of arbitrary rank is straightforward:

Proposition 7. The operators

$$\mathcal{S}_{(p)} = \frac{1}{p!} \sum_{\ell=1}^{p} P_{\pi_\ell} \qquad \text{and} \qquad \mathcal{A}_{(p)} = \frac{1}{p!} \sum_{\ell=1}^{p} \chi(\pi_\ell) P_{\pi_\ell} \tag{33}$$

with $\pi_\ell \in S_p$, $\pi_1 \equiv \pi_e$, and P_{π_ℓ} a definite linear combination of the Kronecker products

$$X_n^{i_1,j_1} \otimes X_n^{i_2,j_2} \otimes \cdots X_n^{i_p,j_p}, \quad i_k, j_k \in \{1, \ldots, n\},$$

produce respectively the symmetrization and anti-symmetrization of the contravariant tensors of rank p.

The Kronecker product of permutation matrices is also compatible with the composition of permutations.

Theorem 4. Let $P_\pi(n)$ and $P_\sigma(m)$ be the n and m-permutation matrices defined by the rules π and σ respectively. The Kronecker product $P_\pi(n) \otimes P_\sigma(m)$ is the nm-permutation matrix $P_\alpha(n,m)$ defined by the rule

$$\alpha(p) = m[\pi(p') - 1] + \sigma(p - mp' + m), \tag{34}$$

with $p' = \lceil \frac{p}{m} \rceil$.

Proof. See Ref. [13], pp 233.

Although the product $X_n^{i,j} \otimes X_m^{k,\ell}$ is non-abelian in general (Proposition 3i), it is possible to arrive at $X_m^{k,\ell} \otimes X_n^{i,j}$ by applying the appropriate permutation of rows and columns in $X_n^{i,j} \otimes X_m^{k,\ell}$. That is, these last operators must be permutation equivalent.

Proposition 8. The Kronecker product $X_n^{i,j} \otimes X_m^{k,\ell}$ is permutation equivalent to $X_m^{k,\ell} \otimes X_n^{i,j}$. That is, there exist P_π, a permutation matrix of order nm, such that

$$P_\pi^T \left(X_n^{i,j} \otimes X_m^{k,\ell} \right) P_\pi = X_m^{k,\ell} \otimes X_n^{i,j}. \tag{35}$$

Proof. See Ref. [13], pp 233.

In general, for any square matrices A and B we have

Theorem 5. Let $A = [a_{i,j}]$ and $B = [b_{k,\ell}]$ be two square matrices of order n and m respectively. The Kronecker product $A \otimes B$ is permutation equivalent to $B \otimes A$.

Proof. From Proposition 8 we know that there exists a permutation matrix P such that (35) is true. Then, by linearity in the conventional matrix product we have

$$\begin{aligned} P^T (A \otimes B) P &= \sum_{i,j}^{n} \sum_{k,\ell}^{m} a_{i,j} b_{k,\ell} \left[P^T \left(X_n^{i,j} \otimes X_m^{k,\ell} \right) P \right] \\ &= \sum_{i,j}^{n} \sum_{k,\ell}^{m} b_{k,\ell} a_{i,j} X_m^{k,\ell} \otimes X_n^{i,j} = B \otimes A. \quad \square \end{aligned}$$

Permutation matrices play an important role in combinatorics and quantum mechanics (see e.g. Ch. 13 of Ref. [18] and Ref. [17]). For example, the Schrödinger equation is invariant under the permutation of electrons since the physical equivalence of all these particles.

4.2. The Haddamard matrix

The n-square matrices with ± 1 entries and having pairwise orthogonal rows are named after Hadamard [19]. In the simplest case ($n = 2$) one has

$$H = \frac{1}{\sqrt{2}} \sum_{i,j=1}^{2} (-1)^{(i-1)(j-1)} X_2^{i,j} = \frac{1}{\sqrt{2}} \begin{pmatrix} 1 & 1 \\ 1 & -1 \end{pmatrix}. \tag{36}$$

This operator is unitary and its action on a vector $|x\rangle \in \mathbb{K}^2$ reads

$$H|x\rangle = \frac{1}{\sqrt{2}} \sum_{i,k=1}^{2} (-1)^{(i-1)(k-1)} x_k |e_i^2\rangle = \frac{1}{\sqrt{2}} \left[(x_1 + x_2)|e_1^2\rangle + (x_1 - x_2)|e_2^2\rangle \right]. \tag{37}$$

The multiplication of H with itself yields

$$H^2 = HH = \sum_{i,\ell=1}^{2} c_{i,\ell} X_2^{i,\ell} = \mathbb{I}_2, \qquad c_{i,\ell} = \frac{1 + (-1)^{i+\ell}}{2} = \delta_{i,\ell}. \tag{38}$$

The Kronecker powers of Hadamard matrices $H^{\otimes k+1}$ are of interest in quantum computing algorithms. Let us give a concrete realization of such powers in terms of Hubbard operators

Proposition 9. Let H be the Hadamard matrix (36), then

$$H^{\otimes k+1} = \frac{1}{\sqrt{2^{k+1}}} \sum_{p,q=1}^{2^{k+1}} (-1)^{\vec{p}\cdot\vec{q}} X^{p,q}_{2^{k+1}}, \quad k \geq 1, \tag{39}$$

with $p_s = \lceil \frac{p}{2^s} \rceil$, $q_s = \lceil \frac{q}{2^s} \rceil$, and

$$\vec{p}\cdot\vec{q} := \sum_{s=0}^{k} (p_s - 1)(q_s - 1). \tag{40}$$

Proof. See Ref. [13], pp 236.

The action of $H^{\otimes k+1}$ on the basis vectors of $\mathbb{K}^{2^{k+1}}$ reads

$$H^{\otimes k+1} |e_j^{2^{k+1}}\rangle = \frac{1}{\sqrt{2^{k+1}}} \sum_{p=1}^{2^{k+1}} (-1)^{\sum_{s=0}^{k}(p_s-1)(j_s-1)} |e_p^{2^{k+1}}\rangle, \tag{41}$$

with $y_s = \lceil \frac{y}{2^s} \rceil$ for $y = p, j$. As an example consider the case $k = 1$ for which we have $H_4 = H \otimes H = H^{\otimes 2}$. Explicitly,

$$H_4 = H^{\otimes 2} = \frac{1}{2} \sum_{p,q=1}^{4} (-1)^{(p-1)(q-1)+(\lceil \frac{p}{2} \rceil -1)(\lceil \frac{q}{2} \rceil -1)} X_4^{p,q} = \frac{1}{2} \begin{pmatrix} 1 & 1 & 1 & 1 \\ 1 & -1 & 1 & -1 \\ 1 & 1 & -1 & -1 \\ 1 & -1 & -1 & 1 \end{pmatrix}.$$

Hence,

$$2H^{\otimes 2}|e_j^4\rangle \to \begin{cases} |e_1^4\rangle + |e_2^4\rangle + |e_3^4\rangle + |e_4^4\rangle, & j = 1 \\ |e_1^4\rangle - |e_2^4\rangle + |e_3^4\rangle - |e_4^4\rangle, & j = 2 \\ |e_1^4\rangle + |e_2^4\rangle - |e_3^4\rangle - |e_4^4\rangle, & j = 3 \\ |e_1^4\rangle - |e_2^4\rangle - |e_3^4\rangle + |e_4^4\rangle, & j = 4 \end{cases}$$

In quantum computing it is usual to express the basis vectors $|e_k^4\rangle$ in the binary form

$$|00\rangle \equiv |e_1^4\rangle, \quad |01\rangle = |e_2^4\rangle, \quad |10\rangle = |e_3^4\rangle, \quad |11\rangle = |e_4^4\rangle.$$

To translate the above results into the binary form, we present the following definition

Definition 3 Consider a positive integer $x \leq 2^{k+1}$ with $k \in \mathbb{N}$. The expansion of x in powers of 2 is defined by the binary coefficients $x_s \in \{0, 1\}$, $s = 0, 1, \ldots, k+1$, as follows

$$x = \sum_{i=0}^{k+1} x_i 2^i. \tag{42}$$

The coefficients of the linear combination (39) can be written in binary form according to the following proposition

Proposition 10 Let p and q be respectively the i-th and j-th powers of 2 with $i, j = 0, 1, \ldots, k+1$, and $k \in \mathbb{N}$. Then

$$(-1)^{\sum_{s=0}^{k}(\lceil \frac{p}{2^s} \rceil -1)(\lceil \frac{q}{2^s} \rceil -1)} = (-1)^{\sum_{s=0}^{k}(p-1)_s (q-1)_s}, \tag{43}$$

where $(p-1)_s$ and $(q-1)_s$ are the s-th binary coefficients of $p-1$ and $q-1$ respectively.

Proof. See Ref. [13], pp 237.

We finally arrive at

$$2H^{\otimes 2}|e_j^4\rangle \to \begin{cases} (-1)^{0\cdot 0+0\cdot 0}|e_1^4\rangle + (-1)^{1\cdot 0+0\cdot 0}|e_2^4\rangle + (-1)^{0\cdot 0+0\cdot 1}|e_3^4\rangle + (-1)^{1\cdot 0+1\cdot 0}|e_4^4\rangle, & j=1 \\ (-1)^{0\cdot 1+0\cdot 0}|e_1^4\rangle + (-1)^{1\cdot 1+0\cdot 0}|e_2^4\rangle + (-1)^{0\cdot 1+1\cdot 0}|e_3^4\rangle + (-1)^{1\cdot 1+1\cdot 0}|e_4^4\rangle, & j=2 \\ (-1)^{0\cdot 0+0\cdot 1}|e_1^4\rangle + (-1)^{1\cdot 0+0\cdot 1}|e_2^4\rangle + (-1)^{0\cdot 0+1\cdot 1}|e_3^4\rangle + (-1)^{0\cdot 1+1\cdot 1}|e_4^4\rangle, & j=3 \\ (-1)^{0\cdot 1+0\cdot 1}|e_1^4\rangle + (-1)^{1\cdot 1+0\cdot 1}|e_2^4\rangle + (-1)^{0\cdot 1+1\cdot 1}|e_3^4\rangle + (-1)^{1\cdot 1+1\cdot 1}|e_4^4\rangle, & j=4 \end{cases}$$

On the other hand, in binary form the basis vectors are given by the rule

$$|e_j^{2^{k+1}}\rangle \to |j-1\rangle_{(k)} := |(j-1)_0, (j-1)_1, \dots, (j-1)_k\rangle,$$

with $(j-1)_s \in \{0,1\}$ the binary coefficients of $j-1$ up to 2^k. Finally, given a vector $|x\rangle$ written in binary notation, the equation (41) reads

$$H^{\otimes n}|x\rangle = \frac{1}{\sqrt{2^n}} \sum_z (-1)^{x\cdot z}|z\rangle,$$

where $|z\rangle$ is also in binary form and $n = 2^{k+1}$.

4.3. The Heisenberg XXX model

In this section we analyze the diagonalization of n-level Hamiltonians of the form

$$H = \sum_{p=1}^{n} \varepsilon_p \, X_n^{p,p} + \sum_{\substack{p,q=1 \\ p\neq q}} V_{p,q} \, X_n^{p,q}, \qquad V_{q,p} = \overline{V}_{p,q}. \tag{44}$$

We use the unitary transformation method introduced in reference [20] (see also [21]). This method is iterative and demands the off-diagonal elements of the transformed Hamiltonian

$$H' = U_{k,m}(\alpha) H U_{k,m}^{\dagger}(\alpha) = \sum_{p=1}^{n} \varepsilon'_p \, X_n^{p,p} + \sum_{\substack{p,q=1 \\ p\neq q}} V'_{p,q} \, X_n^{p,q}, \tag{45}$$

be equal to zero. Here the set of operators

$$U_{k,m}(\alpha) = \exp\left(\alpha X_n^{k,m} - \overline{\alpha} X_n^{m,k}\right), \quad m > k = 1, 2, \dots, n, \quad \alpha = |\alpha| e^{i\mu}, \tag{46}$$

are unitary and α is a complex parameter. Moreover, the coefficients in the transformed Hamiltonian (45) are given by

$$\begin{aligned} \varepsilon'_k &= \tfrac{1}{2}\left[\varepsilon_k + \varepsilon_m + (\varepsilon_k - \varepsilon_m)\cos 2|\alpha| + 2\Re(V_{k,m} e^{-i\mu})\sin 2|\alpha|\right], \\ \varepsilon'_m &= \tfrac{1}{2}\left[\varepsilon_k + \varepsilon_m - (\varepsilon_k - \varepsilon_m)\cos 2|\alpha| - 2\Re(V_{k,m} e^{-i\mu})\sin 2|\alpha|\right], \\ V'_{k,m} e^{i\mu} &= \tfrac{1}{2}\left[\tfrac{1}{2}(\varepsilon_m - \varepsilon_k)\sin 2|\alpha| + V_{k,m} e^{-i\mu}\cos^2|\alpha| - \overline{V}_{k,m} e^{i\mu}\sin^2|\alpha|\right], \\ V'_{k,p} &= V_{k,p}\cos|\alpha| + V_{m,p} e^{i\mu}\sin|\alpha|, \\ V'_{m,p} &= V_{p,m}\cos|\alpha| - V_{p,k} e^{i\mu}\sin|\alpha|, \\ \varepsilon'_p &= \varepsilon_p, \qquad V'_{p,q} = V_{p,q} \qquad p,q \neq k,m. \end{aligned} \tag{47}$$

The condition $V'_{k,m} = 0$ produces

$$\tan 2|\alpha| = \frac{2(-1)^{\kappa+1}|V_{k,m}|}{\varepsilon_m - \varepsilon_k}. \tag{48}$$

Hence

$$\varepsilon'_k = \frac{1}{2} - \sqrt{\frac{1}{4}(\varepsilon_m - \varepsilon_k)^2 + |V_{k,m}|^2}, \quad \varepsilon'_m = \frac{1}{2} + \sqrt{\frac{1}{4}(\varepsilon_m - \varepsilon_k)^2 + |V_{k,m}|^2}. \tag{49}$$

The Hamiltonian (44) is diagonalized by iterating the above procedure as many times as necessary. Now we consider a system of n spin-1/2 particles that interact according to the Hamiltonian

$$H = -\frac{1}{2}\sum_{j=1}^{n}\left(J_x\sigma_j^x\sigma_{j+1}^x + J_y\sigma_j^y\sigma_{j+1}^y + J_z\sigma_j^z\sigma_{j+1}^z\right), \tag{50}$$

$$\sigma_j^k = \underbrace{\mathbb{I}_2 \otimes \ldots \otimes \mathbb{I}_2}_{j-1} \otimes \sigma^k \otimes \mathbb{I}_2 \otimes \ldots \otimes \mathbb{I}_2.$$

Here $\sigma_{n+1}^k := \sigma_1^k$, with σ^k, $k = x, y, z$, standing for the Pauli matrices. Such a Hamiltonian is known as the Heisenberg Hamiltonian [21]. We will analyze the case when $J_x = J_y = J_z = J$ (the XXX-model) and $n = 2$

$$H = -\frac{J}{2}\left(\sigma^x \otimes \sigma^x + \sigma^y \otimes \sigma^y + \sigma^z \otimes \sigma^z\right). \tag{51}$$

In Hubbard representation, the last Hamiltonian reads

$$H = -\frac{J}{4}\left(X_4^{1,1} - X_4^{2,2} - X_4^{3,3} + X_4^{4,4}\right) - 2J\left(X_4^{2,3} + X_4^{3,2}\right). \tag{52}$$

To diagonalize it we make only a single transformation $U_{2,3}(\alpha)$ and from (47) we get $\cos(2|\alpha|) = 0$, $\sin\mu = 0$, and the system

$$\begin{aligned} \varepsilon'_1 &= \varepsilon_1 = -\tfrac{J}{4}, \quad \varepsilon'_2 = \tfrac{1}{2}(\varepsilon_2 + \varepsilon_3) - V_{2,3} = \tfrac{9}{4}J, \\ \varepsilon'_3 &= \tfrac{1}{2}(\varepsilon_2 + \varepsilon_3) + V_{2,3} = -\tfrac{7}{4}J, \quad \varepsilon'_4 = \varepsilon_4 = -\tfrac{J}{4}. \end{aligned} \tag{53}$$

Therefore we arrive at the diagonal Hamiltonian

$$\mathcal{H}' = -\frac{J}{4}\left(X_4^{1,1} - 9X_4^{2,2} + 7X_4^{3,3} + X_4^{4,4}\right). \tag{54}$$

4.4. Atom + Field models of interaction

In this section we consider the problem of the interaction between a two-level atom (qubit) and a radiation field from both the semiclassical and quantized approaches. In both cases, the Hubbard operators are useful in the construction of the time evolution operator.

4.4.1. Semiclassical model. The ground and excited states of a qubit will be denoted as $|-\rangle$ and $|+\rangle$, respectively. The corresponding Hamiltonian is given by

$$H_0 = \hbar\frac{\omega_a}{2}\sigma_3, \tag{55}$$

here ω_a is the transition frequency. On the other hand, the interaction Hamiltonian is given through the dipolar approximation

$$H_I = -\vec{p} \cdot \vec{E}(t). \tag{56}$$

Here $\vec{p}$ is the dipolar momentum of the atom and $\vec{E}(t)$ is a classical radiation field of the form

$$\vec{E}(t) = \mathcal{E}\left(e^{-i\omega_f t} + e^{i\omega_f t}\right)\hat{e}_f, \tag{57}$$

where $\mathcal{E}$ is the amplitude, ω_f the frequency and $\hat{e}_f$ the polarization vector. Thus, the total Hamiltonian is therefore $H = H_0 + H_I$. In the rotating wave approximation this Hamiltonian H can be written in a free-units system as [22]:

$$H = \frac{\Delta}{2}\sigma_3 + g(\sigma_+ + \sigma_-), \qquad \Delta = 1 - \omega_f/\omega_a, \tag{58}$$

with Δ the detuning and g the coupling constant. In Hubbard notation the previous Hamiltonian is written as follows

$$H = H_0 + H_I, \qquad H_0 = \frac{\Delta}{2}\sum_{p=1}^{2} X_2^{p,p}, \quad H_I = g\sum_{p=1}^{2} X_2^{p,3-p}. \tag{59}$$

In the case of exact resonance, the time evolution operator is given by

$$U(t) = e^{iH_I t} = \sum_{p,q=1}^{2} u_{p,q}(t) X_2^{p,q}, \qquad u_{p,q}(t) = e^{i\frac{\pi}{2}|p-q|}\cos\left(gt - \frac{\pi}{2}|p-q|\right) \tag{60}$$

The last expression (60) is useful in considering the evolution of two non-interacting qubits. The time evolution in this case would read

$$U = U_1(t) \otimes U_2(t), \tag{61}$$

where each one of the $U_i(t)$ are written in terms of the X-operators

$$U_i(t) = \sum_{p,q=1}^{2} u_{p,q}(t)\; X_2^{p,q}, \quad i = 1, 2. \tag{62}$$

According to equation (15), the time evolution operator of the whole system reads

$$U(t) = \sum_{p,q=1}^{4} u_{p',q'}(t)\; u_{p+2-2p',q+2-2q'}(t)\; X_4^{p,q}. \tag{63}$$

The time evolution of any state in the corresponding Hilbert space is obtained by the action of this last operator.

4.4.2. Jaynes-Cummings model. The Jaynes-Cummings model describes the interaction between a single atom and a single mode of the quantized electromagnetic field [23]. This considers the Hilbert space associated to the composite system atom+field, so that the vector states of the entire system are spanned by the Kronecker products of the basis vectors belonging to each of the subsystems. In this case, the Hamiltonian includes three parts: the free atom

Hamiltonian given in equation (55), the field alone H_f and the interaction term, given by the Jaynes Cummings Hamiltonian

$$H_I = \gamma(\sigma^+ a + \sigma^- a^\dagger). \tag{64}$$

where a and $a^\dagger$ are the boson ladder operators. Using the Kronecker algebra of the Hubbard operators it is easy to show that $H_0 + H_f$ and H_I are first integrals of the system [24]. The Hamiltonian can be written in terms of the Hubbard operators

$$H_I = \sum_{p=1}^{2} N_p\, X^{p,3-p} \qquad N_p = \sqrt{N+2-p}, \tag{65}$$

where the boson number operator $N = a^\dagger a$ has been promoted to act on the vector space of the entire atom+field system: $N = \mathbb{I}_{at} \otimes N$, with $\mathbb{I}_{at}$ the identity operator in the vector space of the atom. Using the X-operator formalism, it is shown that the time evolution operator reads

$$U(t) = e^{-iH_I t} = \sum_{p,q=1}^{2} u_{p,q}(N_p)\, X^{p,q}, \tag{66}$$

with

$$u_{p,q}(N_p) = e^{i\frac{\pi}{2}|p-q|} \cos\left(\gamma t N_p - \frac{\pi}{2}|p-q|\right). \tag{67}$$

This operator is analogous to the time evolution operator given in Eq. (60). For instance, the dynamics of two non-interacting atoms can be analyzed in terms of the unitary operator

$$U(t) = U_1(t) \otimes U_2(t) \tag{68}$$

where $U_1(t)$ and $U_2(t)$ are expressed in X-operator notation [24]:

$$U_i = \sum_{p,q=1}^{2} u_{p,q}(N_{i;p})\, X_4^{p,q}, \quad i = 1, 2. \tag{69}$$

The operators $N_{i;p}$ are equivalent to the ones defined in (65) for each mode of the field. By virtue of (15), the time evolution operator of the whole system reads

$$U(t) = \sum_{p,q=1}^{4} u_{p',q'}\left(N_{p'}\right)\, u_{p+2-2p',q+2-2q'}\left(N_{q+2-2q'}\right)\, X_4^{p,q}. \tag{70}$$

This last operator is useful in the study of the time evolution of entanglement since the reduced atomic density matrix is straightforwardly computed in this framework [25, 26]. To end up this section we remark that our model could include an interaction like the Heisenberg Hamiltonian (52). In such a case, the involved operators are compatible with the Hubbard representation.

5. Conclusions

The Kronecker product algebra has been revisited in terms of Hubbard operators. This representation allows to reduce complicated calculations involving large matrices or a large number of factors into simple relations of subscripts. In particular, we have shown that the construction of permutation matrices and the construction of symmetrization operators is straightforward in X-representation. Besides, the Hadamard matrix and its tensor powers are nicely dealt in the Hubbard framework. On the other hand, we have analyzed the Hamiltonian of the XXX Heisenberg model for two particles. The diagonalization of this Hamiltonian and the calculation of the time evolution operator have been easily done in the X-representation.

Acknowledgments
The support of CONACyT project 152574 and IPN project SIP-SNIC-2011/04 is acknowledged. M.E. gratefully acknowledges the funding received towards his Ph.D. through the CONACyT Scholarship.

References

[1] Brewer J W 1978 Kronecker products and matrix calculus in system theory *IEEE Trans. Circuits Systems* **25** 772
[2] Steeb W-H and Hardy Y Entanglement 2012 Kronecker Product, Pauli Spin Matrices and a Nonlinear Eigenvalue Problem *Open Syst. Inf. Dyn.* **19** 1250004
[3] Black G R E, King R C and Wybourne B G 1983 Kronecker products for compact semisimple Lie groups *J. Phys. A: Math. Gen.* **16** 1555
[4] Henderson H V, Pukelsheim F and Searle S R 1983 On the history of the Kronecker product *Linear and Multilinear Algebra* **14** 113
[5] Graham A 1981 *Kronecker Products and Matrix Calculus with Applications* (New York: John Wiley and Sons)
[6] Horn R A and Johnson C A 1991 *Topics in Matrix Analysis* (New York: Cambridge University Press)
[7] Steeb W-H and Hardy Y 2011 *Matrix Calculus and Kronecker Product* (Singapore: World Scientific) 2nd ed
[8] Langville A N and Stewart W J 2004 The Kronecker product and stochastic automata networks *J. Comp. Appl. Math.* **167** 429
[9] Van Loan C F 2000 The ubiquitous Kronecker product *J. Comp. Appl. Math.* **123** 85
[10] Peres A 1995 *Quantum Theory: Concepts and Methods* (Dordrecht: Kluwer Academic Publishers)
[11] Nielsen M A and Chuang I L 2000 *Quantum Computation and Quantum Information* (Cambridge: Cambridge University Press)
[12] Mielnik B and Rosas-Ortiz O 2009 Quantum Mechanical Laws *Fundamentals of Physics* ed J L Morán-López, Encyclopedia of Life Support Systems (EOLSS), Developed under the Auspices of the UNESCO (Oxford: Eolss Publishers) 2007/Rev. 2009
[13] Enríquez M and Rosas-Ortiz O 2013 The Kronecker product in terms of Hubbard operators and the Clebsch–Gordan decomposition of SU(2)×SU(2) *Annals of Physics* **339** 218
[14] Hubbard J 1963 Electron correlations in narrow energy bands *Proc. R. Soc. (London)* A **276** 238
[15] Hubbard J 1964 Electron correlations in narrow energy bands II. The degenerate band case *Proc. R. Soc. (London)* A **277** 237
[16] Hubbard J 1965 Electron correlations in narrow energy bands IV. The atomic representation *Proc. R. Soc. (London)* A **285** 542
[17] Bacry H 1977 *Lectures on Group Theory and Particle Theory* (New York: Gordon and Breach)
[18] Wigner E P 1959 *Group Theory and its Application to the Quantum Mechanics of Atomic Spectra* transl J J Griffin (New York: Academic Press)
[19] Hadamard J 1893 Resolution d'une question relative aux determinants *Bull. Sci. Math.* **17** 240
[20] Val'kov V V 1985 Unitary transformations of the group U(N) and diagonalization of multilevel Hamiltonians *Theor. Math. Phys.* **76** 766
[21] Ovchinnikov S G and Val'kov V V 2004 *Hubbard Operators in the Theory of Strongly Correlated Electrons* (London: Imperial College Press)
[22] Enríquez M, Reyes-García R and Rosas-Ortiz R 2013 Time evolution in classical and quantized fields: geometric representation *preprint* Cinvestav
[23] Jaynes E T and Cummings F W 1963 Comparison of quantum and semiclassical radiation theories with applications to the beam maser *Proc. IEEE* **51** 89
[24] Enríquez M and Rosas-Ortiz O 2013 Spectral design to control the time evolution of a two-level atom in single-mode radiation fields *preprint* Cinvestav
[25] Enríquez M and Rosas-Ortiz O 2013 Dynamics of two atoms in a double Jaynes-Cummings model *preprint* Cinvestav
[26] Enríquez M, Quintana-Huerta C and Rosas-Ortiz O 2013 On the entanglement of a pair of two-level atoms in quantized radiation fields *preprint* Cinvestav

Journal of Physics: Conference Series **538** (2014) 012008 doi:10.1088/1742-6596/538/1/012008

Non-exponential decay in Quantum Mechanics and Quantum Field Theory

Francesco Giacosa

Abstract. We describe some salient features as well as some recent developments concerning short-time deviations from the exponential decay law in the context of Quantum Mechanics by using the Lee Hamiltonian approach and Quantum Field Theory by using relativistic Lagrangians. In particular, the case in which two decay channels are present is analyzed: the ratio of decay probability densities, which is a constant equal to the ratio of decay widths in the exponential limit, shows in general sizable fluctuations which persist also at long times.

Institut für Theoretische Physik, Johann Wolfgang Goethe - Universität, Max von Laue-Str. 1, D-60438 Frankfurt, Germany

E-mail: giacosa@th.physik.uni-frankfurt.de

1. Introduction

Decays are ubiquitous in Physics: examples are tunnelling processes in quantum mechanics, very slow decays of nuclei (such as double-β decays, lifetime of about 10^{21} y), as well as very fast decays of hadrons and the Higgs particle (lifetime of about 10^{-22} sec). Although the involved decay times are astonishingly different, the basic phenomenon is the same: a irreversible transition (infinite Poincare' time) of an unstable initial state coupled to a continuum of final states.

Both in the context of Quantum Mechanics (QM) and Quantum Field Theory (QFT), an unstable state $|S\rangle$ is not an eingenstate of the Hamiltonian H of the system, but is described by the so-called energy distribution $d_S(E)$: $d_S(E)dE$ is the probability that the unstable state $|S\rangle$ has an energy between E and $E + dE$. The survival probability amplitude $a(t)$ that the state $|S\rangle$, prepared at the initial time $t = 0$, has not yet decayed at the instant $t > 0$, is given by

$$a(t) = \langle S| e^{-iHt} |S\rangle = \int_{-\infty}^{+\infty} d_S(E)e^{-iEt}dE \; . \qquad (1)$$

The survival probability is given, as usual, by the amplitude squared: $p(t) = |a(t)|^2$. The exponential decay $p(t) = e^{-\Gamma t}$ is obtained when $d_S(E)$ takes the Breit-Wigner (BW) form: $d_S^{\rm BW}(E) = \frac{\Gamma}{2\pi}\left[(E-M)^2 + \Gamma^2/4\right]^{-1}$. This is however an unphysical limiting case, because a physical distribution $d_S(E)$ should fulfill the following two requirements: (i) Existence of an energy threshold, $d_S(E) = 0$ for $E < E_0$, which suffices to prove that long-time deviations exist, usually implying a power-law of the type $p(t) \propto t^{-\alpha}$ [1, 2]. (ii) Finiteness of the mean energy, $\langle E\rangle = \int_{-\infty}^{+\infty} E d_S(E)dE < \infty$, which implies that $p'(0) = 0$ and therefore that deviations at short times occur [2, 3, 4].

Moreover, a non-exponential behavior at short times implies the existence of the quantum Zeno effect, which is the inhibition of the decay (or, more in general, of a quantum transition), due to reiterated frequent measurements [5, 6, 7]. From an experimental standpoint, the Quantum Zeno effect was observed by inhibition of a Rabi oscillation between atomic energy levels [8]. An improved set-up, in which the Zeno effect was verified by using single ions and by means of interaction-free repeated measurements, was shown in Ref. [9].

For what concerns the decay law of a 'genuine' unstable quantum state, the first experimental demonstration of short-time deviations from the exponential law has been observed by studying the tunneling of cold atoms in an optical potential [10]. Later on, the same group [11] could verify also the Zeno (as well as the Anti-Zeno) effect by applying a series of intermediate measurements on this system. Finally, the experimental proof of long-time deviations from the exponential law has been observed by investigating the decay via fluorescence of organic molecules [12].

An experiment performed in the GSI has also found deviations from the exponential decay law in Hydrogen-like ions decaying via electron capture [13]. In Ref. [14] these deviations have been linked to a modification of the BW distribution due to interactions with the measuring apparatus. Other explanations, based on neutrino oscillations or energy level splitting of the initial state, have also been put forward [15].

In this proceeding we review some aspects of the non-exponential decay at short times both in QM and in QFT. In Sec. 2 we present for the QM case a simple but general Hamiltonian to elucidate the main points. Then, we describe two recent developments in QM: the broadening of the emitted spectrum at short times [16] and the case of a decay into two decay channels [4]. In Sec. 3 we turn to the QFT case and we describe the recent results of Refs. [4, 17, 18], in which it has been shown that, contrary to previous claims [19], deviations from the exponential decay law also take place in a genuine QFT framework. Finally, in Sec. 4 we present our conclusions.

2. Non-exponential decay in QM

We introduce a general approach to study decays in QM, which is based on the so-called Lee Hamiltonian $H = H_0 + H_1$ [20]:

$$H_0 = M\left|S\right\rangle\left\langle S\right| + \int_{-\infty}^{+\infty} dk\omega(k)\left|k\right\rangle\left\langle k\right| \ , \ H_1 = \int_{-\infty}^{+\infty} dk\frac{gf(k)}{\sqrt{2\pi}}\left(\left|k\right\rangle\left\langle S\right| + \left|S\right\rangle\left\langle k\right|\right) \ . \quad (2)$$

The quantity g is a coupling constant with the dimension of energy$^{1/2}$. The dimensionless function $f(k)$ describes the interaction of the unstable state $|S\rangle$ with the 'decay-product' state $|k\rangle$; the energy $\omega(k)$ is the energy of $|k\rangle$ in the interaction free case. Similar approaches to Eq. (2) are used in other areas of physics, e.g. in quantum optics with the Jaynes-Cummings Hamiltonian. The spectral function $d_S(E)$ arises as the imaginary part of the propagator, $d_S(E) = \frac{1}{\pi}\,\mathrm{Im}\left[(E - H + i\varepsilon)^{-1}\right]$, and is correctly normalized, $\int_{-\infty}^{+\infty} d_S(E)dE = 1$.

The BW limit is obtained for $f(k) = 1$ and $\omega(k) = k$. The decay width reads $\Gamma = g^2$ and the spectral function $d_S^{\mathrm{BW}}(E) = \frac{\Gamma}{2\pi}\left[(E - M)^2 + \Gamma^2/4\right]^{-1}$ [16]. Note, the choice $\omega(k) = k$ means that the energy is not bounded from below and the choice $f(k) = 1$ means that the unstable state $|S\rangle$ couples to each state of the continuum with the same intensity. For generic functions $f(k)$ and $\omega(k)$ the decay is not exponential (see below), but is usually very well approximated by an exponential decay where the decay width is given by the Fermi golden rule: $\Gamma = g^2 f^2(k_M)/\left|\omega'(k_M)\right|$, where $\omega(k_M) = M$.

Restricting to the BW limit, an interesting feature is found when studying the energy of the emitted spectrum. For definiteness, we consider the case of atomic spontaneous emission: an electron decays from an atomic excited state to the ground state by emitting a photon. This is indeed the original framework in which the decay was studied [21] (see also e.g. Refs. [22, 23] and refs. therein). Denoting M as the energy difference between the excited and the ground state, the

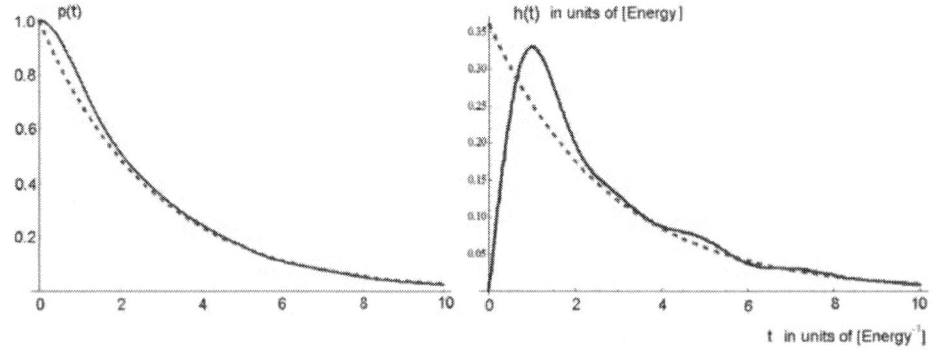

Figure 1. The functions $p(t)$ (left panel, solid line) and $h(t) = -p'(t)$ (right panel, solid line) are shown for $f(k) = \theta(k - E_0)\theta(\Lambda - k)$ with (arbitrary energy units): $E_0 = 0$, $\Lambda = 5$, $M = 2$, $g^2 = 0.36$. The (exponential) dashed lines correspond to the BW limit.

energy distribution of the photon at the time $t > 0$ is given by $\eta(t,\omega) = \frac{\Gamma}{2\pi}\left|\frac{e^{-i\omega t} - e^{-i(M-i\Gamma/2)t}}{\omega - M + i\Gamma/2}\right|^2$. The energy uncertainty of the photon $\delta\omega = \delta\omega(t)$ is given by the width at mid height: $\eta(t,M)/2 = \eta(t, M + \delta\omega/2)$. For $t \to \infty$ one has $\delta\omega \to \Gamma$, which is the natural broadening of the spectral line. For $t \lesssim 3/\Gamma$ the quantity $\delta\omega$ increases as $\delta\omega \simeq 5.56/t$. This means that, if measured at short times after the emission, the photon can have an energy which deviates from the mean value M by several decay widths Γ, see Ref. [16] for details.

We now turn to the a simple example of non-exponential decay. To this end, we choose $f(k) = \theta(k - E_0)\theta(\Lambda - k)$, which means that the state $|S\rangle$ couples *only* to an energy window of final states. A numerical example is shown in Fig. 1, in which the survival probability $p(t)$ and its derivative, $h(t) = -p'(t)$, are plotted (solid lines) and compared with the exponential counterparts (dashed lines, obtained by using the Fermi rule). Deviations from the exponential decay are evident. Note, the function $h(t)$ has a clear physical meaning: $h(t)dt = p(t) - p(t + dt)$ is the probability that the decay occurs in the time-interval $(t, t + dt)$. For this reason, $h(t)$ is also denoted as the density of decay probability.

We now turn to the two-channel case, in which the unstable state $|S\rangle$ couples to two sets of final states $|k,1\rangle$ and $|k,2\rangle$. The Hamiltonian is a straightforward generalization of Eq. (2):

$$H_0 = M\,|S\rangle\langle S| + \sum_{i=1,2}\int_{-\infty}^{+\infty} dk\omega_i(k)\,|k,i\rangle\langle k,i| \ , \ H_1 = \sum_{i=1,2}\int_{-\infty}^{+\infty} dk\frac{g_i f_i(k)}{\sqrt{2\pi}}\left(|S\rangle\langle k,i| + \text{h.c.}\right) \ .$$

In the BW limit $f_i(k) = 1$ and $\omega_i(k) = k + \alpha_i$, where α_1 and α_2 are constants. The partial decay widths are $\Gamma_1 = g_1^2$ and $\Gamma_2 = g_2^2$ and the survival probability reads $p(t) = e^{-\Gamma t}$ with $\Gamma = \Gamma_1 + \Gamma_2$.

In the presence of two decay channels, it is useful to introduce the quantity $h_i(t)$: $h_i(t)dt$ is the probability that the state $|S\rangle$ decays in the i-th channel between t and $t + dt$. In the BW limit $h_i(t) = \Gamma_i e^{-\Gamma t}$ and the ratio $h_1(t)/h_2(t)$ is a constant equal to Γ_1/Γ_2. However, when deviations from the exponential decay are considered, the ratio $h_1(t)/h_2(t)$ is in general not a constant, but shows sizable departures from Γ_1/Γ_2 [4]. This is shown for a particular numerical example in Fig. 2, in which the choices $f_i(k) = \theta(k - E_{0,i})\theta(\Lambda - k)$ and $\alpha_i = 0$ have been used.

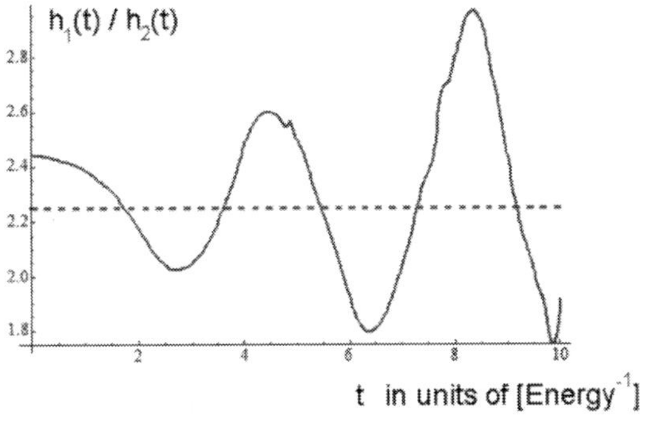

Figure 2. The solid curve is the ratio $h_1(t)/h_2(t)$ plotted as function of t for the numerical choice (arbitrary energy units) $E_{0,1} = 0$, $E_{0,2} = 0.5$, $\Lambda = 5$, $g_1^2 = 0.36$, $g_2^2 = 0.16$. The constant dashed line corresponds to the BW limit Γ_1/Γ_2.

3. Non-exponential decay in QFT

QFT is the theoretical framework in which particles are created and annihilated. It is therefore the most fundamental environment to study the nature of decays. In Refs. [19] it was claimed that in QFT the deviations from the exponential decay law do not take place. These conclusions, however, rely on a perturbative treatment of the decay law. This issue was (re)analyzed in Refs. [4, 17] with the following results:

(i) In Ref. [4, 17, 24, 25] the superrenormalizable QFT interaction Lagrangian $\mathcal{L} = gS\varphi^2$, leading to the decay $S \to \varphi\varphi$, was studied in detail. The spectral function $d_S(E)$ is calculated as the imaginary part of the propagator of S. To this end, it is necessary to perform a resummation of (at least) the one-loop self-energy diagram of the field S. (Namely, a perturbative expression of $d_S(E)$ in a series of g is not a meaningful quantity [4].) The survival probability takes the very same formal expression of Eq. (1)and short-time deviations from the exponential law do occur [4, 17]. Quite remarkably, the time interval in which such deviations take place is independent on the cutoff [17]. This is due to the fact that the energy distribution $d_S(E)$ behaves as E^{-3} for large energies, thus being insensitive on the high energy scale of the theory.
The two-channel case can be easily studied by using the Lagrangian $\mathcal{L} = g_1 S\varphi_1^2 + g_2 S\varphi_2^2$, which implies that the two decay processes $S \to \varphi_1\varphi_1$ and $S \to \varphi_2\varphi_2$ take place. The ratio $h_1(t)/h_2(t)$ shows also in this QFT framework qualitatively similar deviations from the constant BW limit as those in Fig. 2, see Ref. [4] for plots and details.

(ii) In Ref. [18] the renormalizable interaction Lagrangian $\mathcal{L} = gS\bar{\psi}\psi$, where ψ is a fermionic field, has been analyzed. (Note, the Higgs coupling to fermions has the same form.) The energy distribution $d_S(E)$ scales only as E^{-1}, which in turn implies that the presence of a high energy cutoff is necessary for its correct normalization. Thus, the existence of a high energy cutoff is not only a mathematical step, but is a necessary physical requirement. In this framework, the duration of the deviations from the exponential decay at short times lasts only Λ^{-1}, and is thus very small if the high-energy cutoff Λ is large.

4. Summary and outlook

In this proceeding we have discussed the deviations from the exponential decay law in QM and QFT. The decay law in QM has been studied by using the Lee Hamiltonian approach [4, 16] and in QFT by using relativistic QFT Lagrangians [4, 17, 18].

In QM we have discussed a peculiar property of the energy spectrum of the photon in a spontaneous emission process: even in the exponential limit, a broadening of the emitted spectrum is realized if the photon is measured at early times [16]. As a next step we have studied short-time deviations from the exponential law in the simple case in which the unstable state couples to the continuum in a window of energy (see Fig. 1 and Refs. [4, 16]).

A particularly interesting feature in both QM and QFT is that of a decay in which two (or more) decay channels are present. The ratio of decay probability densities is a constant in the BW limit and equals the ratio of decay widths, but show sizable fluctuations on top of this constant limit when deviations from the BW limit are included (see Fig. 2 and Ref. [4]). A systematic study of this ratio for a variety of cases, also including the analysis of realistic physical situations, is an important outlook for the future.

Acknowledgments

The author thanks G. Pagliara and T. Wolkanowski-Gans for useful discussions and the Foundation of the Polytechnical Society of Frankfurt for support through an Educator fellowship.

References

[1] Khalfin L A 1957 Zh. Eksp. Teor. Fiz. **33** 1371 (Engl. trans. Sov. Phys. JETP **6** 1053)
[2] Fonda L, Ghirardi G C and Rimini A 1978 *Rept. Prog. Phys.* **41** 587
[3] Nicolaides C A and Mercouris T 1996 *Journal of Physics* B **29** 1151–67
Urbanowski K 2009 *Central European Journal of Physics* **7** 696–703
Torrontegui E, Muga J G, Martorell J and Sprung D W L 2009 *Phys. Rev.* A **80** 012703
[4] Giacosa F 2012 *Found. Phys.* **42** 1262 [arXiv:1110.5923 [nucl-th]]
[5] Misra B and Sudarshan E C G 1977 *J. Math. Phys.* **18** 756
[6] Facchi P, Nakazato H and Pascazio S 2001 *Phys. Rev. Lett.* **86** 2699–703
Facchi P and Pascazio S 1998 *Phys. Lett.* A **241** 139 [arXiv:quant-ph/9905017]
[7] Koshino K and Shimizu A 2005 *Phys. Rept.* **412** 191 [quant-ph/0411145]
[8] Itano W M, Heinzen D J, Bollinger J J and Wineland D J 1990 *Phys. Rev.* A **41** 2295
[9] Balzer C et al 2002 *Optics Communications* Volume 211 pp 235–41
[10] Wilkinson S R, Bharucha C F, Fischer M C, Madison K W, Morrow P R, Niu Q, Sundaram B and Raizen M G 1997 *Nature* **387** 575
[11] Fischer M C, Gutiérrez-Medina B and Raizen M G 2001 *Phys. Rev. Lett.* **87** 040402
[12] Rothe C, Hintschich S I and Monkman A P 2006 *Phys. Rev. Lett.* **96** 163601
[13] Litvinov Yu A et al *Phys. Lett.* **B664** 162–68 [arXiv:0801.2079 [nucl-ex]]
Kienle P et al. 2013 *Phys. Lett.* B **726** 638 [arXiv:1309.7294 [nucl-ex]]
[14] Giacosa F and Pagliara G 2013 *Quant. Matt.* **2** (2013) 54 [arXiv:1110.1669 [nucl-th]]
Giacosa F and Pagliara G 2012 *PoS BORMIO* **2012** 028 [arXiv:1204.1896 [nucl-th]]
[15] Ivanov A N and Kienle P 2009 *Phys. Rev. Lett.* **103** 062502 [arXiv:0908.0877 [nucl-th]]
Kienle P 2009 *Nucl. Phys.* A **827** 510C
Giunti C 2008 *Phys. Lett.* B **665** 92 [arXiv:0805.0431 [hep-ph]]
Lambiase G, Papini G and Scarpetta G 2013 *Phys. Lett.* B **718** 998 [arXiv:1205.0684 [nucl-th]]
Merle A 2010 *Prog. Part. Nucl. Phys.* **64** 445 [arXiv:1004.2347 [hep-ph]]
Gal A 2010 *Nucl. Phys.* A **842** 102 .
[16] Giacosa F 2013 *Phys. Rev.* A 88, **052131** [arXiv:1305.4467 [quant-ph]]
[17] Giacosa F and Pagliara G 2011 *Mod. Phys. Lett.* **A26** 2247–259 [arXiv:1005.4817 [hep-ph]]
Pagliara G and Giacosa F 2011 *Acta Phys. Polon. Supp.* **4** 753–8 [arXiv:1108.2782 [hep-ph]]
[18] Giacosa F and Pagliara G 2013 *Phys. Rev.* D **88** 025010 [arXiv:1210.4192 [hep-ph]]
[19] Bernardini C, Maiani L and Testa M 1993 *Phys. Rev. Lett.* **71** 2687
Alvarez-Estrada R F and Sanchez-Gomez J L 1999 *Phys. Lett.* A **253** 252
[20] Lee T D 1954 *Phys. Rev.* **95** 1329–34
Chiu C B, Sudarshan E C G and Bhamathi G 1992 *Phys. Rev.* D **46** 3508

[21] Weisskopf V and Wigner E P 1930 *Z. Phys.* **63** 54
Weisskopf V and Wigner E 1930 *Z. Phys.* **65** 18
Breit G 1959 *Handbuch der Physik* **41** 1
[22] Scully M O and Zubairy M S 1997 *Quantum optics* (Cambridge UK: Cambridge University Press)
[23] Facchi P and Pascazio S 2000 *Phys. Rev.* A **62** 023804 [arXiv:quant-ph/9909043]
Berman P R and Ford G W 1994 *Phys. Rev* A **82** 023818
Kofman A G, Kurizki G and Sherman B 1994 *Journal of Modern Optics* **41** 353–84
[24] Achasov N N and Kiselev A V 2004 *Phys. Rev.* D **70** 111901 [hep-ph/0405128]
[25] Giacosa F and Pagliara G 2007 *Phys. Rev.* **C76** 065204 [arXiv:0707.3594 [hep-ph]]
Giacosa F and Wolkanowski T 2012 *Mod. Phys. Lett.* A **27** 1250229 arXiv:1209.2332 [hep-ph]]
Wolkanowski T and Giacosa F 2013 *Acta Phys. Polon. Supp.* **6** 893 [arXiv:1305.2603 [hep-ph]]

Journal of Physics: Conference Series **538** (2014) 012009 doi:10.1088/1742-6596/538/1/012009

Primary classification of symmetries from the solution manifold in Classical Mechanics

J Guerrero[a,b], V Aldaya[b], F Cossío[b], and F F López-Ruiz[c]

[a] Departamento de Matemática Aplicada, Universidad de Murcia, Campus de Espinardo, 30100 Murcia, Spain
[b] Instituto de Astrofísica de Andalucía (IAA - CSIC), Apartado Postal 3004, 18080 Granada, Spain
[c] Departamento de Física Aplicada, Universidad de Cádiz, Campus de Puerto Real, 11510, Puerto Real, Cádiz, Spain

E-mail: juguerre@um.es, valdaya@iaa.es, fcossiop@gmail.com, paco.lopezruiz@uca.es

Abstract. The symmetries of the equations of motion of a classical system are characterized in terms of vector field subalgebras of the whole diffeomorphism algebra of the solution manifold (the space of initial constants endowed with a symplectic structure). Among them, naturally arises the subalgebra of Hamiltonian (contact) vector fields corresponding to (jet-prolongued) point symmetries, those not corresponding to point symmetries and the remaining symmetries being associated with non-Hamiltonian (hence non-symplectic) non-strict contact symmetries.

1. Introduction

In this work we aim at finding symmetries of a general (non-linear) system from the diffeomorphisms of its solution manifold, the manifold of the constants of motion, endowed with a symplectic structure. The interest in finding (basic) symmetries of a system is twofold: classically, the symmetries allow for the reduction of order of the equations of motion (EoM), and even for its complete solution when enough symmetries are known, and at the quantum level symmetries allow for the possibility of applying a non-canonical, group-theoretical quantization method like that of [1].

In 1978 V.I. Arnold [2] introduced a transformation that maps an arbitrary Linear Second Order Ordinary Differential Equation (LSODE) into that of the one-dimensional free particle. This was an explicit realization of a particular case of the transformations introduced by S. Lie, who proved that certain systems given by a non-linear SODE whose EoM are up to cubic in the derivatives can be mapped to the free particle [3].

One of the most important properties of the Arnold transformation (also shared by the more general Lie transformations) is that, being a local diffeomorphism, it maps point symmetries of the EoM (or Lagrangian) of the one-dimensional free particle into point symmetries of the EoM (or Lagrangian) of the LSODE system, resulting in the surprising fact that they have the same point symmetries.

This construction is not easily generalizable to more complex systems not even to higher dimensions (where the Arnold transformation does not exist even in the case of LSODE systems unless the system is isotropic). However, inspired by the Arnold transformation, we shall try to find symmetries of general non-linear systems by importing them from those of a simpler system like the free particle. The rationale under the symmetry *tour* of the Arnold transformation is that the LSODE system has the same solution manifold as the free particle. Thus, we can skip the free-particle step, and import directly the symmetries from the solution manifold. These symmetries will be a suitable subset of the diffeomorphisms of the solution manifold (depending on the structure that we wish to preserve).

The passage to the solution manifold is accomplished by a transformation where the new coordinates and momenta are constants of motion. We shall denote a transformation of this kind a Hamilton-Jacobi (HJ) transformation, since, when the new coordinates and momenta are chosen to be canonical, it is generated by Hamilton's principal function S satisfying the Hamilton-Jacobi equation (see below).

It should be stressed that a HJ transformation provides the general solution to the equations of motion, since it expresses the coordinates and momenta in terms of some (integration) constants and time.

To map the symmetries from the solution manifold to the original system, the Jacobian of this transformation is required. This is the main obstacle to accomplish our task, since for an arbitrary non-linear system the HJ transformation is not available in closed form. However, for the basic symmetries, a power series expansion in terms of the adjoint of the vector field associated with the Hamiltonian is available (see Sec. 5.6).

Note that the existence of the HJ transformation is assured by the *Straightening theorem*, the first step in the induction proof of Frobenius Theorem (see, for instance, [4]). This result states that any vector field can be written, with a suitable choice of local coordinates, as the partial derivative with respect to the first coordinate ($\frac{\partial}{\partial\tau}$ in our case). Applying this to the Cartan formulation of Mechanics, the quotient of the contact manifold by the distribution generated by the vector field associated with Hamilton's EoM leads to a symplectic manifold parametrized by the initial constants (constants of the motion). By an ulterior use of Darboux theorem, a suitable family of constants can be found which are canonical coordinates, thus completing the HJ transformation. This provides an alternative to solving the Hamilton-Jacobi equation for Hamilton's principal function S, which is a type-2 generating function for a canonical transformation where the new coordinates and momenta are constants.

The main purpose of lifting symmetries from the solution manifold is that it provides a simple way of computing all contact symmetries, and these are necessary and sufficient to generate the solution manifold (i.e. expand the tangent space at each point). This is required, for instance, to quantize properly the system. It should be stressed that point symmetries do not provide, except in simple cases like linear systems, enough symmetries to generate the solution manifold (see Sec. 6.3).

Working with contact symmetries release us from the requirement of fixing a given Lagrangian (and considering just variational symmetries of this concrete Lagrangian), since it accounts for all symmetries of all Lagrangian providing the same EoM. Of course, to define a given system, we shall use a particular Lagrangian, but we shall not focus on this Lagrangian and will consider variational and non-variational symmetries of it.

In this paper we shall not restrict ourselves to symmetries generated by jet-prolonged vector fields on the configuration space, but we shall consider in general vector fields on the phase space. We shall not consider other kinds of "exotic" symmetries, such as *non-local* symmetries (see [5]). We shall restrict ourselves to dynamical systems with EoM given by the Euler-Lagrange (EL) equations derived from a (regular) Lagrangian.

The usual notion of symmetry for a dynamical system is that of transforming solutions of the EoM into solutions. More precisely, if M denotes the m-dimensional configuration space, then $\Phi : \mathbb{R} \times M \to \mathbb{R} \times M$ is a symmetry if given a curve $\gamma : I \subset \mathbb{R} \to U \subset M$ solution of the EoM, $\Phi(t, \gamma(t)) = (\tilde{t}, \tilde{\gamma}(\tilde{t}))$ defines a curve $\tilde{\gamma} : \tilde{I} \subset \mathbb{R} \to \tilde{U} \subset M$ which is also a solution of the EoM.

The paper is organized as follows. In Section 2 the point symmetries of the EoM are characterized. In Section 3 variational point symmetries are addressed. Generalized symmetries, in particular contact symmetries are discussed in Section 4. In Section 5 symmetries are lifted from the solution manifold. Some examples are given in Section 6 and finally in Section 7 some outlook is provided.

2. Point Symmetries of the equations of motion

The most elementary notion of symmetry is that which transforms just the variables[1] (t, q). The subsequent transformation on the derivatives q', q'', etc. is the one derived from the transformation of (t, q) (i.e. the derivatives are not considered as independent quantities). These symmetries are usually denoted as *point* or *geometrical*. When the transformation Φ depends smoothly on some parameter ϵ, it can be expanded around the identity up to first order:

$$\tilde{t} = t + \epsilon\,\xi(t, q) + O(\epsilon^2)\,, \qquad \tilde{q}_i = q_i + \epsilon\,\eta_i(t, q) + O(\epsilon^2)\,. \tag{1}$$

Deriving the previous equation with respect to ϵ at $\epsilon = 0$ we obtain the coefficients of the infinitesimal generator of the transformation:

$$X = \xi(t, q)\frac{\partial}{\partial t} + \eta_i(t, q)\frac{\partial}{\partial q_i}\,. \tag{2}$$

Infinitesimal generators of symmetries are usually denoted Lie symmetries, since the Lie bracket of two of them is also an infinitesimal generator of a symmetry (thus closing a Lie algebra).

2.1. Characterization of Lie symmetries

If the EoM are of the form $F_i(t, q, q', q'', \ldots, q^{(n)}) = 0$, then the infinitesimal generator X of the symmetry verifies:

$$X^{[n]}(F_i) = 0\,, \tag{3}$$

on solutions $q(t)$ of the equations of motion, where $X^{[n]}$ is the n-th jet prolongation of X, given by [6]:

$$X^{[n]} = X + X_i^1\frac{\partial}{\partial q_i'} + X_i^2\frac{\partial}{\partial q_i''} + \cdots X_i^n\frac{\partial}{\partial q_i^{(n)}}\,, \tag{4}$$

with

$$X_i^1 = \frac{d\eta_i}{dt} - q_i'\frac{d\xi}{dt}\,, \qquad X_i^k = \frac{dX_i^{k-1}}{dt} - q_i^{(k)}\frac{d\xi}{dt} = \frac{d^k}{dt^k}(\eta_i - q_i'\xi) + q_i^{(k+1)}\xi\,. \tag{5}$$

2.2. Characterization in terms of the total derivative vector field

If the EoM can be written as $q_i^{(n)} = f_i(t, q, q', \ldots, q^{(n-1)})$, then an alternative description can be given in terms of the total derivative (along trajectories) vector field, defined as:

$$D_t = \frac{\partial}{\partial t} + q_i'\frac{\partial}{\partial q_i} + q_i''\frac{\partial}{\partial q_i'} + \cdots + f_i\frac{\partial}{\partial q_i^{(n-1)}}\,. \tag{6}$$

Then

$$X^{[n]}(F) = 0 \quad \Leftrightarrow \quad X^{[n-1]}f_i = X_i^n \quad \Leftrightarrow \quad [X^{[n-1]}, D_t] = -(D_t\xi)D_t\,. \tag{7}$$

on solutions $q(t)$ of the EoM.

[1] To avoid cluttering the notation, we shall denote the vector $(q_1, q_2, \ldots, q_m)$ by q.

2.3. First Integrals

A function $I(t, q, q', \ldots, q^{(n-1)})$ satisfying $D_t I = 0$ is named a first integral of the EoM. First integrals are quantities constant along trajectories which are solutions of the EoM. They take the same value along a single trajectory, although it can take different values in different trajectories. The knowledge of $(n-1)m$ independent first integrals solves completely the EoM.

3. Point Symmetries of the Lagrangian

When the EoM result from the EL equations of a Lagrangian L, determining the extremals of the action functional $\mathcal{L} = \int dt L(t, q, q', \ldots, q^{(n-1)})$, it is natural to wonder about the transformations preserving this action functional (the importance of this is related to Noether theorem and conserved quantities). These symmetries are usually called *variational* symmetries.

The infinitesimal generators $X = \xi(t,q)\frac{\partial}{\partial t} + \eta_i(t,q)\frac{\partial}{\partial q_i}$ of variational symmetries verify:

$$X^{[n-1]}(L) + L D_t \xi = D_t B\,, \tag{8}$$

for some function B. Note that if $B \neq 0$, although the action functional $\mathcal{L}$ is invariant, the Lagrangian is invariant up to the total derivative $D_t B$.

3.1. Noether theorem

By Noether theorem [6, 7, 8], with each variational symmetry a conserved quantity (first integral) can be associated.

For a first-order Lagrangian $L(t, q, q')$, the conserved quantity associated with the variational symmetry $X = \xi(t,q)\frac{\partial}{\partial t} + \eta_i(t,q)\frac{\partial}{\partial q_i}$ is:

$$N = (\eta_i - \xi q_i')\frac{\partial L}{\partial q_i'} + \xi L - B\,. \tag{9}$$

4. Generalized Symmetries and contact structure

A generalized vector field on $\mathbb{R} \times M$ is a vector field of the form:

$$X = \xi(t, q, q', \ldots)\frac{\partial}{\partial t} + \eta_i(t, q, q', \ldots)\frac{\partial}{\partial q_i}\,. \tag{10}$$

A generalized vector field is not a proper vector field in $\mathbb{R} \times M$ in the sense that it does not generate a flow on it (however it is a proper vector field in a suitable jet space). But further computations can be carried out without inconsistencies as though it were a true vector field in $\mathbb{R} \times M$. In particular, formulas for prolongations, invariance of the EoM and of the Lagrangian apply without changes.

4.1. Evolutionary vector fields

Given $f_i(t, q, q', \ldots)$, an evolutionary vector field with characteristic f is defined as:

$$X_f = f_i \frac{\partial}{\partial q_i}\,. \tag{11}$$

To any generalized vector field $X = \xi(t, q, q', \ldots)\frac{\partial}{\partial t} + \eta_i(t, q, q', \ldots)\frac{\partial}{\partial q_i}$, there corresponds an evolutionary vector field X_f with characteristics $f_i = \eta_i - \xi q_i'$, in the sense that they generate the same symmetry transformation. In fact, $X^{[1]} = (X_f)^{[1]} + \xi D_t$, and therefore, according to (7), $X^{[1]}$ is a symmetry iff $(X_f)^{[1]}$ also is, and both vector fields coincide on solutions of the EoM.

4.2. Contact symmetries and contact structure

Generalized symmetries of first order are also known as contact symmetries since they preserve the contact 1-form, also known as Poincaré-Cartan 1-form,

$$\theta_L = \frac{\partial L}{\partial q_i'}(dq_i - q_i' dt) + L dt\,, \tag{12}$$

in the sense that $\mathbb{L}_X \theta_L = \theta_{L'}$, with $\theta_{L'}$ a contact 1-form for the Lagrangian $L' = X^{[1]}(L) + L D_t \xi$ with the same EL EoM (see [9]).

Note that by (8) (extended to generalized symmetries), if X is a variational (non-point, in general) symmetry, $L' = D_t B$, and $\theta_{L'} = \mathbb{L}_X \theta_L = df$, for some function f.

The Poincaré-Cartan 1-form defines a contact structure on the odd dimensional manifold $\mathbb{R} \times T^*M$, where it is written as

$$\theta_H = p_i dq^i - H\, dt\,, \tag{13}$$

where $H = p_i q^{i\prime} - L$ is the Legendre transform of L and $p_i = \frac{\partial L}{\partial q^{i\prime}}$. θ_H is also known as the Poincaré-Cartan 1-form associated with H.

5. Symmetries from the solution manifold

5.1. Hamiltonian formulation

Given a Poincaré-Cartan 1-form θ_H, its differential $\omega_H = d\theta_H = dp_i \wedge dq^i - dH \wedge dt$, known as the Poincaré-Cartan 2-form, has maximal rank. Its radical is generated, as a module, by a vector field $\bar{X}_H$ such that $i_{\bar{X}_H} \omega_H = 0$. This vector field can be written as:

$$\bar{X}_H = \frac{\partial}{\partial t} + \frac{\partial H}{\partial p_i}\frac{\partial}{\partial q^i} - \frac{\partial H}{\partial q^i}\frac{\partial}{\partial p_i}\,. \tag{14}$$

The flow of $\bar{X}_H$ is given by Hamilton's EoM:

$$\frac{dt}{ds} = 1\,, \qquad \frac{dq^i}{ds} = X_H^{q^i} = \frac{\partial H}{\partial p_i}\,, \qquad \frac{dp_i}{ds} = X_H^{p_i} = -\frac{\partial H}{\partial q^i}\,. \tag{15}$$

The time evolution is given in terms of the Poisson bracket in T^*M:

$$\bar{X}_H F = \frac{d}{dt}F = \frac{\partial F}{\partial t} + \{F, H\}\,. \tag{16}$$

5.2. The solution manifold

Taking quotient on the contact manifold by Hamiltons's equations (i.e. by the distribution generated by $\bar{X}_H$), we obtain a symplectic manifold $\mathbb{S}$ parametrized by, say, the constants of motion, with symplectic form $\omega = \omega_H$ (i.e. ω_H simply "falls down" to the quotient).

On Darboux coordinates, $\omega = dP_i \wedge dQ^i$, where $P_i\,, Q^i$ are canonically conjugated constants of motion, and the Poisson bracket is given by:

$$\{F, G\} = \frac{\partial F}{\partial P_i}\frac{\partial G}{\partial Q^i} - \frac{\partial F}{\partial Q^i}\frac{\partial G}{\partial P_i}\,. \tag{17}$$

Given a function F on $\mathbb{S}$, the Hamiltonian vector field X_F associated with F is given by:

$$X_F = \{F, \cdot\} = \frac{\partial F}{\partial P_i}\frac{\partial}{\partial Q^i} - \frac{\partial F}{\partial Q^i}\frac{\partial}{\partial P_i}\,. \tag{18}$$

5.3. Hamilton-Jacobi transformation

The passage from the contact manifold parametrized by (q^i, p_j, t) to the solution manifold parametrized by (Q^i, P_j, τ) is given by the Hamilton-Jacobi transformation:

$$q_i = q_i(Q, P, \tau), \qquad p^i = p^i(Q, P, \tau), \qquad t = \tau. \tag{19}$$

Note that a trivial change in time, $t = \tau$, has been introduced, in order to render the transformation invertible (in the solution manifold the time evolution is *frozen* since it is parametrized by constants of the motion, therefore τ *decouples* everywhere and can be safely removed). In this way the associated change in the partial derivatives, necessary to compute the transformation of vector fields, can be obtained:

$$\begin{aligned} \frac{\partial}{\partial Q^i} &= \frac{\partial q^j}{\partial Q^i}\frac{\partial}{\partial q^j} + \frac{\partial p_j}{\partial Q^i}\frac{\partial}{\partial p_j} \\ \frac{\partial}{\partial P_i} &= \frac{\partial q^j}{\partial P_i}\frac{\partial}{\partial q^j} + \frac{\partial p_j}{\partial P_i}\frac{\partial}{\partial p_j} \\ \frac{\partial}{\partial \tau} &= \frac{\partial}{\partial t} + \frac{\partial q^j}{\partial \tau}\frac{\partial}{\partial q^j} + \frac{\partial p_j}{\partial \tau}\frac{\partial}{\partial p_j} = \frac{d}{dt}. \end{aligned} \tag{20}$$

Under the Hamilton-Jacobi transformation the Poincaré-Cartan 1-form is written as

$$\theta_H = p_i dq^i - H dt = P_i dQ^i + d(S - P_i Q^i), \tag{21}$$

where $S = S(q, P, t)$ is Hamilton's Principal function, which coincides with the action integral $\int dt L(t, q, q')$ (on solutions) up to a constant.

Note that we recover that $\omega_H = d\theta_H = dP_i \wedge dQ^i = \omega$, therefore the Poincaré-Cartan 2-form "falls down" to the Solution Manifold.

It should be stressed that the trivial change of variables $\tau = t$ in eq. (19) and the fact that the constants of motion (Q^i, P_j) are canonically conjugated variables have been chosen for simplicity. In general we can take $\tau = \tau(t, q, p)$ and (Q^i, P_j) not canonically conjugated, but in this case the HJ transformation does not coincide with the one generated by Hamilton's Principal function S. The expressions appearing in Secs. 5.1, 5.2 and 5.3 should be modified accordingly, but the main conclusions still hold (see [10] for details)[2].

5.4. Noether theorem in the Solution Manifold

All Hamiltonian vector fields X_F generate symmetries since:

$$i_{X_F}\omega = -dF \quad \Rightarrow \quad \mathbb{L}_{X_F}\omega = 0 \quad \Rightarrow \quad \mathbb{L}_{X_F}\omega_H = 0, \tag{22}$$

and this implies $\mathbb{L}_{X_F}\theta_H = df$, therefore X_F (when lifted to the contact manifold) is a variational contact symmetry, and it might be (the evolutionary form of) a point symmetry.

In particular $\frac{\partial}{\partial Q^i}$ and $\frac{\partial}{\partial P_i}$, when lifted by the inverse HJ transformation, i.e. the expressions (20), are (basic) variational symmetries.

It should be stressed that in the expression of X_F there is no "dynamical" information. It is only when lifted by the inverse HJ transformation that the information on the Hamiltonian or the Lagrangian appears.

[2] This freedom, far from being a drawback of the method, is an important advantage, allowing to exploit the natural (and non-canonical) symplectic structure of the Solution Manifold when it is a co-adjoint orbit of a Lie group.

5.5. non-Noether symmetries in the Solution Manifold

If $X \neq X_F$ in $\mathbb{S}$, then

$$\mathbb{L}_X \omega = i_X d\omega + d(i_X \omega) = d(i_X \omega) = \omega' , \tag{23}$$

with ω' a closed 2-form: $\omega' = d\theta'$ with $\theta' = i_X \omega$. If $\theta' = \theta_{H'}$ with H' the Legendre transform of L' providing the same EL EoM as L, then X, when lifted by the inverse HJ transformation, is a (non-Noether) contact symmetry.

Note that in one dimension, since ω is a volume form, $\omega' = f\omega$ for some function f on $\mathbb{S}$ (in fact $f = div(X)$).

5.6. Lifting symmetries from the solution manifold

In the solution manifold $\mathbb{S}$ there are, obviously, infinitely many symmetries. The problem is lifting these symmetries to the contact manifold by the inverse HJ transformation.

If the system is conservative[3] ($L \neq L(t)$, i.e $H \neq H(t)$), a formal solution can be obtained through the exponential map:

$$\begin{aligned} q &= e^{\tau X_H} Q \\ p &= e^{\tau X_H} P \\ t &= \tau . \end{aligned} \tag{24}$$

The partial derivatives necessary to compute (20) can be obtained through ($Y = Q^i$, P_i):

$$\frac{\partial}{\partial Y} e^{\tau X_H} = e^{\tau X_H} e^{-\tau ad_{X_H}} (\frac{\partial}{\partial Y}) = e^{\tau X_H} \sum_{n=0}^{\infty} \frac{(-1)^n}{n!} \tau^n ad^n_{X_H} (\frac{\partial}{\partial Y}) . \tag{25}$$

In the previous formulas

$$\begin{aligned} ad_{X_H}(\frac{\partial}{\partial Y}) &= [X_H, \frac{\partial}{\partial Y}] \\ ad^n_{X_H}(\frac{\partial}{\partial Y}) &= [X_H, ad^{n-1}_{X_H}(\frac{\partial}{\partial Y})] . \end{aligned} \tag{26}$$

This provides a series expansion in powers of τ for $q^i(Q, P, \tau), p_j(Q, P, \tau)$ and their partial derivatives with respect to Q^i, P_j.

This way of obtaining symmetries from the solution manifold is, to the best of our knowledge, new, and the formal series appearing in Eqs. (25)-(26) can be easily summed up when the vector fields $(\frac{\partial}{\partial Q^i}, \frac{\partial}{\partial P_j}, X_H)$ close a finite dimensional Lie algebra[4], since the terms in the expansion can be arranged in a finite number of expansions (each one proportional to an element of a basis of the Lie algebra) and these are numerical series that can be easily shown to be absolutely convergent.

If $(\frac{\partial}{\partial Q^i}, \frac{\partial}{\partial P_j}, X_H)$ do not close a finite-dimensional Lie algebra, but the infinite-dimensional Lie algebra generated from them is of *finite (polynomial) growth* [12], in the sense that the successive Lie brackets appearing in eq. (25) are proportional to increasing powers of a small perturbative parameter λ, in such a way that the vector space generated at each order of λ is finite, then the formal series can be arranged in terms proportional to powers of λ (and each of these terms is of the type considered before and therefore absolutely convergent). Note that this is the situation when a conserved Hamiltonian is of the form $H = H_0 + \lambda H_I$, where H_0 represents an exactly solvable system, H_I an interaction Hamiltonian (non-solvable) and λ is a small coupling constant.

[3] For time-dependent Hamiltonians, Dyson or Magnus expansion should be used instead, see [11, 10].

[4] For this purpose a proper choice of the variables (Q^i, P_j) can be crucial, see the comment before Sec. 5.4.

6. Examples

In this section we shall provide some examples to illustrate the construction of symmetries from the solution manifold. In all these cases the corresponding HJ transformation is known, and therefore explicit formulas can be provided. In general, a construction like the one given in Sec. 5.6 should be used to obtain approximate solutions and approximated symmetries. Note however that for many purposes (for instance when the system is a small perturbation of an exactly solvable Hamiltonian, see [10]) this is enough.

6.1. Symmetries of the free one-dimensional particle

For the free one-dimensional particle $\ddot{q}=0$, $L=\frac{1}{2}m\dot{q}^2$ and $H=\frac{p^2}{2m}$, where $p=m\dot{q}$. The HJ transformation is $q=Q+\frac{P}{m}\tau$ and $p=P$, providing the general solution to the EoM. Its points symmetries are:

$$\begin{array}{rclcl}
G_1 & = & \frac{\partial}{\partial q} & = & \frac{\partial}{\partial Q} \\
G_2 & = & t\frac{\partial}{\partial q}+\frac{\partial}{\partial \dot{q}} & = & m\frac{\partial}{\partial P} \\
G_3 & = & \frac{\partial}{\partial t} & = & \frac{\partial}{\partial \tau}-\frac{P}{m}\frac{\partial}{\partial Q} \\
G_4 & = & t^2\frac{\partial}{\partial t}+qt\frac{\partial}{\partial q}+(q-\dot{q}t)\frac{\partial}{\partial \dot{q}} & = & \tau^2\frac{\partial}{\partial \tau}+mQ\frac{\partial}{\partial P} \\
G_5 & = & t\frac{\partial}{\partial t}+\frac{q}{2}\frac{\partial}{\partial q}-\frac{\dot{q}}{2}\frac{\partial}{\partial \dot{q}} & = & \tau\frac{\partial}{\partial \tau}+\frac{Q}{2}\frac{\partial}{\partial Q}-\frac{P}{2}\frac{\partial}{\partial P} \\
G_6 & = & q\frac{\partial}{\partial q}+\dot{q}\frac{\partial}{\partial \dot{q}} & = & Q\frac{\partial}{\partial Q}+P\frac{\partial}{\partial P} \\
G_7 & = & qt\frac{\partial}{\partial t}+q^2\frac{\partial}{\partial q}+\dot{q}(q-\dot{q}t)\frac{\partial}{\partial \dot{q}} & = & (Q+\frac{P}{m}\tau)\tau\frac{\partial}{\partial \tau} \\
 & & & & +Q^2\frac{\partial}{\partial Q}+QP\frac{\partial}{\partial P} \\
G_8 & = & q\frac{\partial}{\partial t}-\dot{q}^2\frac{\partial}{\partial \dot{q}} & = & (Q+\frac{P}{m}\tau)\frac{\partial}{\partial \tau} \\
 & & & & -\frac{QP}{m}\frac{\partial}{\partial Q}-\frac{P^2}{m}\frac{\partial}{\partial P}\,.
\end{array} \tag{27}$$

In the previous equation $G_1,\ldots,G_5$ are variational point symmetries, whereas G_6, G_7, G_8 are non-variational point symmetries. The first expression corresponds to the first jet prolongation in M, and the second expression corresponds to the solution manifold (extended with τ). Note that in the second expression the terms in $\frac{\partial}{\partial \tau}$ can be safely removed (the evolutionary form is then obtained) since its objective was rendering the transformation geometrical.

General contact symmetries are obtained by:

$$\begin{aligned}
X=f(Q,P)\frac{\partial}{\partial Q}+g(Q,P)\frac{\partial}{\partial P} &= \left[f(q-\dot{q}t,\dot{q})+\frac{t}{m}g(q-\dot{q}t,\dot{q})\right]\frac{\partial}{\partial q} \\
&\quad +\frac{1}{m}g(q-\dot{q}t,\dot{q})\frac{\partial}{\partial \dot{q}}\,,
\end{aligned} \tag{28}$$

where this expression is in evolutionary form. The term in $\frac{\partial}{\partial \dot{q}}$ is a prolongation since $D_t(f)=D_t(g)=0$ (f and g are first integrals).

If div $X=0$, then X is a Hamiltonian vector field providing, when lifted to the contact manifold, a (non-geometrical, in general) variational symmetry. This is the case of G_1 ($f=1$, $g=0$), G_2 ($f=0$, $g=m$), G_3 ($f=-P/m$, $g=0$), G_4 ($f=0$, $g=mQ$) and G_5 ($f=Q/2$, $g=-P/2$). In this case these symmetries are rendered geometrical[5] by the addition of a suitable term in $\frac{\partial}{\partial \tau}$.

If div $X\neq 0$, then X is not a Hamiltonian vector field, therefore when lifted to the contact manifold it provides a non-variational (non-geometrical in general) symmetry. This is the case

[5] Note that in general this is not possible and for non-linear systems only a few symmetries turn out to be geometrical.

of G_6 ($f = Q$, $g = P$), G_7 ($f = Q^2$, $g = QP$), and G_8 ($f = -QP/m$, $g = -P^2/m$). Again, these symmetries are rendered geometrical by the addition of a suitable term in $\frac{\partial}{\partial\tau}$.

Equation (27) provides all point symmetries of the free particle (either variational or not), whereas (28) (with the addition of an arbitrary term in $\frac{\partial}{\partial\tau}$) provides all contact symmetries of the free particle, either variational, non variational, geometrical or non geometrical.

6.2. The free relativistic particle

For the free relativistic particle in one dimension, $L = -mc^2\sqrt{1-\frac{\dot{q}^2}{c^2}} \equiv -mc^2/\gamma$, $H = p_0 c$, where $p = m\gamma\dot{q}$ and $p_0 = \sqrt{m^2c^2+p^2} = mc\gamma$. The HJ transformation is: $p = P$, $q = Q + \frac{P}{P_0}ct$, where $P_0 = \sqrt{m^2c^2+P^2}$, providing the general solution to the EoM.

The basic symmetries in this case are:

$$\begin{aligned} \frac{\partial}{\partial Q} &= \frac{\partial}{\partial q}, \\ \frac{\partial}{\partial P} &= \frac{\partial}{\partial p} + \frac{m^2c^3t}{p_0^3}\frac{\partial}{\partial q} = \frac{1}{m\gamma^3}\left(t\frac{\partial}{\partial q} + \frac{\partial}{\partial \dot{q}}\right). \end{aligned}$$

Note that $\frac{\partial}{\partial Q}$, when lifted to the solution manifold, is a variational point symmetry (reflecting the translation invariance of the Lagrangian). However $\frac{\partial}{\partial P}$ is not a point symmetry, and cannot be rendered a point symmetry by just adding of a term in $\frac{\partial}{\partial\tau}$.

If we consider the vector field $\frac{Po^3}{m^2c^3}\frac{\partial}{\partial P} = t\frac{\partial}{\partial q} + \frac{\partial}{\partial\dot{q}}$, we recover the point (variational) symmetry G_2 of the free non-relativistic particle, which is however a non-variational symmetry for the relativistic particle. A variational symmetry obtained from $\frac{\partial}{\partial P}$ is $\frac{Po^2}{mc^2}\frac{\partial}{\partial P} = \frac{1}{\gamma}\left(t\frac{\partial}{\partial q} + \frac{\partial}{\partial\dot{q}}\right)$, but this is non geometrical.

To obtain a variational point symmetry from $\frac{\partial}{\partial P}$, different expressions for the solutions should be employed, using a non-canonical HJ transformation involving a change in time (i.e. $\tau \neq t$). In this way the expression of the relativistic boost generator can be recovered by lifting $\frac{Po}{c}\frac{\partial}{\partial P}$ (see [10]). Note that this is not possible with the conventional canonical transformation generated by the solution of the Hamilton-Jacobi equation.

6.3. The Pöschl-Teller potential

As a non-linear example, we shall consider a particle in a Pöschl-Teller potential, with Lagrangian given by $L = \frac{1}{2}m\dot{q}^2 - \frac{D}{\cos(\alpha q)^2}$, and Hamiltonian $H = \frac{p^2}{2m} + \frac{D}{\cos(\alpha q)^2}$. Here D is the energy of the minimum of the potential, and α is related to the width of the potential.

Written in terms of $\xi = \frac{1}{\alpha}\sin(\alpha q)$ and $p_\xi = \frac{\partial L}{\partial\dot{\xi}}$,

$$H = (1-\alpha^2\xi^2)\frac{p_\xi^2}{2m} + \frac{D}{1-\alpha^2\xi^2} = \frac{1}{2}m\dot{\xi}^2 + \frac{1}{2}m\omega(E)^2\xi^2 + D, \tag{29}$$

where $\omega(E) = \sqrt{\frac{2E}{D}}\alpha$ is an energy-dependent frequency.

Note that written in this coordinates the Pöschl-Teller potential looks like a harmonic oscillator. However, the frequency of the oscillations depends on the energy E, and therefore on the particular trajectory considered. Therefore they are different systems, although the Pöschl-Teller potential can be interpreted as an harmonic oscillator in the circle (or the n-sphere in higher dimensions).

The solutions of the EoM, providing the HJ transformation, are (see [13]):

$$q = \frac{1}{\alpha}\arcsin\left[\sin(\alpha Q)\cos(\omega(E)t) + \frac{\alpha}{m\omega(E)}P\cos(\alpha Q)\sin(\omega(E)t)\right] \quad (30)$$

$$p = m\dot{q} = \frac{1}{\cos(\alpha q)}\left[\cos(\alpha Q)\cos(\omega(E)t) - \frac{\alpha}{m\omega(E)}P\sin(\alpha Q)\sin(\omega(E)t)\right]. \quad (31)$$

In these solutions, the energy appearing in the frequency should be written in (Q, P) variables. Also, in the denominator of the *rhs* of the second line, q should be substituted by the *rhs* of the first line.

The inverse of the HJ transformation can also be computed, resulting in:

$$Q = \frac{1}{\alpha}\arcsin\left[\sin(\alpha q)\cos(\omega(E)t) - \frac{\alpha}{m\omega(E)}p\cos(\alpha q)\sin(\omega(E)t)\right] \quad (32)$$

$$P = \frac{1}{\cos(\alpha Q)}\left[\cos(\alpha q)\cos(\omega(E)t) + \frac{\alpha}{m\omega(E)}p\sin(\alpha q)\sin(\omega(E)t)\right]. \quad (33)$$

In these expressions, the energy appearing in the frequency should now be expressed in (q, p) variables. As before, in the denominator of the *rhs* of the second line, Q should be substituted by the *rhs* of the first line.

Basic symmetries in the Solution Manifold lifted by the inverse HJ transformation can be computed, although their expressions are rather involved. They can be obtained from the Jacobian appearing in eq. (20), which is given by:

$$\begin{aligned}
\frac{\partial q}{\partial Q} &= (A + Bt)\cos(\omega t) + (C + Dt)\sin(\omega t) \\
A &= \frac{\cos(\alpha Q)}{\cos(\alpha q)} \\
B &= \frac{2\alpha^3 DP\tan(\alpha Q)}{m^2\omega^2\cos(\alpha Q)\cos(\alpha q)} \\
C &= -\frac{4\alpha^3 DP\tan(\alpha Q) + m\alpha P\omega^2\sin(2\alpha Q)}{2m^2\omega^3\cos(\alpha Q)\cos(\alpha q)} \\
D &= -\frac{2\alpha^2 D\sin^2(\alpha Q)}{m\omega\cos(\alpha Q)^3\cos(\alpha q)}
\end{aligned}$$

$$\begin{aligned}
\frac{\partial p}{\partial Q} &= (A + Bt)\cos(\omega t) + (C + Dt)\sin(\omega t) \\
A &= \frac{\alpha\left(m^2p\omega^2\tan(\alpha q)\cos(\alpha Q) - m^2P\omega^2\sin(\alpha Q)\right)}{m^2\omega^2\cos(\alpha q)} \\
B &= \frac{\alpha\left(\alpha^3 DpP\tan(\alpha q)\sin(2\alpha Q) - 2\alpha Dm^2\omega^2\sin^2(\alpha Q)\right)}{m^2\omega^2\cos(\alpha q)\cos(\alpha Q)^3} \\
C &= -\frac{2\alpha^2 D\sin^2(\alpha Q) + m\omega^2\cos^4(\alpha Q)}{\omega\cos(\alpha Q)^3\cos(\alpha q)} \\
&\quad + \frac{\alpha^2 pP\tan(\alpha q)\left(\alpha^2 D\sin(2\alpha Q) + m\omega^2\sin(\alpha Q)\cos^3(\alpha Q)\right)}{m^2\omega^3\cos(\alpha Q)^3\cos(\alpha q)} \\
D &= \frac{-\alpha^3 DmP\omega^2\sin(2\alpha Q) - 2\alpha^3 Dmp\omega^2\tan(\alpha q)\sin^2(\alpha Q)}{m^2\omega^3\cos(\alpha Q)^3\cos(\alpha q)}
\end{aligned}$$

$$
\begin{aligned}
\frac{\partial q}{\partial P} &= (A + Bt)\cos(\omega t) + (C + Dt)\sin(\omega t) \\
A &= 0 \\
B &= \frac{m\omega^2 \cos^2(\alpha Q) - 2\alpha^2 D}{m^2\omega^2 \cos(Q\alpha)\cos(\alpha q)} \\
C &= \frac{2\alpha^2 D}{m^2\omega^3 \cos(\alpha Q)\cos(\alpha q)} \\
D &= -\frac{\alpha P \tan(\alpha Q)\cos(Q\alpha)}{m^2\omega \cos(\alpha q)}
\end{aligned}
$$

$$
\begin{aligned}
\frac{\partial p}{\partial P} &= (A + Bt)\cos(\omega t) + (C + Dt)\sin(\omega t) \\
A &= \frac{\cos(\alpha Q)}{\cos(\alpha q)} \\
B &= \frac{-4\alpha^3 Dp\tan(\alpha q) + 2\alpha m p\omega^2 \tan(\alpha q)\cos^2(\alpha Q) - \alpha m P\omega^2 \sin(2\alpha Q)}{2m^2\omega^2 \cos(\alpha Q)\cos(\alpha q)} \\
C &= -\frac{\alpha m P\omega^2 \sin(2\alpha Q) - 4\alpha^3 Dp\tan(\alpha q)}{2m^2\omega^3 \cos(\alpha Q)\cos(\alpha q)} \\
D &= -\frac{-4\alpha^2 Dm\omega^2 + 2m^2\omega^4 \cos^2(\alpha Q) + \alpha^2 pP\omega^2 \tan(\alpha q)\sin(2\alpha Q)}{2m^2\omega^3 \cos(\alpha Q)\cos(\alpha q)}
\end{aligned}
$$

In the previous formulas we have introduced $\omega \equiv \omega(E)$, and at any occurrence of (Q, P) in the *rhs* of an equation, equations (32) and (33) should be used (we write the expressions in such a way in order to make them more compact). As before, the expression of the energy in terms of (q, p) inside ω should be used.

These basic symmetries are non geometrical. It is possible to obtain point symmetries for the Pöschl-Teller potential (apart from the trivial one $\frac{\partial}{\partial t}$) by lifting certain functions on the solution manifold closing an $sl(2, \mathbb{R})$ subalgebra of the Poisson algebra (see [13] for details).

6.4. Other examples

The method of lifting symmetries from the solution manifold can be applied to many other examples, either in exact or in approximate form. Among them, the symmetries of the Kepler problem, the λq^4 potential, the particle in the circle, the Sine-Gordon model or Sigma models can be addressed (see [10]).

7. Comments and outlook

The idea of importing symmetries from the solution manifold *à la Arnold* is, to the best of our knowledge, new, and we feel that deserves further study. A more complete discussion, along with a thoughtful classification of symmetries will be addressed in [10]. Let us comment only on the extension of this classification to the quantum level.

The role of the solution manifold is played by the Hilbert space $\mathcal{H}_0$ of solutions of the time dependent Schrödinger equation, which can be realized by fixing a value of time t_0 (for instance $t_0 = 0$). The role of the Hamilton-Jacobi transformation is played by the inverse of the time evolution operator $\hat{U}(t, t_0)$. The main difference is that, although we have $\hat{Q}$ and $\hat{P}$ operators, wavefunctions depend only on one of the variables Q or P (or a combination thereof). In the context of Geometric Quantization, this means that they are polarized. The EoM is the Schrödinger equation (which is also polarized).

The consequences of all this is that only variational symmetries survive. This can be easily understood since the Schrödinger equation involves the Hamiltonian. Therefore, transformations changing the Lagrangian to an alternative Lagrangian (leading to the same EL equations) are not allowed at the quantum level.

A different approach to the symmetries of the quantum problem is that of Lie symmetries of the Schrödinger equation, see [14, 15].

Acknowledgments
Work partially supported by the Fundación Séneca, Spanish MICINN and Junta de Andalucía under projects 08814/PI/08, FIS2011-29813-C02-01 and FQM219-FQM1951, respectively.

References

[1] Aldaya V and de Azcárraga J A 1982 *J. Math. Phys.* **23** 1297
[2] Arnold V 1978 *Supplementary Chapters to the Theory of Ordinary Differential Equations* (Moscow: Nauka) Geometrical Methods in the Theory of Ordinary Differential Equations (New York: Springer) (1983) (Engl.transl.)
[3] Lie S 1883 *Arch. Math.* **VIII,IX** 187
[4] Abraham R and Marsden J E 1978 *Foundations of Mechanics* (London: Benjamin-Cummings)
[5] Govinder K S and Leach P G L 1995 *J. Phys. A* **28** 5349–5359
[6] Olver J P 1993 *Applications of Lie Groups to Differential Equations* (Springer)
[7] Stephani H 1989 *Differential equations: their solution using Symmetries* (Cambridge Univ. Press)
[8] Ibragimov N H 1985 *Transformations Groups Applied to Mathematical Physics* (D. Reidel Publishing Company)
[9] Lutzky M 1982 *J. Phys. A: Math. Gen.* **15** L87–L91
[10] Aldaya V, Cossío F, Guerrero J and López-Ruiz F F in preparation
[11] Blanes S, Casas F, Oteo J A and Ros J 2009 *Phys. Rep.* **470** 151
[12] Kac V G 1968 *Math. USSR Izv. 2 1271* **2** 1271
[13] Guerrero J, López-Ruiz F F, Calixto M and Aldaya V 2009 *Rep. Math. Phys.* **64** 329–340
[14] Nucci M C 2012 *Journal of Physics: Conference Series* **380** 012008 (12 pages)
[15] Nucci M and Leach P 2013 *J. Math. Anal. Appl.* **406** 219–228

Journal of Physics: Conference Series **538** (2014) 012010 doi:10.1088/1742-6596/538/1/012010

An Introduction to Geometric Algebra with some Preliminary Thoughts on the Geometric Meaning of Quantum Mechanics

Martin Erik Horn

bbw University of Applied Sciences Berlin-Brandenburg,
Leibnizstraße 11 – 13, D – 10625 Berlin, Germany

E-mail: mail@martinerikhorn.de

Abstract. It is still a great riddle to me why Wolfgang Pauli and P.A.M. Dirac had not fully grasped the meaning of their own mathematical constructions. They invented magnificent, fantastic and very important mathematical features of modern physics, but they only delivered half of the interpretations of their own inventions. Of course, Pauli matrices and Dirac matrices represent operators, which Pauli and Dirac discussed in length. But this is only part of the true meaning behind them, as the non-commutative ideas of Grassmann, Clifford, Hamilton and Cartan allow a second, very far reaching interpretation of Pauli and Dirac matrices.
An introduction to this alternative interpretation will be discussed. Some applications of this view on Pauli and Dirac matrices are given, e.g. a geometric algebra picture of the plane wave solution of the Maxwell equation, a geometric algebra picture of special relativity, a toy model of SU(3) symmetry, and some only very preliminary thoughts about a possible geometric meaning of quantum mechanics.

1. Aim of this paper

"The neglect of exterior algebra is the mathematical tragedy of this century. (…) meanwhile, we have to bear with mathematicians who are exterior algebra-blind," tells us Gian-Carlo Rota [1]. Consequently mathematics teachers and their students at schools and universities are exterior algebra-blind, too. Even today nobody tells them how to deal with exterior algebraic structures. It is the aim of this paper, to change this sad truth a little bit.

The remedy Hestenes and other scientists doing physics and mathematics education research propose against exterior algebra-blindness is called *Geometric Algebra*. Thus this paper gives first and foremost a short introduction to Geometric Algebra. But as the author is spelled by the aesthetical and conceptual strength of Geometric Algebra, some more speculative thoughts about Geometric Algebra will be discussed in the second part of this paper.

2. The space we all live in

Obviously, we live in three-dimensional space: We can go to the left or to the right (x direction), forwards or backwards (y direction), and upwards or downwards (z direction). There is no forth spatial direction we can move into. When moving around or moving our arms we feel and experience just three directions.

But there is a severe problem with the construction of our sensory system. Most of our perceptions we use to characterize and interpret the world around us are visual perceptions. We see with our eyes.

But the light, which carries the information about our three-dimensional surroundings, is absorbed by the retina, which is a two-dimensional, curved surface. Thus the three-dimensional objects outside us are represented as two-dimensional images in our eyes. Neurons then conduct this two-dimensional information into our brain which tries to reconstruct the three-dimensional objects around us. But as only information about a two-dimensional distorted picture reaches our brain, the visual model our brain constructs might be incomplete, unsuited for some situations or even faulty. And if the visual models our brain delivers are not correct, the mathematical models we build on these visual models might be partly wrong too.

There are strong hints that the mathematical models of sighted persons are indeed deficient. As knot theorist Alexei Sossinsky remarks, the most complicated knots were invented by blind mathematicians [2]. And "almost all blind mathematicians are (or were) geometers. (...) A blind person's spatial intuition is primarily the result of tactile and operational experience. It is deeper – in the literal as well as the metaphorical sense" [2].

Children are not able to interpret two-dimensional pictures in a way we do as grown-ups. They can not distinguish properly between illuminated convex or concave objects [3]. They have to learn it. In a similar way we all are not able to interpret rotations correctly. We have not learned it. Usually we all suppose objects to be in the same topological state after a rotation of 2π. But this is wrong. An object attached to its surroundings by some strings exhibits a rotation symmetry of 4π which we do not re-

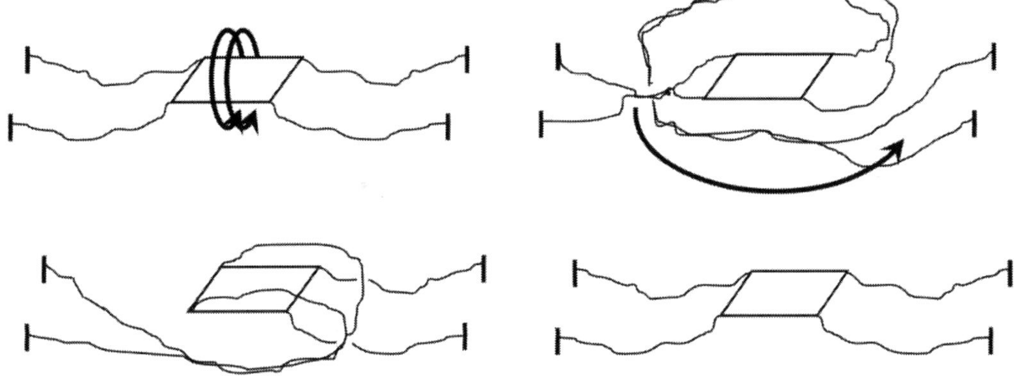

Figure 1. Visualization of the Dirac belt trick.

cognize due to our visual deficiencies. Thus such an object has a spin of ½. Dirac's belt trick [4], [5], [6], [7], [8] clearly demonstrates this fact (see figure 1). We therefore have to draw two important conclusions:

- 4π symmetries are nothing exclusively quantum mechanical.
- 4π symmetries are an essential part of our everyday world (our space) we live in.
- We need an appropriate mathematical language to describe 4π symmetries.

This language is Clifford Algebra. When applied to didactical situations in physics or other sciences Clifford Algebra $C\ell_{3,0}$ of three-dimensional Euclidean space is called Geometric Algebra [9], and Clifford Algebra $C\ell_{1,3}$ of four-dimensional pseudo-Euclidean spacetime is called Spacetime Algebra [10], [11].

3. Geometric algebra is simple

"Much of Clifford algebra is quite simple" [12] or even "simple-minded" (see page XIII at the introduction of [7]). It is even so simple and easy to understand, that Geometric Algebra "can be explained to the first person you meet in the street" (page 820 of [13]). Its basic features are described in many introductory books like [14], [15], [16], [17], [18], [19], [20] from a more or less university level perspective. But to my astonishment even university professors who should be competent with modeling geometric situations mathematically confess that they were "finding it so difficult to understand GA" (see page VII at the introduction of [17]). Therefore I decided to present the basic foundations of Geometric Algebra in this paper in a way suitable for school lessons. I myself taught Geometric Algebra and Spacetime Algebra at physics lessons of upper secondary level in school [21], [22]. And the results of the examinations were indeed encouraging [23].

The first step at these lessons has been to introduce oriented line elements (vectors) and oriented area elements (bivectors) with the aim to immediately hit the algebraic heart of Geometric Algebra: anti-commutativity. A simple discussion of the geometrical situation directly reveals the central idea of Geometric Algebra (see figure 2).

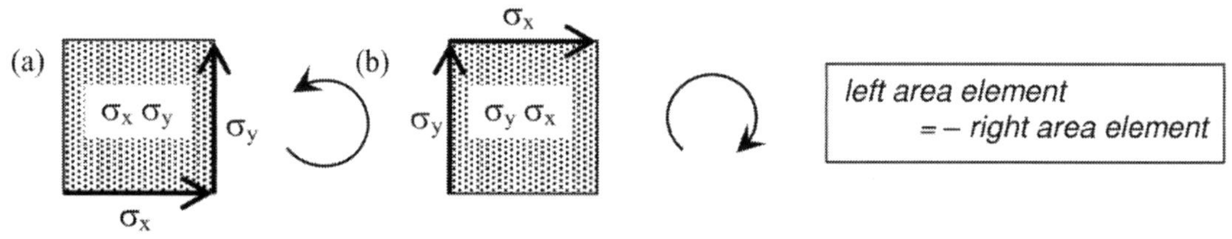

Figure 2. Visualization of the geometric product of two base vectors as base bivector.

If two different vectors are multiplied, we get an oriented area element. This multiplication is called geometric product (or Clifford product if you like) and contains the complete geometric information about the relation of the two vectors [11], [15]. Thus the product of two base vectors results in a base bivector, which represents an oriented unit square. The orientation then of course depends on the order of the multiplication. If we first make a step into the direction of the x-axis and then a step into the direction of the y-axis (figure 2a), we will get a positive or counter-clockwise orientation (which equals the direction of a traffic circle on the continent). If we first make a step into the direction of the y-axis instead and then a step into the direction of the x-axis (figure 2b), we will get a negative or clockwise orientation (which equals the direction of a ghost driver in a traffic circle). This change of orientation is codified in the negative signs of equations (1).

$$\sigma_x\sigma_y = -\sigma_y\sigma_x \qquad \sigma_y\sigma_z = -\sigma_z\sigma_y \qquad \sigma_z\sigma_x = -\sigma_x\sigma_z \tag{1}$$

Together with the condition, that unit vectors square to one

$$\sigma_x^2 = \sigma_y^2 = \sigma_z^2 = 1 \tag{2}$$

these equations form the algebraic core of Geometric Algebra. This is indeed rather simple, and Parra Serra concludes in his didactical analysis of Geometric Algebra: "The only rules to remember are that different orthogonal generating units (vectors) anticommute and that their square is +1…"[1] [13].

When I read for the first time about these rules in a paper from Cambridge, a bizarre announcement followed: "We have now reached the point which is liable to cause the greatest intellectual shock" (see page 1184 of [24]). And it was indeed a severe shock which influences my scientific work till today: These rules define Pauli algebra. Therefore we are forced to conclude: **Pauli matrices represent base vectors of three-dimensional Euclidean space.**

And with the help of the oriented unit volume element (or base trivector or pseudoscalar) I

$$I = \sigma_x\sigma_y\sigma_z \qquad I^2 = (\sigma_x\sigma_y\sigma_z)^2 = -1\,, \tag{3}$$

which acts as an imaginary unit, the Pauli relations can be rediscovered in standard form:

$$\sigma_x\sigma_y = I\,\sigma_z \qquad \sigma_y\sigma_z = I\,\sigma_x \qquad \sigma_z\sigma_x = I\,\sigma_y\,. \tag{4}$$

[1] …or in spacetime – 1 for timelike vectors or 0 in the case of light rays, which will be discussed later.

Every vector r can then be written as linear combination of the Pauli matrices σ_x, σ_y, and σ_z, a fact already noticed by Cartan (see page 43 of [25]):

$$r = x\,\sigma_x + y\,\sigma_y + z\,\sigma_z\,. \tag{5}$$

Following the idea to describe all "geometric objects as real linear combinations of the generating units and their geometric products" (see page 823 of [13]) the most general mathematical object possible in Pauli algebra can be written as multivector

$$\begin{gathered} M = k + x\,\sigma_x + y\,\sigma_y + z\,\sigma_z + A_{xy}\,\sigma_x\sigma_y + A_{yz}\,\sigma_y\sigma_z + A_{zx}\,\sigma_z\sigma_x + V\,\sigma_x\sigma_y\sigma_z \\ M = (k + V\,I) + (x + A_{yz}I)\,\sigma_x + (y + A_{zx}I)\,\sigma_y + (z + A_{xy}I)\,\sigma_z\,. \end{gathered} \tag{6}$$

The second version of multivector (6) is called paravector [26], because the real coefficients $k, x, y, z, A_{xy}, A_{yz}, A_{zx}, V \in \mathbb{R}$ are grouped to form a complex-like vectorial structure.

4. The double nature of Pauli matrices

Pauli invented Pauli matrices to model the strange behaviour of electrons in 1927 (page 608 of [27]). He never considered them to represent base vectors. Pauli always interpreted his matrices as operators (or as a substantial part of an operator) and used the sandwich product $\mathbf{P} = \mathbf{S}\,\mathbf{p}\,\mathbf{S}^{-1}$ (page 611 of [27]) in his calculations.

Of course this is right, too. Pauli matrices possess a double nature: They can be interpreted as operands (vectors, on which is acted on) as well as operators which act on vectors. As $\sigma_i = \sigma_i^{-1}$ the sandwich product of different Pauli matrices with a vector r results in:

$$r_1 = \sigma_x\,r\,\sigma_x = \ \ x\,\sigma_x - y\,\sigma_y - z\,\sigma_z \tag{7}$$

$$r_2 = \sigma_y\,r\,\sigma_y = -x\,\sigma_x + y\,\sigma_y - z\,\sigma_z \tag{8}$$

$$r_3 = \sigma_z\,r\,\sigma_z = -x\,\sigma_x - y\,\sigma_y + z\,\sigma_z\,. \tag{9}$$

These are reflections: vector r_3 of equation (9) equals the original vector r reflected at the z-axis. If we multiply a vector r from the left and from the right by a base vector, the vector will be reflected at the axis which points in the direction of the base vector. **Pauli matrices represent base reflection in three-dimensional Euclidean space.**

This can be generalized for arbitrary directions. If we multiply a vector r from the left and from the right by a unit vector n, the vector r will be reflected at the axis which points in the direction of the unit vector n:

$$r_{ref} = n\,r\,n \qquad \text{with} \quad n^2 = (n_x\,\sigma_x + n_y\,\sigma_y + n_z\,\sigma_z)^2 = 1\,. \tag{10}$$

The two interpretations of figure 3 form the geometric core of Geometric Algebra.

Pauli matrices ⇔ base vectors of three-dimensional Euclidean space	**Pauli matrices ⇔ base reflections in three-dimensional Euclidean space**

Figure 3. Central ideas of Geometric Algebra.

It is well known that two succeeding reflections always result in a rotation. Or the other way round: every rotation can be split into two (or more even numbered) reflections (see pages 1137/1138 of [5] and pages 45/46 of [25]). Therefore rotations can be modeled by two reflections:

$$r_{rot} = m\,r_{ref}\,m = m\,n\,r\,n\,m \qquad \text{with} \quad m^2 = n^2 = 1\,. \tag{11}$$

If we multiply a vector r from the left and from the right first by a unit vector n and then multiply the result from the left and from the right by another unit vector m, the vector r will be rotated in the plane spanned by the two unit vectors n and m through twice the angle between the two unit vectors n and m (see page 111 of [9]).

This is extraordinary. Please notice the many exclamation marks Doran and Lasenby use when discussing these points: "This is starting to look extremely simple!" (page 43 of [15]). And: "The rule also works for any grade of multivector!" (page 44 of [15]). Modeling rotations in Geometric Algebra is indeed a complete conceptual highlight and much easier than in other mathematical systems.

5. Historical background: a missed opportunity

Geometric Algebra goes back to the theory of extensions, first formulated by Grassmann in 1844. Unfortunately, Grassmann's theory was written in an obscure mathematical language nearly nobody understood at this time. It seems that only two eminent mathematicians really noticed, what Grassmann had written: Hamilton, the inventor of quaternion theory, and Charles Sanders Peirce.

Dyson describes the reason for this tremendous ignorance of nearly all mathematicians: "Grassmann was an obscure high-school teacher in Stettin, while Hamilton was the world-famous mathematician whose official titles occupy six lines of print after his name (…). So it is regrettable, but not surprising, that quaternions were hailed as a great discovery, while Grassmann had to wait 23 years before his work received any recognition at all from professional mathematicians" (see page 644 of [28]). And Peirce, fully aware of the importance of Grassmann's contribution, translated Grassmann's ideas into his own mathematical language [29], which again nearly nobody understood at this time – except Clifford, who finally succeeded in writing down [30] all these ideas in a language comprehensible by his contemporaries.

In his short, but important paper Peirce clearly states that "in truth, Grassmann has got hold (though he did not say so) of an eightfold algebra" [29], just the algebra of the multivector (6). So it was totally clear in 1877 that the basic entities of Grassmann's algebra are "three rectangular vectors …, three rectangular planes …, one solid …, unity" [29].

In his Josiah Willard Gibbs Lecture [28] Dyson identifies the delay in the reception of Grassmann's ideas as one of several missed opportunities in the history of mathematics. According to him the central question should have been: "How can it happen that the properties of three-dimensional space are represented equally well by two quite different and incompatible algebraic structures?" (page 644 of [28]). Unfortunately the answer Wolfgang Pauli and P.A.M. Dirac gave to this question partly neglected the geometrical meaning. They never fully grasped the meaning of their own mathematical constructions, and they only delivered half of the interpretations of their inventions. In his Oersted Medal Lecture David Hestenes urgently points to this failure: "The most basic of these misconceptions is that the Pauli matrices are intrinsically related to spin. On the contrary, I claim that their physical significance is derived solely from their correspondence with orthogonal directions in space" (p. 115 of [9]).

This is especially sad in the case of the great Dirac, because it seems that he knew about these ideas and renounced them due to aesthetical considerations. He writes: "In this way the physical world is put into correspondence with the scheme of bi-quaternions, instead of with the scheme of quaternions. Now the scheme of bi-quaternions is not of any special interest in mathematical theory" (page 261 of [31]). This lack of special mathematical interest convinced Dirac to follow another route and construct a very interesting, but also very complicated and sort of conformal connection of quaternions with physical reality [31]. And he writes: "Also (bi-quaternions) have eight components, which is rather too many for a simple scheme for describing quantities in space-time" (page 261 of [31]). Here Dirac failed twice: Eight components are not too many but too few for a scheme for describing quantities in four-dimensional spacetime with necessarily $2^4 = 16$ geometric base elements. And eight components are just the right number of components to describe three-dimensional space, as already Peirce has noticed.

Even today the missed opportunity Dyson identified has an effect on the work of scientists. For example field medalist Michael Atiyah writes: "The geometric significance of spinors is still very mysterious. Unlike differential forms, which are related to areas and volumes, spinors have no such simple explanation. They appear out of some slick algebra, but the geometrical meaning is obscure" (see pages 113/114 of [32]).

These words are astonishing, and we should compare them with the conclusion David Hestenes published 17 years earlier: "Thus … every spinor uniquely determines a rotation dilatation. Therein lies the operational geometric significance of spinors. The spinors form a subalgebra of Geometric Algebra. The algebra of spinors was discovered independent of the full Geometric Algebra by Hamilton, who gave it the name quaternion algebra. Some readers may want to say, rather, that we have here two isomorphic algebras, but there is no call for any such distinction. A quaternion is a spinor. The identification of quaternions with spinors is fully justified not only because they have equivalent algebraic properties, but more important, because they have the same geometric significance" (see page 1022 of [33]).

An example of the ignorance with respect to such a picture of spinors can be found when solving the vacuum Maxwell equation. Even before the re-formulation of Geometric Algebra by David Hestenes in 1966 [34] it was known (but widely ignored) that the spinorial plane wave solution of the Maxwell equations results in circular polarized electromagnetic waves (see page 14 of [35]). This is shown in the following section.

6. Geometric Algebra calculus & plane electromagnetic waves

In three dimensions geometric calculus builds on the geometric derivative

$$\nabla = \sigma_x \frac{\partial}{\partial x} + \sigma_y \frac{\partial}{\partial y} + \sigma_z \frac{\partial}{\partial z} \,. \tag{12}$$

This operator acts on multivectors. In electrodynamics we consider especially the multivector which represents the electromagnetic field strength. As usual the electric field E and the magnetic field B can be combined with the help of an imaginary unit (see page 114 of [9], page 1025 of [33], and page 509 of [36]) into the "complex Faraday vector" F

$$F = E + I\,B = E + \sigma_x\sigma_y\sigma_z\,B \,. \tag{13}$$

But this electromagnetic field strength F should not be looked upon with algebraic eyes only, as the term "complex vector" totally disguises the geometric meaning behind electric and magnetic fields. In Geometric Algebra the electric field strength E is considered as a vector while the magnetic field strength is considered as a bivector I B. Thus the electromagnetic field strength F is always a sum of a vector and an oriented area element.

These two basic ingredients (12) and (13) are all the ingredients needed to dramatically condense the four Maxwell equations into one single Maxwell equation (see again [9], [33], [36])

$$\left(\frac{\partial}{c\,\partial t} + \nabla\right) F = \rho - \frac{1}{c}\,J \qquad \text{or in empty space:} \qquad \left(\frac{\partial}{c\,\partial t} + \nabla\right) F = 0 \,. \tag{14}$$

Instead of four rather unconnected algebraic equations we now possess one equation which expresses all the relevant physical information geometrically as scalar part, vector part, bivector part and trivector part. The great didactical advantage of this geometric interpretation becomes clear when we substitute the purely oscillatory trial solution

$$F = f\,e^{I(\omega t - k z)} \tag{15}$$

for a monochromatic plane wave (with frequency $\omega > 0$ travelling parallel to the z-axis) into the vacuum Maxwell equation (14) and solve it for f. The constant factor f must not be a scalar. According to equation (13) f has again to be the sum of a vector and a bivector, which indicates the electric and the magnetic fields of the origin at time $t = 0$.

Inserting equation (15) into equation (14) results after some straightforward mathematical steps in the following simple intermediate equation (see page 1026 of [33]):

$$F = \sigma_z F \qquad \text{or} \qquad E + \sigma_x\sigma_y\sigma_z B = \sigma_z E + \sigma_x\sigma_y B \,. \tag{16}$$

The different terms of this equation can now be split into its geometrical constituents with the help of the inner and the outer products. The inner product of two vectors a and b or of a bivector **a** and a vector b is defined as:

$$a \bullet b = \frac{1}{2}(ab + ba) \qquad \mathbf{a} \bullet b = \frac{1}{2}(\mathbf{a}b - b\mathbf{a}) \,. \tag{17}$$

It reduces the geometrical grades of the constituents: While a and b are vectors, $a \bullet b$ is a scalar (see left equation of (17)). While **a** is a bivector and b is a vector, $\mathbf{a} \bullet b$ is a vector (see right equation of (17)). The outer product of two vectors a and b or of a bivector **a** and a vector b is defined as:

$$a \wedge b = \frac{1}{2}(ab - ba) \qquad \mathbf{a} \wedge b = \frac{1}{2}(\mathbf{a}b + b\mathbf{a}) \,. \tag{18}$$

It increases the geometrical grades of the constituents: While a and b are vectors, $a \wedge b$ is a bivector (left equation of (18)). While **a** is a bivector and b is a vector, $\mathbf{a} \wedge b$ is a trivector or pseudoscalar (right equation of (18)). And please note the interesting variation of signs in equations (17) and (18). It really depends on the geometrical grades of the factors whether a product is symmetric or antisymmetric!

The sum of inner and outer products now form the geometric product

$$ab = a \bullet b + a \wedge b \qquad \mathbf{a}b = \mathbf{a} \bullet b + \mathbf{a} \wedge b \tag{19}$$

This is indeed strong stuff, and Sobczyk still today remembers the moment, when he first heard about these products: "I remember my sense of amazement when he (David Hestenes) wrote down the basic identity for the geometric multiplication of vectors ... Why hadn't I ever heard of this striking product, and why hadn't I ever heard of a bivector or directed plane segment, since it was the natural generalization of a vector. Twenty-five years later I still find myself asking these same questions" (see page 1291 of [37]).

Now we are able to evaluate equation (16):

$$\underset{\text{vector}}{\underset{\uparrow}{E}} + \underset{\text{bivector}}{\underset{\uparrow}{\sigma_x\sigma_y\sigma_z B}} = \underset{\text{scalar}}{\underset{\uparrow}{\sigma_z \bullet E}} + \underset{\text{bivector}}{\underset{\uparrow}{\sigma_z \wedge E}} + \underset{\text{vector}}{\underset{\uparrow}{\sigma_x\sigma_y \bullet B}} + \underset{\text{trivector}}{\underset{\uparrow}{\sigma_x\sigma_y \wedge B}} \tag{20}$$

The scalar part of this equation

$$\sigma_z \bullet E = 0 \tag{21}$$

tells us that the direction of propagation σ_z and the direction of the electric field vector E are perpendicular. The trivector part of this equation

$$\sigma_x\sigma_y \wedge B = 0 \tag{22}$$

tells us that the conventional magnetic field vector B is parallel to the xy-plane. Therefore the magnetic bivector I B is perpendicular to this plane. The vector and bivector equations

$$E = \sigma_x\sigma_y \bullet B \qquad \text{and} \qquad \sigma_x\sigma_y\sigma_z B = \sigma_z \wedge E \tag{23}$$

both allow us to eliminate the magnetic field because they both result in

$$\sigma_z E = \sigma_x\sigma_y\sigma_z B = I B \tag{24}$$

and therefore

$$F = (1 + \sigma_z) E = E (1 - \sigma_z) \,. \tag{25}$$

If the initial condition $E = E_0\, \sigma_x$ holds at $z = 0$ and $t = 0$, then the plane wave solution of the Maxwell equation in empty space is

$$F_{(t,z)} = E_0\, \sigma_x\, (1 + \sigma_z)\, e^{I(\omega t - k z)} = E_0\, (\sigma_x - \sigma_z \sigma_x)\, [\cos(\omega t - k z) + I \sin(\omega t - k z)] \qquad (26)$$

which is shown in figure 4 (see also figure 2 at page 510 of [36]).

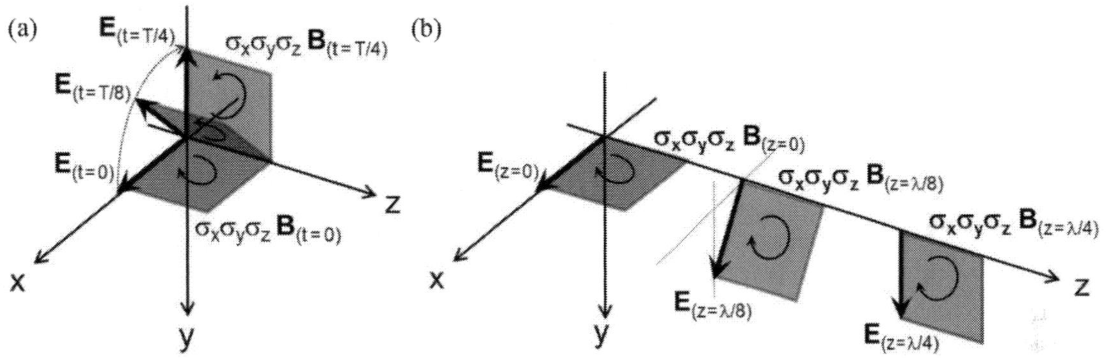

Figure 4. Graphic representations of the plane wave solution of Maxwell's equation in empty space (a) at $z = 0$ and (b) at $t = 0$.

The plane wave solution given in standard textbooks showing linear polarized electromagnetic waves is thus an artificial algebraic solution which neglects the geometric structure of the space we live in. It can only be constructed by a superposition of left and right polarized electromagnetic waves (see also page 2027 of [33]).

7. Four-dimensional spacetimes & special relativity

Obviously, we live in a three-dimensional spatial world. But we only lived there till the invention of special relativity by Einstein in 1905. Since 1905 we live in a four-dimensional spacetime world. This world can be modeled in Geometric Algebra by increasing the number of base vectors – and by changing their quality.

Starting at the origin of a coordinate system light travels the distance x during the time interval ct which can be expressed by the spacetime interval

$$+ (c\,t)^2 - x^2 = 0\,. \qquad (27)$$

This is a disturbing equation! The theory of relativity claims that space and time can be treated structurally in the same way which according to Minkowski calls for an "identical treatment of the four coordinates x, y, z, t" (page 83 of [38]). But space and time are not treated identically here. There is a positive sign in front of the time interval square and a negative sign in front of the space interval square in equation (27). Of course we cannot get rid of these different signs. But we can hide them. To do this, we write equation (27) with the use of two base vectors:

$$+ (c\,t\, \gamma_t)^2 + (x\, \gamma_x)^2 = 0\,. \qquad (28)$$

Now we have identical signs in front of space and time interval squares, because we define the time-like base vector γ_t and the three space-like base vectors γ_x, γ_y, γ_z as

$$\gamma_t^2 = 1 \qquad \gamma_x^2 = \gamma_y^2 = \gamma_z^2 = -1\,. \qquad (29)$$

Again different base vectors anti-commute:

$$\gamma_i\, \gamma_k = -\gamma_k\, \gamma_i \qquad \text{with} \quad i \neq k \quad \text{and} \quad i, k \in \{t, x, y, z\}\,. \qquad (30)$$

And we have now again "reached the point which is liable to cause the greatest intellectual shock" (page 1184 of [24]): These rules define Dirac algebra. Therefore we are forced to conclude: **Dirac matrices represent base vectors of four-dimensional spacetime.**

Every spacetime vector r can then be written as linear combination of the Dirac matrices γ_t, γ_x, γ_y, and γ_z, a fact already noticed by Cartan (page 133 of [25]):

$$r = ct\,\gamma_t + x\,\gamma_x + y\,\gamma_y + z\,\gamma_z\,. \tag{31}$$

Hestenes does not hide his astonishment about the path the history of mathematics has taken here: "The Dirac matrices are no more and no less than matrix representations of an orthonormal frame of spacetime vectors and thereby they characterize spacetime geometry. But how can this be? Dirac never said any such thing!" (page 694 of [10])

Dirac invented Dirac matrices to model the strange behaviour of relativistic electrons in 1928. He never considered them to represent spacetime base vectors. Dirac and Pauli always interpreted their matrices as operators (or as a substantial part of an operator). This is really surprising: clearly all the mathematics was there, and Pauli and Dirac were masters of it. But instead of following the idea of Cartan to use the spacetime vector (31), Pauli tried to model physics with the complete spacetime multivector (see lemma 5 at page 113 of [39])

$$X = \sum_{A=1}^{16} x_A \gamma_A \qquad \text{(with } \gamma_A \text{ every possible product of several Dirac matrices)} \tag{32}$$

as the most general mathematical object possible in Dirac algebra. And of course again he used sandwich products like $\gamma_\mu' = \mathbf{S}\,\gamma_\mu\,\mathbf{S}^{-1}$ (see equation 1 at page 110 of [39]) or $\gamma_\mu X \gamma_\mu = \ldots$ (page 114 of [39]) in his calculations. This makes sense, as Dirac matrices possess also a double nature: They can be interpreted as operands as well as operators which act on operands.

The sandwich products of the Dirac matrices with a spacetime vector r are:

$$r_0 = \gamma_t\, r\, \gamma_t = ct\,\gamma_t - x\,\gamma_x - y\,\gamma_y - z\,\gamma_z \tag{33}$$

$$r_1 = -\gamma_x\, r\, \gamma_x = -ct\,\gamma_t + x\,\gamma_x - y\,\gamma_y - z\,\gamma_z \tag{34}$$

$$r_2 = -\gamma_y\, r\, \gamma_y = -ct\,\gamma_t - x\,\gamma_x + y\,\gamma_y - z\,\gamma_z \tag{35}$$

$$r_3 = -\gamma_z\, r\, \gamma_z = -ct\,\gamma_t - x\,\gamma_x - y\,\gamma_y + z\,\gamma_z \tag{36}$$

These are spacetime reflections. In equation (33) the vector r is reflected at the time axis, and in the other equations (34) to (36) the vector r is reflected at the spatial axis. **Dirac matrices represent base reflections in four-dimensional spacetime.**

This can be generalized for arbitrary directions. If we multiply a vector r from the left and from the right by a unit vector n, the vector r will be reflected at the axis which points in the direction of the unit vector n:

$$r_{ref} = n\, r\, n^{-1} = \pm\, n\, r\, n\,. \tag{37}$$

with $n^2 = 1$ if n is a time-like unit vector ($n^{-1} = +n$) and $n^2 = -1$ if n is a space-like unit vector ($n^{-1} = -n$). The two interpretations of figure 5 form the geometric core of Spacetime Algebra.

Dirac matrices ⇔ base vectors of four-dimensional spacetime	Dirac matrices ⇔ base reflections in four-dimensional spacetime

Figure 5. Central ideas of Spacetime Algebra.

With all that in mind, Lorentz transformations are "starting to look extremely simple" (page 43 of [15]), because they are nothing else than spacetime rotations or two successive spacetime reflections. Therefore they can be modeled with the following very easy to apply formula:

$$r_{rot} = m\, r_{ref}\, m^{-1} = \pm\, m\, n\, r\, n\, m \qquad \text{with} \quad m, n \text{ spacetime unit vectors.} \tag{38}$$

Some Geometric Algebra examples for Lorentz transformations, special relativistic length contractions or time dilations, and other special relativistic effects which can be used for secondary school or introductory university courses can be found in [19], [21], [22], [23], [40].

8. Algebraic and geometric relations between Pauli and Dirac matrices

Usually the three Pauli matrices are considered as (2 x 2) matrices (page 44 of [25]):

$$\sigma_x = \begin{pmatrix} 0 & 1 \\ 1 & 0 \end{pmatrix} \qquad \sigma_y = \begin{pmatrix} 0 & -i \\ i & 0 \end{pmatrix} \qquad \sigma_z = \begin{pmatrix} 1 & 0 \\ 0 & -1 \end{pmatrix} \tag{39}$$

In contrast to that Dirac matrices are considered as (4 x 4) matrices (page 278 of [15]):

$$\gamma_t = \begin{pmatrix} \mathbf{1} & 0 \\ 0 & -\mathbf{1} \end{pmatrix} \qquad \gamma_x = \begin{pmatrix} 0 & -\sigma_x \\ \sigma_x & 0 \end{pmatrix} \qquad \gamma_y = \begin{pmatrix} 0 & -\sigma_y \\ \sigma_y & 0 \end{pmatrix} \qquad \gamma_z = \begin{pmatrix} 0 & -\sigma_z \\ \sigma_z & 0 \end{pmatrix} \tag{40}$$

Applying the direct product of Zehfuss and Kronecker [41], [42], [43], Dirac matrices can be composed of Pauli matrices. They then have the signature (+, –, –, –)[2]:

$$\gamma_t = \sigma_z \otimes \mathbf{1} \qquad \gamma_x = -(\sigma_z\sigma_x) \otimes \sigma_x \qquad \gamma_y = -(\sigma_z\sigma_x) \otimes \sigma_y \qquad \gamma_z = -(\sigma_z\sigma_x) \otimes \sigma_z \tag{41}$$

The main algebraic message is: Two (2 x 2) matrices directly multiplied result in one (4 x 4) matrix. A Dirac matrix is algebraically composed of two Pauli matrices (or Pauli matrix products). Or colloquially: There are two Pauli matrices inside one Dirac matrix.

But there is a deep geometric relationship between Pauli algebra and Dirac algebra which contrasts this sort of superficial algebraic picture: The spatial base vectors of Pauli algebra have to be constructed not only by the spatial base vectors of Dirac algebra, but also by its time-like base vector, as the base vectors of an observer in three-dimensional space depend on the time direction of this observer in four-dimensional spacetime. This base-vector of time is relevant for the three base vectors of space we construct.

This can be seen, if we compare the geometric derivative of three-dimensional space (12) with the geometric derivative of four-dimensional spacetime

$$\Box = \gamma_t \frac{\partial}{c\,\partial t} + \gamma_x \frac{\partial}{\partial x} + \gamma_y \frac{\partial}{\partial y} + \gamma_z \frac{\partial}{\partial z}\,. \tag{42}$$

Obviously the two geometric derivatives of a multivector are not equal, if we add the time direction according to the successful procedure of equation (14):

$$\left(\frac{\partial}{c\,\partial t} + \nabla\right) M \neq \Box\, M \qquad \text{but:} \quad \left(\frac{\partial}{c\,\partial t} + \nabla\right) M = \Box\, \gamma_t\, M\,. \tag{43}$$

At the left equation of (43) the time derivatives have different grades. To adjust this, the geometric derivative (also called Dirac operator) has to be post-multiplied by the time-like base vector γ_t. This results in the balanced equation (44).

[2] Of course other systems of (4 x 4) matrices comply with Dirac algebra, too. For instance, Cartan proposed the following matrices as spacetime base vectors (see page 133 in connection with page 82 of [25]) with signature (+, +, +, –): $A_1 = \sigma_z \otimes \sigma_x$, $A_2 = -\sigma_z \otimes \sigma_y$, $A_3 = \sigma_x \otimes \mathbf{1}$, and $A_4 = c\,(\sigma_z\sigma_x) \otimes \mathbf{1}$

$$\left(\frac{\partial}{c\,\partial t}+\sigma_x\frac{\partial}{\partial x}+\sigma_y\frac{\partial}{\partial y}+\sigma_z\frac{\partial}{\partial z}\right)M=\left(\frac{\partial}{c\,\partial t}+\gamma_x\gamma_t\frac{\partial}{\partial x}+\gamma_y\gamma_t\frac{\partial}{\partial y}+\gamma_z\gamma_t\frac{\partial}{\partial z}\right)M \quad (44)$$

The relation between Pauli and Dirac matrices can be found now by comparing the base vectors in front of the spatial derivatives:

$$\sigma_x=\gamma_x\gamma_t \qquad \sigma_y=\gamma_y\gamma_t \qquad \sigma_z=\gamma_z\gamma_t \quad (45)$$

Geometry tells us, that there are not (colloquially spoken) two Pauli matrices inside one Dirac matrix. There are two Dirac matrices inside every Pauli matrix instead! Of course, the Pauli matrices of equation (45) are written as (4 x 4) matrices. Nevertheless, they totally follow the rules of Pauli algebra[3].

"Consequently, a vector in the Pauli algebra of space is a bivector in the Dirac algebra of space-time" (page 1292 of [37]). Whenever we move along a spatial line, we actually move along a spacetime area element. And this fact is so important, that Sobczyk included it as special highlight in his birthday address for David Hestenes [37].

9. A toy model of SU(3) symmetry

In a sandwich-like product of Geometric Algebra more or less the same factors have to be pre- and post-multiplied. In this section products will be discussed which have different pre- and post-multiplication factors. As an example we try to translate the eight Gell-Mann matrices

$$\lambda_1=\begin{pmatrix}0&1&0\\1&0&0\\0&0&0\end{pmatrix}\quad \lambda_2=\begin{pmatrix}0&-i&0\\i&0&0\\0&0&0\end{pmatrix}\quad \lambda_3=\begin{pmatrix}1&0&0\\0&-1&0\\0&0&0\end{pmatrix}\quad \lambda_4=\begin{pmatrix}0&0&1\\0&0&0\\1&0&0\end{pmatrix}$$
$$\lambda_5=\begin{pmatrix}0&0&-i\\0&0&0\\i&0&0\end{pmatrix}\quad \lambda_6=\begin{pmatrix}0&0&0\\0&0&1\\0&1&0\end{pmatrix}\quad \lambda_7=\begin{pmatrix}0&0&0\\0&0&-i\\0&i&0\end{pmatrix}\quad \lambda_8=\begin{pmatrix}1&0&0\\0&1&0\\0&0&-2\end{pmatrix} \quad (46)$$

of SU(3) symmetry into a toy model of Geometric Algebra. To mimic the multiplication of the first Gell-Mann matrix λ_1 by a complex column vector

$$\lambda_1=\begin{pmatrix}0&1&0\\1&0&0\\0&0&0\end{pmatrix}\begin{pmatrix}x+i\,A_{yz}\\y+i\,A_{zx}\\z+i\,A_{xy}\end{pmatrix}=\begin{pmatrix}y+i\,A_{zx}\\x+i\,A_{yz}\\0\end{pmatrix} \quad (47)$$

we have to find a Geometric Algebra transformation of a multivector M which gets rid of the third paravector component $z+A_{xy}\,I$. And of course the scalar part should disappear too, because it is not part of the resulting column vector of equation (47). This can be achieved by adding the following equations:

$$\begin{aligned}\sigma_x M \sigma_y &= \sigma_x\left((k+VI)+(x+A_{yz}I)\,\sigma_x+(y+A_{zx}I)\,\sigma_y+(z+A_{xy}I)\,\sigma_z\right)\sigma_y\\ &= (k+VI)\,\sigma_x\sigma_y+(x+A_{yz}I)\,\sigma_y+(y+A_{zx}I)\,\sigma_x-(z+A_{xy}I)\,\sigma_x\sigma_y\sigma_z\end{aligned} \quad (48)$$

$$\sigma_y M \sigma_x = -(k+VI)\,\sigma_x\sigma_y+(x+A_{yz}I)\,\sigma_y+(y+A_{zx}I)\,\sigma_x+(z+A_{xy}I)\,\sigma_x\sigma_y\sigma_z \quad (49)$$

[3] In five-dimensional spacetimes or spacetimevelocities this works in a similar way. Then Pauli matrices (base vectors of three-dimensional space) can be identified as products of three Dirac matrices (base vectors of five-dimensional spacetimes or spacetimevelocities, see equations 24-26 of [44]) as $\sigma_x' = \gamma_x\gamma_t\gamma_v$, $\sigma_y' = \gamma_y\gamma_t\gamma_v$, $\sigma_z' = \gamma_z\gamma_t\gamma_v$, if together with the Dirac matrices (41) the time-like Dirac matrix $\gamma_v = \sigma_x \otimes \mathbf{1}$ (see equation 45 at page 497 of [45]) is used as additional base vector for the direction of the fifth coordinate.

Journal of Physics: Conference Series **538** (2014) 012010 doi:10.1088/1742-6596/538/1/012010

This can be done for the other Gell-Mann matrices in a similar way. The results are in three-dimensional space (see following equations on the left side) and according to transformation (45) in four-dimensional spacetime (equations on the right side):

$$\lambda_1(M) = \frac{1}{2}\,(\sigma_x\, M\, \sigma_y + \sigma_y\, M\, \sigma_x) \qquad \lambda_1(M) = -\frac{1}{2}\,\gamma_t\,(\gamma_x\, M\, \gamma_y + \gamma_y\, M\, \gamma_x)\,\gamma_t \tag{50}$$

$$\lambda_2(M) = \frac{1}{2}\,(\sigma_z\, M\, \mathbf{1} - \mathbf{1}\, M\, \sigma_z) \qquad \lambda_2(M) = -\frac{1}{2}\,\gamma_t\,(\gamma_z\, M\, \gamma_t - \gamma_t\, M\, \gamma_z)\,\gamma_t \tag{51}$$

$$\lambda_3(M) = \frac{1}{2}\,(\sigma_x\, M\, \sigma_x - \sigma_y\, M\, \sigma_y) \qquad \lambda_3(M) = -\frac{1}{2}\,\gamma_t\,(\gamma_x\, M\, \gamma_x - \gamma_y\, M\, \gamma_y)\,\gamma_t \tag{52}$$

$$\lambda_4(M) = \frac{1}{2}\,(\sigma_z\, M\, \sigma_x + \sigma_x\, M\, \sigma_z) \qquad \lambda_4(M) = -\frac{1}{2}\,\gamma_t\,(\gamma_z\, M\, \gamma_x + \gamma_x\, M\, \gamma_z)\,\gamma_t \tag{53}$$

$$\lambda_5(M) = \frac{1}{2}\,(\mathbf{1}\, M\, \sigma_y - \sigma_y\, M\, \mathbf{1}) \qquad \lambda_5(M) = -\frac{1}{2}\,\gamma_t\,(\gamma_t\, M\, \gamma_y - \gamma_y\, M\, \gamma_t)\,\gamma_t \tag{54}$$

$$\lambda_6(M) = \frac{1}{2}\,(\sigma_y\, M\, \sigma_z + \sigma_z\, M\, \sigma_y) \qquad \lambda_6(M) = -\frac{1}{2}\,\gamma_t\,(\gamma_y\, M\, \gamma_z + \gamma_z\, M\, \gamma_y)\,\gamma_t \tag{55}$$

$$\lambda_7(M) = \frac{1}{2}\,(\sigma_x\, M\, \mathbf{1} - \mathbf{1}\, M\, \sigma_x) \qquad \lambda_7(M) = -\frac{1}{2}\,\gamma_t\,(\gamma_x\, M\, \gamma_t - \gamma_t\, M\, \gamma_x)\,\gamma_t \tag{56}$$

$$\lambda_8(M) = \frac{1}{2\sqrt{3}}(\sigma_x M \sigma_x + \sigma_y M \sigma_y - 2\sigma_z M \sigma_z) \qquad \lambda_8(M) = -\frac{1}{2\sqrt{3}}\gamma_t(\gamma_x M \gamma_x + \gamma_y M \gamma_y - 2\gamma_z M \gamma_z)\gamma_t \tag{57}$$

These results are nearly identical to results which can be found in the literature. Only the transformations for the diagonal Gell-Mann matrices $\lambda_3(M)$ and $\lambda_8(M)$ slightly differ from these results, see table 1 at page 168 of [46] for the equations given above on the left side and equation 26 a-h at page 7 of [47] for the equations on the right side.

10. Preliminary Thoughts on the Geometric Meaning of Quantum Mechanics

The introduction to Geometric Algebra given in the preceding sections follows ideas published by many other authors. All these things described in these sections are not new, but already known since decades – or even longer.

The main focus of my paper might be a little bit different compared to standard presentations, because for didactical reasons I didn't emphasize coordinate-free methods in the way other authors would expect [48]. But the mathematical and physical interpretations are surely identical.

This changes when we reach quantum mechanics. Very probably no human being till now has actually understood quantum mechanics in a way we understand physical laws from other domains. "I don't understand it. Nobody does," Feynman tells about quantum mechanics (page 9 of [49]).

Therefore the interpretation of equations describing quantum mechanical situations is unclear and controversial. But these equations obviously describe experiments to a very high accuracy. The algebra of quantum mechanics is understood. It's the geometry of quantum mechanics which accounts for nearly all open questions. For this reason a mathematical language which uniquely combines algebra and geometry might be suited to shed more light on the meaning of quantum mechanics.

Then the basic question is: What is the geometry of the world we live in? We can describe this geometry in the non-relativistic case by the 16 basic elements of Pauli algebra: One base scalar which equals the (2 x 2) unit matrix, three base vectors σ_x, σ_y, σ_z, given in equation (39), three base bivectors or oriented unit area elements $\sigma_x\sigma_y$, $\sigma_y\sigma_z$, $\sigma_z\sigma_x$ and one base trivector or oriented unit volume element

$\sigma_x \sigma_y \sigma_z$. But then we are stuck: There are two possibilities for an oriented volume element in nature. In the preceding sections we always decided to take $I = \sigma_x \sigma_y \sigma_z$ as oriented volume element. But does nature really decides for a positive sign if all base vectors are multiplied? This positive sign is nothing else than the consequence of an arbitrary decision we as human beings make. But there is another possibility: It might as well be that nature decides for a negatively oriented unit volume element $-I$.

Or in other words: Do we live in a left-handed or a right-handed coordinate system designed by nature? Obviously we do not know. And the more interesting question is: What will happen, if even nature does not know?

A possible answer to this question could be, that at every measurement there will be a probability $\rho(1)$, that the sign is positive. And there will be a probability $\rho(-1) = 1 - \rho(1)$, that the sign is negative. The consequences are clear: Whenever we measure a physical entity U which should distinguish between the different possible signs of the oriented volume element (for example spin, magnetic moment) we will get a statistical result which represents the mean value of U:

$$\langle U \rangle = (-1)\,\rho(-1) + (1)\,\rho(1) \tag{58}$$

Therefore the hypothesis is: **It is not the spin of the electron, which is uncertain in quantum mechanics, but the orientation or handedness of the space we live in.**

As said, the algebra of quantum mechanics is understood and is completely available for all of us. For instance you can find the algebra of equation (58) in Jordan's book (page 73 of [50]), where this equation is the starting point for his discussion of quantum mechanics. Jordan develops an interpretation of quantum mechanics, in which "all quantities (are) made from spin" (chapter 10 of [58]). We should try to follow his mathematical route, which really offers an "easy way to look rather deeply into quantum mechanics" (pager VIII of [58]). But at the same time we should also try to change the interpretative frame of Jordan's presentation. Quantum mechanical quantities are made from geometry and might be a consequence of uncertainty in the handedness of the space[4].

11. Final remarks: we should teach Geometric Algebra

The following two points, David Hestenes again and again emphasized (see pages 120/121 of [9]):

- Geometric Algebra provides a unified language for the whole of physics that is conceptually and computationally superior to alternative mathematical systems in every application domain,
- Reforming the mathematical language of physics is the single most essential step toward simplifying physics education at all levels from high school to graduate school,

should have become clear by this paper. We therefore should not hesitate to promote and enlarge the knowledge about Geometric Algebra. This surely is no complicated task, because Seneca is still right today: "Non quia difficilia sunt, non audemus, sed quia non audemus, difficilia sunt" [51] – It is not because Geometric Algebra is difficult that we do not dare, but because we do not dare, Geometric algebra is difficult.

---

[4] Perhaps this is not the right place to admit that I have a severe problem with general relativity: It delivers far too much solutions. In our eyes today, most solutions of Einstein's field equation seem to be irrelevant to physics, and especially solutions with closed time-like curves are considered as suspicious. Therefore one of three alternatives should be correct: 1. Einstein's field equation is wrong. The correct field equations should only give us solutions which describe situations of physical reality. 2. Einstein's field equation is incomplete. There should be an additional equation which tells us whether a solution of Einstein's field equation describes a situation possible in physical reality or not. 3. Einstein's field equation is correct and all solutions of the field equation describe situations possible in physics. Then solutions with closed time-like curves are just solutions which describe quantum mechanical phenomena. An elementary particle following a time-like curve has a probability $\rho(1)$, that its coordinate system is left-handed. And there will be a probability $\rho(-1) = 1 - \rho(1)$, that its coordinate system is right-handed, if the timelike curve changes the orientation of space within the elementary particles time travel like the orientation of an ant moving along a Moebius strip. This would explain quantum mechanics as a general relativistic effect according to equation (58).

References

[1] Rota G-C 1997 *Indiscrete Thoughts* (Boston: Birkhäuser) p 232–233
[2] Sossinsky A 2002 *Knots. Mathematics with a Twist* (Cambridge: Harvard University Press) p 12–13
[3] Stone J and Pascalis O 2010 Footprints Sticking Out of the Sand (Part I) – Children's Perception of Naturalistic and Embossed Symbol Stimuli. *Perception* **39** (9) pp 1254–1260
[4] Penrose R and Rindler W 1984 *Spinors and Space-Time. Vol. 1: Two-Spinor Calculus and Relativistic Fields* (Cambridge: Cambridge University Press) p 43
[5] Misner C W, Thorne K S and Wheeler J A 1973 *Gravitation* (New York: Freeman) p 1149
[6] Feynman R and Weinberg S 1987 *Elementary Particles and the Laws of Physics. The 1986 Dirac Memorial Lectures* (Cambridge: Cambridge University Press) pp 29, 30, 58
[7] Snygg J 1997 *Clifford Algebra. A Computational Tool for Physicists* (New York: Oxford University Press) p 12
[8] Staley M 2010 Understanding Quaternions and the Dirac Belt Trick. *European Journal of Physics* **31** (3) pp 467–478
[9] Hestenes D 2003 Oersted Medal Lecture 2002: Reforming the Mathematical Language of Physics. *American Journal of Phyics* **71** (2) pp 104–121
[10] Hestenes D 2003 Spacetime Physics with Geometric Algebra. *American Journal of Physics* **71** (7) pp 691–714
[11] Jung F 2006 *Geometrische Algebra und die Rolle des Clifford-Produkts in der Klassischen und Quantenmechanik*. Diploma thesis, written at the Institute of Physics, University of Mainz
[12] Lounesto P 1998 Book review of [7]. *Foundations of Physics* **28** (6) p 1021
[13] Parra Serra J M 2009 Clifford Algebra and the Didactics of Mathematics. *Advances in Applied Clifford Algebras* **19** (3/4) pp 819–834
[14] Hestenes D 2002 *New Foundations for Classical Mechanics*. 2nd ed. (New York: Kluwer Academic Publishers)
[15] Doran C and Lasenby A 2003 *Geometric Algebra for Physicists* (Cambridge: Cambridge University Press)
[16] Chisolm E 2012 *Geomertric Algebra*. arXiv:1205.5935v1 [math-ph] May 27th, 2012
[17] Vince J 2008 *Geometric Algebra for Computer Graphics* (London: Springer) p VII
[18] Hildenbrand D 2013 *Foundations of Geometric Algebra Computing* (Berlin: Springer)
[19] Horn M E 2012 *Pauli-Algebra und S_3-Permutationsalgebra – Eine algebraische und geometrische Einführung*. Published at www.bookboon.com/de (London: Ventus Publishing ApS)
[20] González Calvet R 2001 *Treatise of Plane Geometry Through Geometric Algebra*. Published at http://campus.uab.es/~PC00018 (Barcelona: University of Barcelona)
[21] Horn M E 2009 Die Spezielle Relativitätstheorie im Kontext der Raumzeit-Algebra & Arbeitsbögen zur Geometrischen Algebra und zur Raumzeit-Algebra. Published in Nordmeier V and Grötzebauch H (Eds) *Didaktik der Physik*. CD-ROM of the Spring Meeting of the German Physical Society in Bochum, paper 30.40 (Berlin: LOB – Lehmanns Media)
[22] Horn M E 2011 Grassmann, Pauli, Dirac: Special Relativity in the Schoolroom. Published in Petsche H-J, Lewis A C, Liesen J and Russ S (Eds) *From Past to Future – Graßmann's Work in Context*, pp 435–450 (Basel: Birkhäuser)
[23] Horn M E 2010 Die Raumzeit-Algebra im Abitur. *PhyDid B – Didaktik der Physik*, paper 28.4
[24] Gull S, Lasenby A and Doran C 1993 Imaginary Numbers are not Real – The Geometric Algebra of Spacetime. *Foundations of Physics* **23** (9) pp 1175–1201
[25] Cartan É 1981 *The Theory of Spinors*. Unabridged republication of the complete English translation first published in 1966 (New York: Dover Publications)
[26] Baylis W E 2002 *Electrodynamics. A Modern Geometric Approach* (Boston: Birkhäuser) p 22
[27] Pauli W 1927 Zur Quantenmechanik des magnetischen Elektrons. *Zeitschrift für Physik* **43** (9/10) pp 601–623
[28] Dyson F J 1972 Missed Oppertunities. Printed version of the Josiah Willard Gibbs Lecture of Jan 17th, 1972. *Bulletin of the American Mathematical Society* **78** (5) pp 635–652

[29] Peirce C S 1877 Note on Grassmann's Calculus of Extension. *Proceedings of the American Academy of Arts and Sciences* **13** pp 115–116
[30] Clifford W K 1878 Applications of Grassmann's Extensive Algebra. *American Journal of Mathematics* **1** (4) pp 350–358
[31] Dirac P A M 1945 Applications of Quaternions to Lorentz Transformations. *Proceedings of the Royal Irish Academy*, Vol L, Sect A, pp 261–270
[32] Atiyah M F 1998 The Dirac Equation and Geometry. Published in Goddard P (Ed) *Paul Dirac. The Man and His Work.* Chap. 4, pp 108–124 (Cambridge: Cambridge University Press)
[33] Hestenes D 1971 Vectors, Spinors, and Complex Numbers in Classical and Quantum Physics. *American Journal of Physics* **39** (9) pp 1013–1027
[34] Hestenes D 1966 *Space-Time Algebra.* (New York: Gordon & Breach)
[35] Moses H E 1958 *The Solution of Maxwell's Equations in Terms of a Spinor Notation. Part I: The Initial Value Problem in Terms of Field Strengths and the Inverse Problem.* New York University, Institute of Mathematical Sciences, Research Report No. EM-114
[36] Vold T 1993 An Introduction to Geometric Calculus and its Application to Electrodynamics. *American Journal of Physics* **61** (6) pp 505–513
[37] Sobczyk G 1993 David Hestenes: The Early Years. *Foundations of Physics* **23** (10) pp 1291–1293
[38] Minkowski H 1909 Space and Time. English version published in Lorentz H A, Einstein A, Minkowski H and Weyl H 1923 *The Principle of Relativity.* Chap 5, pp 73–91 (New York: Dover Publications)
[39] Pauli W 1936 Contributions Mathématique à la Théorie des Matrices de Dirac. *Annales de l'Institut Henri Poincaré* **6** (2) pp 109–136
[40] Horn M E 2010 Die Spezielle Relativitätstheorie in der Mathematikerausbildung & Pauli-Algebra und Dirac-Algebra. OH-Folien zur Einführung in die Spezielle Relativitätstheorie im Rahmen eines Kurses zur Physik für Mathematiker. *PhyDid B – Didaktik der Physik*, paper 19.35 & attachment 19.35, URL: http://www.phydid.de [30 Nov 2013]
[41] Zehfuss J G 1858 Über eine gewisse Determinante. *Zeitschrift für Mathematik und Physik* **3** pp 298–301
[42] Steeb W-H 1991 *Kronecker Product of Matrices and Applications* (Mannheim, Wien, Zürich: Bibliographisches Institut/Wissenschaftsverlag)
[43] Henderson H V, Pukelsheim F and Searle S R 1983 On the History of the Kronecker Product. *Linear and Multilinear Algebra* **14** (2) pp 113–120
[44] Horn M E 2011 Die fünfdimensionale Raumzeit-Algebra am Beispiel der Kosmologischen Relativität. *PhyDid B – Didaktik der Physik*, paper 17.1
[45] Horn M E 2012 Translating Cosmological Special Relativity into Geometric Algebra. Published in Sivasundaram S (Ed) *Proceedings of the 9th International Conference on Mathematical Problems in Engineering, Aerospace and Sciences in Vienna (ICNPAA).* American Institute of Physics, AIP Conference Proceedings, **1493** pp 492–498
[46] Hestenes D 1982 Space-Time Structure of Weak and Electromagnetic Interactions. *Foundations of Physics* **12** (2) pp 153–168
[47] Vroegindewey P G 1989 *Gauge Theories and Space-Time Algebra.* Reports on Applied and Numerical Analysis (RANA 89-06). Department of Mathematics and Computing Science, Eindhoven University of Technology, May 1989.
[48] Hestenes D 1992 Mathematical Viruses. Published in Micali A, Boudet R and Helmstetter J (Eds) *Clifford Algebras and Their Applications in Mathematical Physics*, pp 3–16 (Dordrecht: Kluwer Academic Publishers)
[49] Feynman R P 2008 *QED – The Strange Theory of Light and Matter.* First Princeton paperback printing (Princeton, New Jersey: Princeton University Press)
[50] Jordan T 2005 *Quantum Mechanics in Simple Matrix Form* (New York: Dover Publications)
[51] Seneca *Letter to Lucilius*, letter 104, section 26

Embedded Random Matrix Ensembles with Lie Symmetries: Results from $U(\Omega)$ Wigner-Racah algebra

V.K.B. Kota[1,*] **and Manan Vyas**[2,3,†]
[1]Physical Research Laboratory, Ahmedabad 380 009, India
[2]Department of Physics and Astronomy, Washington State University, Pullman, Washington 99164-2814, USA
[3]Department of Physics, Yeshiva University, New York, New York 10016, USA
E-mail: *vkbkota@prl.res.in, †manan.vyas@yu.edu

Abstract.
Random matrix ensembles for a system of m number of fermions or bosons in Ω number of single particle levels each r-fold degenerate and interacting with two-body forces are considered. The spectrum generating algebra for these systems is $U(r\Omega)$ and a subalgebra of interest is $U(r\Omega) \supset U(\Omega) \otimes SU(r)$ algebra. Now, for random two-body interactions preserving $SU(r)$ symmetry, one can introduce embedded Gaussian unitary ensemble of random matrices with $U(\Omega) \otimes SU(r)$ embedding and this class of ensembles are denoted by EGUE(2)-$SU(r)$. Ensembles with $r = 1$, 2 and 4 for fermions correspond to spinless fermions, fermions with spin and fermions with Wigner's spin-isospin $SU(4)$ symmetry respectively. Similarly, for bosons $r = 1$, 2 and 3 correspond to spinless bosons, two species boson systems and bosons with spin one respectively. The distinction between fermions and bosons is in the $U(\Omega)$ irreducible representations. General formulation based on Wigner-Racah algebra for lower order moments of the one- and two-point functions in eigenvalues generated by EGUE(2)- $SU(r)$ is briefly reviewed. The final formulas for the moments involve only $SU(\Omega)$ Racah coefficients. For the fourth moment of the one-point function for $r > 1$ and for the higher order (> 4) bivariate moments of the two-point function for $r \geq 1$, formulas are not available for the $SU(\Omega)$ Racah coefficients that are needed. It is necessary to derive analytical formulas for these or develop methods that give asymptotic results (an example for this is given in the paper) or develop methods that allow for their numerical evaluation. This important open problem is discussed in some detail.

1. Introduction

Wigner introduced random matrix theory (RMT) in physics in 1955 and as Wigner stated: *The assumption is that the Hamiltonian which governs the behavior of a complicated system is a random symmetric matrix, with no special properties except for its symmetric nature.* Further, if 'complicated'had been replaced by 'non-integrable', it would have led to the foundations of what is now well known as 'Quantum Chaos'. Depending on the global symmetry properties of the Hamiltonian of a quantum system, namely rotational symmetry and time-reversal symmetry, we have Dyson's tripartite classification of random matrices giving the classical random matrix ensembles, the Gaussian orthogonal (GOE), unitary (GUE) and symplectic (GSE) ensembles. For example, given the matrix elements $a_{ij} = b_{ij} + ic_{ij}$ of a GUE matrix with $a_{ji} = (a_{ij})^\dagger$, the b_{ij}, $i \geq j$ and $c_{ij}, i > j$ are independent Gaussian random variables with zero center and variance $v^2(1 + \delta_{ij})$ where v is a free parameter. Moreover, $\overline{a_{ij}a_{kl}} = 2v^2(\delta_{il}\delta_{jk})$ where the 'overline'indicates ensemble average. Here after we call this GUE(v^2). In the last three decades, RMT has found applications not only in all branches of quantum physics but also in many other disciplines such as Econophysics, Wireless communication, information theory, multivariate statistics, number theory, neural and biological networks and so on [1, 2, 3, 4].

As stated by H.A. Weidenmüller [5]: *although used with increasing frequency in many branches of Physics, random matrix ensembles sometimes are too unspecific to account for important features of the physical system at hand. One refinement which retains the basic stochastic approach but allows for such features consists in the use of embedded ensembles.* Finite quantum systems such as nuclei, atoms, quantum dots, small metallic grains, interacting spin systems modeling quantum computing core and ultra-cold atoms, share one common property - their constituents (predominantly) interact via two-particle interactions. Therefore, it is more appropriate to represent an isolated finite interacting quantum system by random matrix models generated by random two-body interactions. Then we have random matrix ensembles in m-particle spaces - these ensembles are defined by representing the two-particle Hamiltonian (H) by GOE/GUE/GSE and then the m particle H matrix is generated by the m-particle Hilbert space geometry. The key element here is the recognition that there is a Lie algebra that transports the information in the two-particle spaces to many-particle spaces. As a GOE/GUE/GSE random matrix ensemble in two-particle spaces is embedded in the m-particle H matrix, these ensembles are more generically called embedded ensembles (EE). It is important to note that the classical GOE/GUE/GSE, now almost universally regarded as models for the corresponding chaotic systems [2], are ensemble of multi-body, not two-body interactions. This difference shows up both in one-point (density of states) and two-point (fluctuations and transition strengths) functions as seen first in nuclear shell model examples [6, 7].

Embedded random matrix ensembles ensembles or simply embedded ensembles (EE), though introduced in 1970 and analyzed in detail first in 1975, have received new emphasis beginning from 1996 and since then a wide variety of EGE have been introduced in literature, both for fermion and boson systems [7, 8, 9]. Motivation to study EE in detail has come from: (i) many-body quantum chaos; (ii) statistical spectroscopy (in nuclei and atoms); (iii) generic electron-electron interaction effects on transport properties of mesoscopic systems; (iv) regular structures generated by random interactions in quantum many-body systems; (v) possible applications to quantum gases and in quantum information science (thermalization, entanglement and other aspects). Since 1997, a variety of EE have been introduced in literature and almost all these arise from different Lie algebras that define the embedding. A major hurdle in deriving statistical properties generated by EE is that they are largely intractable analytically. Although numerical experiments have given some interesting and valuable results [7, 8, 9, 10, 11, 12], perturbation theory, trace propagation methods and the so-called binary correlation method are used to derive some analytical results. However, one general formulation for deriving analytical results is to use the Wigner-Racah algebra of the embedding Lie algebra. Though this is indicated in some early papers [13, 14, 15, 16], it is clearly established only in 2005 [17] and applied to a variety of EGUEs thereafter [17, 18, 19, 20]. Our purpose here is to review the progress made in applying Wigner-Racah algebra to a class of EE, the embedded Gaussian unitary ensembles with $U(\Omega) \otimes SU(r)$ embedding and generated by random two-body interactions with $SU(r)$ symmetry. These are called EGUE(2)-$SU(r)$ ensembles. Secondly, the purpose is also to clearly bring out the outstanding unsolved problems. Now we will give a preview.

In Section 2, introduced and defined are EGUE(2)-$SU(r)$ ensembles. Section 3 gives a brief discussion of the Wigner-Racah algebra formulation for the moments of the one- and two-point functions in eigenvalues generated by these ensembles. In Section 4, the formulation for the fourth moment of the density of eigenvalues is presented and discussed in detail the difficulties associated with the Wigner-Racah formulation. In Section 5, as an example of asymptotic methods, results obtained using an extended binary correlation method for the moments of a bivariate transition strength density are presented. Finally, Section 6 gives future outlook.

2. EGUE(2)-$SU(r)$ ensembles: Definition

Let us begin with m spinless fermions in Ω number of single particle (sp) states (labeled by $|\nu_i\rangle$, $i = 1, 2, \ldots, \Omega$) and the Hamiltonian for the system is say two-body preserving particle number m. Now, each distribution of the m fermions in the Ω number of sp states defines a basis state. The number of basis states $d_f(\Omega, m)$ is

$$d_f(\Omega, m) = \binom{\Omega}{m} . \tag{1}$$

For example, $d_f(12, 6) = 924$ and $d_f(16, 8) = 12870$. General form of a two-body Hamiltonian is,

$$H(2) = \sum_{i>j,k>\ell} H_{ijk\ell}\, a^\dagger_{\nu_k} a^\dagger_{\nu_\ell} a_{\nu_j} a_{\nu_i}\ ; \quad H_{ijk\ell} = \langle \nu_k \nu_\ell \mid H(2) \mid \nu_i \nu_j \rangle\ . \tag{2}$$

Here, $a^\dagger$ are creation operators for a single fermion and a are annihilation operators. It is easy to write formulas for the matrix elements of H in the m fermion basis states and hence it is easy to construct the m-fermion H matrix. Assuming GUE(1) representation for H in two-particle space will give EGUE(2) in m particle spaces. The embedding algebra here is $U(\Omega)$. Also, this corresponds to $r = 1$ in EGUE(2)-$SU(r)$ with $U(\Omega)$ irreducible representation (irrep) being $\{1^m\}$. See Fig. 1 for the Young tableaux representation of irreps and this is used throughout this paper. Just as above, it is possible to define BEGUE(2) for m bosons in Ω number of sp states. Here, the number of basis sates is given by

$$d_b(\Omega, m) = \binom{\Omega + m - 1}{m} . \tag{3}$$

For example, $d_b(5, 10) = 1001$ and $d_b(8, 20) = 888030$. Similarly, the two-body Hamiltonian is

$$H(2) = \sum_{i\geq j,k\geq\ell} \frac{H_{ijk\ell}}{\sqrt{(1+\delta_{ij})(1+\delta_{k\ell})}}\, b^\dagger_{\nu_k} b^\dagger_{\nu_\ell} b_{\nu_j} b_{\nu_i}\ ; \quad H_{ijk\ell} = \langle \nu_k \nu_\ell \mid H(2) \mid \nu_i \nu_j \rangle\ . \tag{4}$$

Assuming GUE(1) representation for H in two-particle spaces will give BEGUE(2) in m particle spaces. The embedding algebra here also is $U(\Omega)$ and $r = 1$ in EGUE(2)-$SU(r)$ but the $U(\Omega)$ irrep is $\{m\}$. Now we will consider generalization to any $r > 1$.

Consider m number of fermions or bosons in Ω number of sp levels each r-fold degenerate and then the total number of sp states is $N = r\Omega$. Also we assume that the Hamiltonian for these systems is a $SU(r)$ scalar. Then the spectrum generating algebra is $SU(r\Omega) \supset U(\Omega) \otimes SU(r)$. All the m-particle states can be classified according to the irreps of $SU(r\Omega)$, $U(\Omega)$ and $SU(r)$. These in turn can be labeled by the Young tableaux $\{f\}$ with no more than k parts for a given $U(k)$ and $\sum_{i=1}^{k} f_i = m$. Moreover, all m particle states for fermion systems will transform as the $SU(r\Omega)$ irrep $\{1^m\}$ (antisymmetric irrep shown in Fig. 1) and for bosons as the symmetric irrep $\{m\}$. Due to this fact, with respect to $SU(\Omega)$ if a m particle state transforms as $\{f\}$, then with respect to $U(r)$ it will transform as $\{\tilde{f}\}$ irrep for fermions and simply as $\{f\}$ for bosons. Note that we are not making a distinction between $U(k)$ and $SU(k)$ although the $SU(k)$ irreps $\{g\}$ that corresponds to a $U(k)$ irrep $\{f\}$ will be such that $g_i = f_i - f_k$. From now on we will denote m particle $U(\Omega)$ irreps by f_m. Then the corresponding $SU(r)$ irrep for bosons is f_m and for fermions it is $\tilde{f}_m$. Therefore, the $SU(r)$ irrep need not be specified in denoting the symmetry defined basis states. In addition, we will denote the irrep labels that belong to the subalgebras of $U(\Omega)$ by v_m and those that belong to the subalgebras of $SU(r)$ by β_m. With all these, m particle states with $U(\Omega) \otimes SU(r)$ symmetry can be labeled as $|f_m, v_m, \beta_m\rangle$. Then, a two-body Hamiltonian that is $SU(r)$ scalar can be written as

$$H(2) = \sum_{f_2, v_2^i, v_2^f, \beta_2; f_2=\{2\},\{1^2\}} H_{f_2 v_2^i v_2^f} \, A^\dagger(f_2 v_2^f \beta_2)\, A(f_2 v_2^i \beta_2) \;; \qquad (5)$$

$$H_{f_2 v_2^i v_2^f} = \left\langle f_2, v_2^f, \beta_2 \mid H(2) \mid f_2, v_2^i, \beta_2 \right\rangle .$$

Note that $H_{f_2 v_2^i v_2^f}(2)$ is independent of β_2 and this ensures that the $H(2)$ operator is $SU(r)$ scalar. Also $A^\dagger$ and A are normalized two-particle creation and annihilation operators. Similarly, the matrix elements of $H(2)$ in m particle space,

$$H_{f_m v_m^i v_m^f} = \left\langle f_m, v_m^f, \beta_m \mid H(2) \mid f_m, v_m^i, \beta_m \right\rangle \qquad (6)$$

are also independent of β_m. Now we are in a position to define the EGUE(2)-$SU(r)$ ensemble.

Firstly, the $H(2)$ matrices in two-particle spaces are taken to be independent GUEs, one for each f_2, i.e. real and imaginary parts of the two-particle matrix elements $H_{f_2 v_2^i v_2^f}$ are independent Gaussian variables with zero center and variance given by (with bar representing ensemble average),

$$\overline{H_{f_2 v_2^1 v_2^2}\; H_{f_2' v_2^3 v_2^4}} = \delta_{f_2 f_2'} \delta_{v_2^1 v_2^4} \delta_{v_2^2 v_2^3} \left(\lambda_{f_2}\right)^2 . \qquad (7)$$

Using this random $H(2)$ operator, we can construct H matrix in m particle spaces with fixed f_m. Then, we have EGUE(2)-$SU(r)$ for each f_m.

In RMT, one is interested in deriving the form for the ensemble averaged eigenvalue density or the one-point function,

$$\overline{\rho^{m,f_m}(E)} = \overline{\langle \delta(H(2) - E) \rangle^{m,f_m}} , \qquad (8)$$

and it is defined by its moments M_p,

$$M_p = \overline{\langle [H(2)]^p \rangle^{m,f_m}} . \qquad (9)$$

Similarly, the two-point function that gives information about fluctuations is

$$S^{mf_m:m'f_{m'}}(E,E') = \overline{\rho^{m,f_m}(E)\rho^{m',f_{m'}}(E')} - \overline{\rho^{m,f_m}(E)}\;\; \overline{\rho^{m',f_{m'}}(E')} \qquad (10)$$

and it is defined by the bivariate moments $\Sigma_{pq}(m, f_m : m', f_{m'})$,

$$\Sigma_{pq}(m, f_m : m', f_{m'}) = \overline{\langle [H(2)]^p \rangle^{m,f_m} \;\; \langle [H(2)]^q \rangle^{m',f_{m'}}} - \overline{\langle [H(2)]^p \rangle^{m,f_m}}\;\; \overline{\langle [H(2)]^q \rangle^{m',f_{m'}}} . \qquad (11)$$

Then, the scaled or reduced bivariate moments are,

$$\hat{\Sigma}_{pq}(m, f_m : m', f_{m'}) = \frac{\Sigma_{pq}(m, f_m : m', f_{m'})}{\left[\overline{\langle [H(2)]^2 \rangle^{m,f_m}}\right]^{p/2} \left[\overline{\langle [H(2)]^2 \rangle^{m',f_{m'}}}\right]^{q/2}} . \qquad (12)$$

Thus, in general we need the following ensemble averaged bivariate moments

$$M_{pq}(m, f_m : m', f_{m'}) = \overline{\langle [H(2)]^p \rangle^{m,f_m} \;\; \langle [H(2)]^q \rangle^{m',f_{m'}}} . \qquad (13)$$

It is useful to note that $M_p = M_{p0}$ or M_{0p}.

In addition to the correlation functions in eigenvalues, one is also interested in transition strength densities and fluctuations in transition strengths generated by the action of a transition operator on eigenstates. We will discuss these with an example in Section 5.

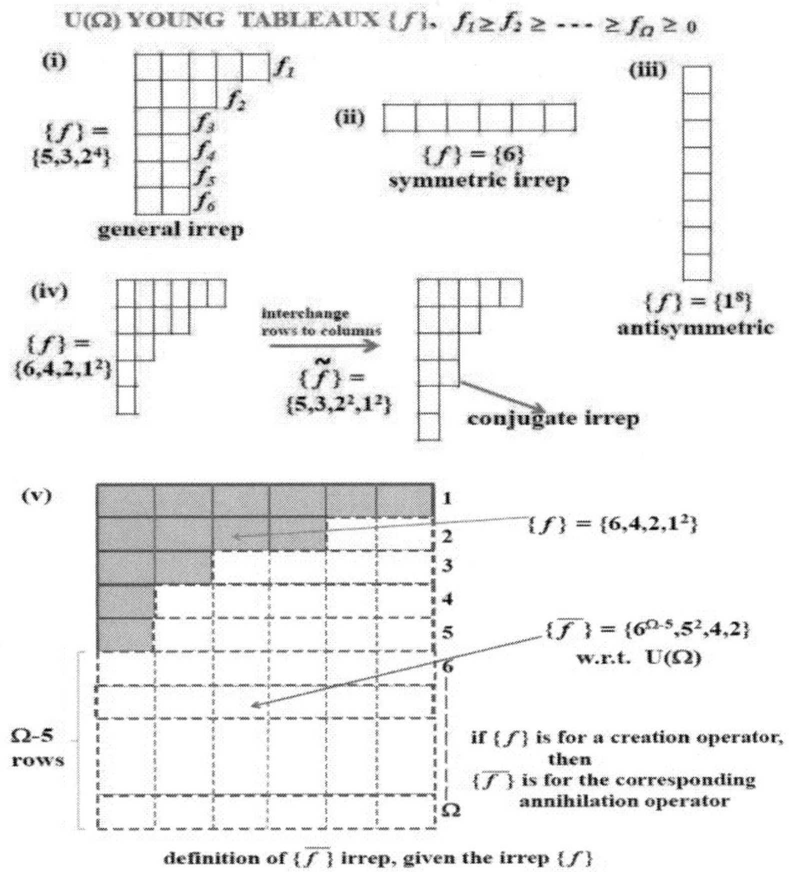

Figure 1. Young tableaux representation of the irreps of $U(\Omega)$. Shown are: (i) a general irrep $\{f\}$; (ii) symmetric irrep $\{m\}$; (iii) antisymmetric irrep $\{1^m\}$; (iv) conjugate irrep $\{\tilde{f}\}$ that corresponds to a given $\{f\}$; (v) irrep $\overline{\{f\}}$ that corresponds to a given $\{f\}$. Note the importance of Ω in defining $\overline{\{f\}}$.

3. Wigner-Racah algebra for EGUE(2) with SU(r) symmetry

Wigner-Racah algebra of the $U(\Omega) \otimes SU(r)$ algebra can be used to derive formulas for the moments M_p and Σ_{pq}. As we shall see ahead, they will lead to tractable formulas involving only $SU(\Omega)$ Racah coefficients for the lower order moments. However, for moments of higher order, evaluation of the needed Racah coefficients will be challenging. We will discuss in this Section, the formulation for lower order moments. First key point in deriving analytical results is tensorial decomposition of $H(2)$ with respect to (w.r.t.) $U(\Omega) \otimes SU(r)$. As $H(2)$ is a $SU(r)$ scalar, it will transform as the irrep $\{0\}$ w.r.t. $SU(r)$. As a result, w.r.t. $SU(\Omega)$ the operator $H(2)$ transforms as the irrep F_ν where $f_2 \times \overline{f_2} \to F_\nu$; the multiplication here is the Kronecker multiplication. Note that $f_2 = \{2\}$ or $\{1^2\}$ and see Fig. 1 for the definition of $\overline{f_2}$ irrep that corresponds to a given f_2 irrep. For $f_2 = \{2\}$, we have $F_\nu = \{0\}$, $\{21^{\Omega-2}\}$ and $\{42^{\Omega-2}\}$. Similarly, for $f_2 = \{1^2\}$, we have $F_\nu = \{0\}$, $\{21^{\Omega-2}\}$ and $\{2^2 1^{\Omega-4}\}$. Unit tensors B's can be defined as,

$$B(f_2F_\nu\omega_\nu) = \sqrt{\frac{1}{d_r(f_2^{(r)})}} \sum_{v_2^i, v_2^f, \beta_2} A^\dagger(f_2 v_2^f \beta_2)\, A(f_2 v_2^i \beta_2) \left\langle f_2 v_2^f\; \overline{f_2}\, \overline{v_2^i} \mid F_\nu \omega_\nu \right\rangle . \tag{14}$$

In Eq. (14), $\langle - - - | \rangle$ is a $U(\Omega)$ Wigner coefficient and $d_r(f_2^{(r)})$ is dimension of the irrep $f_2^{(r)}$ w.r.t. $SU(r)$. Note that for boson systems $f_2^{(r)} = f_2$ and for fermions $f_2^{(r)} = \tilde{f}_2$. It is possible to expand $H(2)$ defined by Eq. (5) in terms of the B's. This gives,

$$H(2) = \sum_{f_2, F_\nu, \omega_\nu} W(f_2F_\nu\omega_\nu)\, B(f_2F_\nu\omega_\nu) \tag{15}$$

and the expansion coefficients W's will be zero centered independent Gaussian variables with

$$\overline{W(f_2F_\nu\omega_\nu)W(f_2'F_\nu'\omega_\nu')} = \delta_{f_2f_2'}\delta_{F_\nu F_\nu'}\delta_{\omega_\nu\omega_\nu'}\,(\lambda_{f_2})^2 d_r(f_2^{(r)}) . \tag{16}$$

Eqs. (15) and (16) are used to derive formulas for M_{pq} and thereby M_p and $\hat{\Sigma}_{pq}$.

Firstly, general form of $M_{pq}(m_1, f_{m_1} : m_2, f_{m_2})$ is [with avg$\{- - -\}$ indicating ensemble average of $\{- - -\}$ and using H for $H(2)$],

$$\begin{aligned} M_{pq}(m_1, f_{m_1} : m_2, f_{m_2}) &= \text{avg}\left\{ \langle H^p \rangle^{m_1, f_{m_1}} \langle H^q \rangle^{m_2, f_{m_2}} \right\} \\ &= \sum_{\alpha_1, \alpha_2, \ldots, \alpha_p : \gamma_1, \gamma_2, \ldots, \gamma_q} \text{avg}\left\{ \langle f_{m_1} v_{m_1}^{\alpha_1} \beta_{m_1}^{\alpha_1} \mid H \mid f_{m_1} v_{m_1}^{\alpha_2} \beta_{m_1}^{\alpha_2} \rangle \right. \\ &\times \langle f_{m_1} v_{m_1}^{\alpha_2} \beta_{m_1}^{\alpha_2} \mid H \mid f_{m_1} v_{m_1}^{\alpha_3} \beta_{m_1}^{\alpha_3} \rangle \ldots \langle f_{m_1} v_{m_1}^{\alpha_p} \beta_{m_1}^{\alpha_p} \mid H \mid f_{m_1} v_{m_1}^{\alpha_1} \beta_{m_1}^{\alpha_1} \rangle \\ &\times \langle f_{m_2} v_{m_2}^{\gamma_1} \beta_{m_2}^{\gamma_1} \mid H \mid f_{m_2} v_{m_2}^{\gamma_2} \beta_{m_2}^{\gamma_2} \rangle \langle f_{m_2} v_{m_2}^{\gamma_2} \beta_{m_2}^{\gamma_2} \mid H \mid f_{m_2} v_{m_2}^{\gamma_3} \beta_{m_2}^{\gamma_3} \rangle \\ &\times \left. \ldots \langle f_{m_2} v_{m_2}^{\gamma_q} \beta_{m_2}^{\gamma_q} \mid H \mid f_{m_2} v_{m_2}^{\gamma_1} \beta_{m_2}^{\gamma_1} \rangle \right\} . \end{aligned} \tag{17}$$

Now, substituting for H the expansion in terms of the unit tensor operators B, the M_{pq} will reduce into an ensemble average involving only the W coefficients and matrix elements of B in (m_1, m_2) spaces. The ensemble average can be carried out using Eq. (16). Similarly, each matrix elements of B can be written, using Wigner-Eckart theorem, as a product of a $SU(\Omega)$ Wigner coefficient involving the labels of the subalgebras of $SU(\Omega)$ and a reduced matrix element that depends only on $SU(\Omega)$ irreps and the f_2 and F_ν of B. The reduced matrix elements (denoted by triple barred matrix elements) in turn can be written in terms of $SU(\Omega)$ Racah (or U-) coefficients. This result follows by combining the form of the operator B given by Eq. (14), Wigner-Eckart theorem and sum rules involving Wigner coefficients and U-coefficients; see for example [21]. The final formula is,

$$\langle f_m \mid\mid\mid B(f_2F_\nu) \mid\mid\mid f_m \rangle_\rho = \sum_{f_{m-2}} \frac{-m(m-1)}{2} \frac{\mathcal{N}_{f_{m-2}}}{\mathcal{N}_{f_m}} \frac{U(f_m\overline{f_2}f_mf_2; f_{m-2}F_\nu)_\rho}{U(f_m\overline{f_2}f_mf_2; f_{m-2}\{0\})} . \tag{18}$$

Note that, $\mathcal{N}_{f_k}$ is the dimension of the irrep f_k of k particles with respect to the symmetric group S_k. Similarly, ρ is the multiplicity label for the occurrence of f_m more than once in the Kronecker product $f_m \times F_\nu \to f_m$. With all these, finally M_{pq} will involve a sum of terms with each term containing a product of Wigner and Racah coefficients of $SU(\Omega)$. As the ensemble average over W's will give several delta-functions [see Eq. (16)], formulas for M_{pq} will simplify into sums involving terms with products of U-coefficients.

The procedure described above has been carried out for M_2, M_{11}, M_{22}, M_{13} and M_{31} for any r; see [20] and references therein for details. Here, we will discuss briefly some of these results. It should be noted that all M_{pq} with $p+q$ odd are zero by the definition of EGUE(2)-$SU(r)$ ensemble. Thus, the first non-trivial moments are of second order ($p+q=2$). With $p=q=1$ we have M_{11} and the formula for this involves only the dimensions $\mathcal{N}_{f_k}$. Going further, with $p=2, q=0$ or $p=0, q=2$ we have M_2 and formula for this is,

$$\begin{aligned} M_2(m,f_m) = \overline{\langle H^2\rangle^{m,f_m}} &= \sum_{f_2} \frac{(\lambda_{f_2})^2}{d_\Omega(f_2)} \sum_{\nu=0,1,2} \mathcal{Q}^\nu(f_2:m,f_m)\;; \\ \mathcal{Q}^\nu(f_2:m,f_m) &= [m(m-1)/2]^2 \sum_{f_{m-2},f'_{m-2}} \frac{\mathcal{N}_{f_{m-2}}}{\mathcal{N}_{f_m}}\frac{\mathcal{N}_{f'_{m-2}}}{\mathcal{N}_{f_m}} X_{UU}(f_2;f_{m-2},f'_{m-2};F_\nu)\;. \end{aligned} \tag{19}$$

The X_{UU} function involves $SU(\Omega)$ Racah coefficients,

$$X_{UU}(f_2;f_{m-2},f'_{m-2};F_\nu) = \sum_\rho \frac{U(f_m,\overline{f_2},f_m,f_2;f_{m-2},F_\nu)_\rho U(f_m,\overline{f_2},f_m,f_2;f'_{m-2},F_\nu)_\rho}{U(f_m,\overline{f_2},f_m,f_2;f_{m-2},\{0\})U(f_m,\overline{f_2},f_m,f_2;f'_{m-2},\{0\})}\;. \tag{20}$$

Proceeding to $p+q=4$, we have the moments M_{22}, M_{31}, M_{13}, M_{40} and M_{04}. Firstly, it is possible to write a formula for M_{22} and this involves X_{UU} given above and also a Y_{UU} function,

$$Y_{UU}(f_{m-2},f'_{m-2};F_\nu) = \sum_\rho \frac{U(f_m,\{1^{\Omega-2}\},f_m,\{1^2\};f_{m-2},F_\nu)_\rho\, U(f_m,\{2^{\Omega-1}\},f_m,\{2\};f'_{m-2},F_\nu)_\rho}{U(f_m,\{1^{\Omega-2}\},f_m,\{1^2\};f_{m-2},\{0\})\,U(f_m,\{2^{\Omega-1}\},f_m,\{2\};f'_{m-2},\{0\})}\;. \tag{21}$$

In addition, it is easy to prove that

$$\begin{aligned} M_{31}(m,f_m:m',f_{m'}) &= 3M_{11}(m,f_m:m',f_{m'})M_2(m,f_m)\;, \\ M_{13}(m,f_m:m',f_{m'}) &= 3M_{11}(m,f_m:m',f_{m'})M_2(m,f_{m'})\;. \end{aligned} \tag{22}$$

Using the above results and the tables for products of U-coefficients given in [19, 20, 22], M_2 (giving spectral variances), $\hat{\Sigma}_{11}$ (giving variance of the fluctuations in eigenvalue centroids for $m=m'$ and cross correlations for $m \neq m'$ and/or $f_m \neq f_{m'}$) and $\hat{\Sigma}_{22}$ (giving variance of the fluctuations in spectral variances for $m=m'$ and cross correlations for $m \neq m'$ and/or $f_m \neq f_{m'}$) are calculated in many examples. Using the exact formulas, obtained are also results in the asymptotic limit (note that $\hat{\Sigma}_{13} = \hat{\Sigma}_{31} = 3\hat{\Sigma}_{11}$). Interesting asymptotic limit for fermions is the dilute limit defined by $m \to \infty$, $N \to \infty$ and $m/N \to 0$ and for bosons it is the dense limit defined by $m \to \infty$, $N \to \infty$ and $m/N \to \infty$. The exact and asymptotic formulas have given many new insights into the statistical properties of interacting fermion and boson systems [17, 18, 19, 20].

With $p=4, q=0$ or $p=0, q=4$, we have $M_4(f_m)$. Except for $r=1$ as given in detail in [14, 15, 16, 17], analytical formulas for M_4 are not available for $r>1$. Thus, with $p+q=4$, we are left with the unsolved problem of M_4 and we will now discuss what is unknown here.

4. Fourth moment in terms of Racah coefficients: Open questions

Hamiltonian operator H for EGUE(2)-$SU(r)$ is a sum of Hamiltonian operators in $f_2=\{2\}$ and $\{1^2\}$ spaces. Let us denote these two parts as H_{f_2}. Then,

$$M_4(m,f_m) = \overline{\langle H^4\rangle^{m,f_m}} = \overline{\langle (H_{\{2\}} + H_{\{1^2\}})^4\rangle^{m,f_m}}\;. \tag{23}$$

Journal of Physics: Conference Series **538** (2014) 012011 doi:10.1088/1742-6596/538/1/012011

Using the cyclic invariance of the averages and applying the property that terms with odd powers of H_{f_2} will vanish will give,

$$\begin{aligned}\overline{\langle H^4\rangle^{m,f_m}} &= \overline{\langle (H_{\{2\}})^4\rangle^{m,f_m}} + \overline{\langle (H_{\{1^2\}})^4\rangle^{m,f_m}} + 4\overline{\langle (H_{\{2\}})^2(H_{\{1^2\}})^2\rangle^{m,f_m}} \\ &+2\overline{\langle H_{\{2\}}H_{\{1^2\}}H_{\{2\}}H_{\{1^2\}}\rangle^{m,f_m}}\,.\end{aligned} \tag{24}$$

Writing H in terms of the unit tensors B's will give,

$$\begin{aligned}\overline{\left\langle (H_{f_2})^4\right\rangle^{m,f_m}} &= [d_\Omega(f_m)]^{-1} \sum_{v_i,F_{\nu_i},\omega_{\nu_i}\,;\,i=1}^{4} \langle f_m v_1 \mid B(f_2F_{\nu_1}\omega_{\nu_1}) \mid f_m v_2\rangle \\ &\times \langle f_m v_2 \mid B(f_2F_{\nu_2}\omega_{\nu_2}) \mid f_m v_3\rangle \langle f_m v_3 \mid B(f_2F_{\nu_3}\omega_{\nu_3}) \mid f_m v_4\rangle \langle f_m v_4 \mid B(f_2F_{\nu_4}\omega_{\nu_4}) \mid f_m v_1\rangle \\ &\times \overline{W(f_2F_{\nu_1}\omega_{\nu_1})W(f_2F_{\nu_2}\omega_{\nu_2})W(f_2F_{\nu_3}\omega_{\nu_3})W(f_2F_{\nu_4}\omega_{\nu_4})}\,.\end{aligned} \tag{25}$$

Now, using the property that W's are independent Gaussian variables will lead to

$$\begin{aligned}\overline{\left\langle (H_{f_2})^4\right\rangle^{m,f_m}} &= 2\left[\overline{\left\langle (H_{f_2})^2\right\rangle^{m,f_m}}\right]^2 + \lambda_{f_2}^4 \left[d_r(f_2^{(r)})\right]^2 d_\Omega(f_m) \\ &\times \sum_{F_{\nu_1},F_{\nu_2},\rho_1,\rho_2,\rho_3,\rho_4} \frac{1}{\sqrt{d_\Omega(F_{\nu_1})d_\Omega(F_{\nu_2})}} U(f_m\overline{f_m}f_mf_m;(F_{\nu_1})_{\rho_1\rho_3}(F_{\nu_2})_{\rho_2\rho_4}) \\ &\times \langle f_m \mid\mid B(f_2F_{\nu_1}) \mid\mid f_m\rangle_{\rho_1} \langle f_m \mid\mid B(f_2F_{\nu_2}) \mid\mid f_m\rangle_{\rho_2} \\ &\times \langle f_m \mid\mid B(f_2F_{\nu_1}) \mid\mid f_m\rangle_{\rho_3} \langle f_m \mid\mid B(f_2F_{\nu_2}) \mid\mid f_m\rangle_{\rho_4}\,.\end{aligned} \tag{26}$$

Similarly, we have

$$\overline{\left\langle H^2_{\{2\}}H^2_{\{1^2\}}\right\rangle^{m,f_m}} = \left\{\overline{\left\langle H^2_{\{2\}}\right\rangle^{m,f_m}}\right\} \left\{\overline{\left\langle H^2_{\{1^2\}}\right\rangle^{m,f_m}}\right\}, \tag{27}$$

and

$$\begin{aligned}\overline{\langle H_{\{2\}}H_{\{1^2\}}H_{\{2\}}H_{\{1^2\}}\rangle^{m,f_m}} &= \lambda^2_{\{2\}}\lambda^2_{\{1^2\}}d_r(\{2\})d_r(\{1^2\})d_\Omega(f_m) \\ &\times \sum_{F_{\nu_1},F_{\nu_2},\rho_1,\rho_2,\rho_3,\rho_4} \frac{1}{\sqrt{d_\Omega(F_{\nu_1})d_\Omega(F_{\nu_2})}} U(f_m\overline{f_m}f_mf_m;(F_{\nu_1})_{\rho_1\rho_3}(F_{\nu_2})_{\rho_2\rho_4}) \\ &\times \langle f_m \mid\mid B(\{2\}F_{\nu_1}) \mid\mid f_m\rangle_{\rho_1} \langle f_m \mid\mid B(\{1^2\}F_{\nu_2}) \mid\mid f_m\rangle_{\rho_2} \\ &\times \langle f_m \mid\mid B(\{2\}F_{\nu_1}) \mid\mid f_m\rangle_{\rho_3} \langle f_m \mid\mid B(\{1^2\}F_{\nu_2}) \mid\mid f_m\rangle_{\rho_4}\,.\end{aligned} \tag{28}$$

Substituting the results in Eqs. (25)-(28) in Eqs. (24) will give $\overline{\langle H^4\rangle^{m,f_m}}$. This involves $SU(\Omega)$ Racah coefficients with multiplicity labels and evaluation of these is in general complicated. Similarly, evaluation of the reduced matrix elements is also complicated. It is important to note that

$$\gamma_2(m,f_m) = \frac{M_4(m,f_m)}{[M_2(m,f_m)]^2} - 3 \tag{29}$$

gives the excess parameter and if it is close to zero (asymptotically or otherwise), then we have Gaussian form for the eigenvalue density. This result was established only for spinless EGUE(2) but not yet for EGUE(2)-$SU(r)$ with $r > 1$ as the formulas for the reduced matrix elements and the U-coefficients appearing in Eqs. (26) and (28) are not available for $r > 1$.

One simple situation where there is some hope of deriving formulas for M_4 is for the f_m irreps that give multiplicity labels all unity. We denote the $U(\Omega)$ irreps that satisfy this as $f_m^{(g)}$ and we have verified that one of these irreps is $\{r^k\}$ for fermions and $\{k^r\}$ for bosons where $m = kr$ and others are $\{m\}$ and $\{1^m\}$. For these irreps, the expression for γ_2 is

$$\begin{aligned}
\left[\gamma_2(m, f_m^{(g)}) + 1\right] &= \left[\overline{\langle H^2 \rangle^{m, f_m^{(g)}}}\right]^{-2} \\
&\times \left\{ \sum_{f_a, f_b = \{2\}, \{1^2\}} \frac{\lambda_{f_a}^2 \lambda_{f_b}^2}{d_\Omega(f_a) d_\Omega(f_b)} \sum_{F_{\nu_1}, F_{\nu_2}} \frac{d_\Omega(f_m^{(g)})}{\sqrt{d_\Omega(F_{\nu_1}) d_\Omega(F_{\nu_2})}} \right. \\
&\times \left. U(f_m^{(g)} \overline{f_m^{(g)}} f_m^{(g)} f_m^{(g)}; F_{\nu_1} F_{\nu_2})\, \mathcal{Q}^{\nu_1}(f_a : m, f_m^{(g)})\, \mathcal{Q}^{\nu_2}(f_b : m, f_m^{(g)}) \right\} .
\end{aligned} \tag{30}$$

The $\mathcal{Q}^\nu(f_2 : m, f_m)$ in Eq. (30) are defined by Eq. (19) and they can be calculated using X_{UU} discussed before. Therefore, the only unknown in Eq. (30) is the $SU(\Omega)$ Racah coefficient $U(f_m^{(g)} \overline{f_m^{(g)}} f_m^{(g)} f_m^{(g)}; F_{\nu_1} F_{\nu_2})$. To the best of our knowledge, formulas or a viable numerical procedure for evaluating this type of Racah coefficients is not available in literature. As understanding the shape of the eigenvalue density for EGUE(2)-$SU(r)$ for any $r > 1$ rests on these Racah coefficients, certainly efforts should be made to derive formulas (exact or asymptotic or purely a numerical procedure) for these. This is one of the most important open problem in the subject of EE.

5. Binary Correlation Approximation

Bivariate moments Σ_{pq} for $p + q$ up to a large value are needed to derive the form of the two-point function for eigenvalues and thereby determine properties of level fluctuations. Formulas for these higher order moments are not available for any r. It is unlikely that exact formulas for U-coefficients will help in solving this problem and it appears that asymptotic methods (methods giving directly results valid in the asymptotic limit) need to be developed for further analysis of EGUE(2)-$SU(r)$. One asymptotic method originally used by Wigner for GOE and found to be useful [13] for spinless EGOE(2) [and also for EGOE(k), $k \geq 2$] is the so called binary correlation approximation (BCA). As an example for asymptotic results, we will briefly discuss here some results for the bivariate moments of the transition strength density $I_{biv:O} = I(E_f) | \langle E_f \mid O \mid E_i \rangle |^2 I(E_i)$, generated by a transition operator O, for a system of spinless fermions in two different orbits with the Hamiltonian preserving the number of fermions in each orbit [with a suitable interpretation, this has applications to beta decay and neutrinoless double-beta decay nuclear transition matrix elements (NTME)]. Note that $I(E)$ are eigenvalue densities. We are hopeful that the results presented in this Section will prompt development of asymptotic methods for EGUE and EGOE with Lie symmetries.

Let us consider $N_1 + N_2$ number of single particle (sp) states forming two orbits with #1 orbit having N_1 sp states and #2 orbit has N_2 sp states. Say there are m_1 number of spinless fermions in the N_1 sp states and m_2 number of spinless fermions in the N_2 sp states. Also, say the Hamiltonian $H(k_H)$ is a k_H-body operator and the transition operator $O(k_O)$ changes k_O

number of fermions from space #2 to fermions in space #1. Then,

$$H(k_H) = \sum_{i+j=k_H;\alpha,\beta,\gamma,\delta} \left[v_H^{\alpha\beta\gamma\delta}(i,j)\right] \alpha_1^\dagger(i)\beta_1(i)\gamma_2^\dagger(j)\delta_2(j)\,,\quad O(k_O) = \sum_{\gamma,\delta} v_O^{\gamma\delta}(k_O)\gamma_1^\dagger(k_O)\delta_2(k_O)\,. \tag{31}$$

In general $\alpha_k^\dagger(l)$ is normalized l-particle creation operator in space $\#k$ and $\alpha_k(l)$ is the corresponding annihilation operator. It is important to note that $H(k_H)$ preserves (m_1, m_2) but $O(k_O)$ does not; $O(k_O)\,|m_1, m_2\rangle = |m_1 + k_O, m_2 - k_O\rangle$. The bivariate moments of $I_{biv:O}$ are defined by

$$\widetilde{M}_{PQ}(m_1, m_2) = \left\langle O^\dagger(k_O)[H(k_H)]^Q O(k_O)[H(k_H)]^P\right\rangle^{m_1,m_2}\,. \tag{32}$$

To proceed further, $H(k_H)$ and $O(k_O)$ in (m_1, m_2) spaces are represented by independent EGOEs. Then, the v_H and the v_O matrix elements in Eq. (31) are independent zero centered Gaussian variables with

$$\overline{\left[v_H^{\alpha\beta\gamma\delta}(i,j)\right]^2} = v_H^2(i,j)\,,\quad \overline{\left[v_O^{\gamma\delta}\right]^2} = v_O^2\,. \tag{33}$$

It is useful to note that the EGOE ensemble for H carries $U(N_1 + N_2) \supset U(N_1) \oplus U(N_2)$ direct sum subalgebra symmetry ($\oplus$ denotes direct sum) with $U(N_1)$ and $U(N_2)$ generating m_1 and m_2 respectively. Also, the $H(k_H)$ matrix in a given (m_1, m_2) space is a matrix with dimension $\binom{N_1}{m_1}\binom{N_2}{m_2}$. However, the O matrix will be a rectangular matrix with $\langle m_1 + k_O, m_2 - k_O, \beta \mid O \mid m_1, m_2, \alpha\rangle$ matrix elements (α and β are additional labels needed for specifying the states completely). Asymptotic formulas for the ensemble average (average over both H and O ensembles) of $\widetilde{M}_{PQ}(m_1, m_2)$ are derived using BCA (assumed is $k_H, k_O << m_1$ and $k_H, k_O << m_2$). Substituting in Eq. (32), the forms for H and O from Eq. (31), we will have terms containing ensemble averages of products of v_H^{----} and V_O^{--} and (m_1, m_2) space matrix elements of products of creation and annihilation operators. In BCA, one retains terms that involve only squares of the defining space matrix elements v_H^{----} and V_O^{--}. Simplifying these terms using the independent Gaussian property of these variables and Eq. (33), formulas for $\widetilde{M}_{PQ}(m_1, m_2)$ with $P + Q \leq 4$ are derived. Final results are [23, 24],

$$\begin{aligned}
\widetilde{M}_{00}(m_1, m_2) &= \overline{\langle O^\dagger(k_O)O(k_O)\rangle^{m_1,m_2}} = v_O^2\binom{\widetilde{m}_1}{k_O}\binom{m_2}{k_O}\,,\\
\widetilde{M}_{11}(m_1, m_2) &= v_O^2 \sum_{i+j=k_H} v_H^2(i,j)\binom{\widetilde{m}_1 - i}{k_O}\binom{m_2 - j}{k_O} T(m_1, N_1, i)T(m_2, N_2, j)\,,\\
\widetilde{M}_{20}(m_1, m_2) &= \widetilde{M}_{00}(m_1, m_2)\overline{\langle H^2(k_H)\rangle^{m_1,m_2}}\,,\\
\widetilde{M}_{02}(m_1, m_2) &= \widetilde{M}_{00}(m_1, m_2)\overline{\langle H^2(k_H)\rangle^{m_1+k_O,m_2-k_O}}\,,\\
\widetilde{M}_{40}(m_1, m_2) &= \widetilde{M}_{00}(m_1, m_2)\overline{\langle H^4(k_H)\rangle^{m_1,m_2}}\,,\\
\widetilde{M}_{04}(m_1, m_2) &= \widetilde{M}_{00}(m_1, m_2)\overline{\langle H^4(k_H)\rangle^{m_1+k_O,m_2-k_O}}\,,\\
\widetilde{M}_{31}(m_1, m_2) &= 2\overline{\langle H^2(k_H)\rangle^{m_1,m_2}}\widetilde{M}_{11}(m_1, m_2) + v_O^2 \sum_{i+j=k_H, t+u=k_H} v_H^2(i,j)v_H^2(t,u)\\
&\times \binom{m_2-j}{k_O}\binom{\widetilde{m}_1-i}{k_O} F(m_1, N_1, i, t)F(m_2, N_2, j, u)\,,
\end{aligned}$$

$$
\begin{aligned}
\widetilde{M}_{13}(m_1,m_2) &= 2\overline{\langle H^2(k_H)\rangle}^{m_1+k_O,m_2-k_O}\widetilde{M}_{11}(m_1,m_2)+v_O^2\sum_{i+j=k_H,t+u=k_H} v_H^2(i,j)v_H^2(t,u) \\
&\times\; G(t,u)\binom{\widetilde{m}_1-k_O-t+i}{i}\binom{m_1+k_O-t}{i}\binom{\widetilde{m}_2-u+k_O+j}{j}\binom{m_2-k_O-u}{j}\;, \\
\widetilde{M}_{22}(m_1,m_2) &= \widetilde{M}_{00}(m_1,m_2)\overline{\langle H^2(k_H)\rangle}^{m_1+k_O,m_2-k_O}\;\overline{\langle H^2(k_H)\rangle}^{m_1,m_2} \\
&+\; v_O^2\sum_{i+j=k_H,t+u=k_H} v_H^2(i,j)v_H^2(t,u)\binom{\widetilde{m}_1-i-t}{k_O}\binom{m_2-u-j}{k_O} \\
&\times\; [F(m_1,N_1,i,t)F(m_2,N_2,j,u) \\
&+\; T(m_1,N_1,i)T(m_2,N_2,j)T(m_1,N_1,t)T(m_2,N_2,u)]\;,
\end{aligned} \tag{34}
$$

with,

$$
\begin{aligned}
T(m,N,k_H) &= \binom{m}{k_H}\left[\binom{\widetilde{m}+k_H}{k_H}+1\right]\;, \\
F(m,N,k_H,k_G) &= \binom{m-k_H}{k_G}\binom{m}{k_H}\binom{N}{k_H}\binom{N}{k_G}\;, \\
G(t,u) &= \binom{\widetilde{m}_1-t}{k_O}\binom{m_2-u}{k_O}T(m_1,N_1,t)T(m_2,N_2,u)\;, \\
\overline{\langle H^2(k_H)\rangle}^{m_1,m_2} &= \sum_{i+j=k_H} v_H^2(i,j)T(m_1,N_1,i)T(m_2,N_2,j)\;, \\
\overline{\langle H^4(k_H)\rangle}^{m_1,m_2} &= 2\left[\sum_{i+j=k_H} v_H^2(i,j)T(m_1,N_1,i)T(m_2,N_2,j)\right]^2 \\
&+\; \sum_{i+j=k_H,t+u=k_H} v_H^2(i,j)v_H^2(t,u)F(m_1,N_1,i,t)F(m_2,N_2,j,u)\;, \\
\widetilde{m} &= N-m,\quad \widetilde{m}_1=N_1-m_1,\quad \widetilde{m}_2=N_2-m_2\;.
\end{aligned} \tag{35}
$$

These formulas are applied to some nuclei and it is seen that the bivariate transition strength density generated by a two-body Hamiltonian ($k_H = 2$) approaches bivariate Gaussian form for the transition operators appropriate for beta decay ($k_O = 1$) and neutrinoless double-beta decay ($k_O = 2$) [23, 24]. Finally, Eqs. (34) and (35) also show that the one-point function $\rho^{m_1,m_2}(E) = \langle\delta(H-E)\rangle^{m_1,m_2}$ will be asymptotically a Gaussian for $H = H(2)$ and the corresponding EE here is EGOE(2)-[$U(N_1)\oplus U(N_2)$].

6. Future outlook: Further open questions

Although some advances are made in deriving analytical results for certain quantities using group theory and BCA, mathematical tractability of embedded ensembles remains a challenge. We are hopeful that the present paper will prompt new efforts in developing further Wigner-Racah algebra for general $U(\Omega)$ algebras so that one may obtain a better understanding of one- and two-point functions for EE with Lie symmetries. Besides this, another important future direction is to develop further BCA or develop some other appropriate asymptotic method for

EGEs with symmetries such as $U(\Omega) \otimes SU(r)$. In this quest, the so called p-expansion method sketched by French [25] may need re-examination and further exploration. Certainly, asymptotic results for fermions with spin, for spinless boson systems and for boson systems with spin (single boson spin $\mathbf{s} = \frac{1}{2}$ and 1) will be interesting and may prove to be useful in mesoscopic physics and in the studies of ultra-cold atoms.

Acknowledgments

Thanks are due to Profs. F. Iachello, H.A. Weidenmüller, K.T. Hecht and P.H. Butler for many useful correspondence.

References

[1] Mehta M L 2004 *Random Matrices* (Netherlands: Elsevier B.V.)
[2] Haake F 2010 *Quantum signatures of Chaos* (Heidelberg: Springer-Verlag)
[3] Akemann G, Baik J and Di Francesco P 2011 *The Oxford Handbook of Random Matrix Theory* (Oxford: Oxford University Press)
[4] Couillet R and Debbah M 2012 *Random Matrix Methods for Wireless Communications* (New York: Cambridge University Press)
[5] Weidenmüller H A, private communication to V.K.B. Kota
[6] Brody T A, Flores J, French J B, Mello P A, Pandey A and Wong S S M 1981 *Rev. Mod. Phys.* **53** 385
[7] Kota V K B 2001 *Phys. Rep.* **347** 223
[8] Papenbrock T and Weidenmüller H A 2007 *Rev. Mod. Phys.* **79** 997
[9] Gómez J M G, Kar K, Kota V K B, Molina R A, Relaño A and Retamosa J 2011 *Phys. Rep.* **499** 103
[10] Vyas Manan, Kota V K B and Chavda N D 2010 Phys. Rev. E **81** 036212
[11] Vyas Manan, Chavda N D, Kota V K B and Potbhare V 2012 *J. Phys. A: Math. Theor.* **45** 265203
[12] Deota H, Chavda N D, Kota V K B, Potbhare V and Vyas Manan 2013 *Phys. Rev. E* **88** 022130
[13] Mon K K and French J B 1975 *Ann. Phys. (N.Y.)* **95** 90
[14] Benet L, Rupp T and Weidenmüller H A 2001 *Ann. Phys. (N.Y.)* **292** 67
[15] Agasa T, Benet L, Rupp T and Weidenmüller H A 2002 *Ann. Phys. (N.Y.)* **298** 229
[16] Pluhař Z and Weidenmüller H A 2002 *Ann. Phys. (N.Y)* **297** 344
[17] Kota V K B 2005 *J. Math. Phys.* **46** 033514
[18] Kota V K B 2007 *J. Math. Phys.* **48** 053304
[19] Vyas Manan and Kota V K B 2010 *Ann. Phys. (N.Y.)* **325** 2451
[20] Vyas Manan and Kota V K B 2012 *J. Math. Phys.* **53** 123303
[21] Hecht K T and Draayer J P 1974 *Nucl. Phys.* A **223** 285
[22] Hecht K T 1974 *J. Math. Phys.* **15** 2148
[23] Vyas Manan and Kota V K B (2011) *Preprint* nucl-th/1106.0395
[24] Vyas Manan 2011 *Some studies on two-body random matrix ensembles* (Ph. D. thesis, M. S. University of Baroda, Vadodara, India)
[25] French J B 1980 *Theory and Applications of Moment Methods in Many Fermion Systems*, eds B J Dalton, S M Grimes, J P Vary and S A Williams (New York: Plenum) pp 91

Order, chaos and quasi symmetries in a first-order quantum phase transition

A Leviatan and M Macek
Racah Institute of Physics, The Hebrew University, Jerusalem 91904, Israel
E-mail: ami@phys.huji.ac.il, mmacek@phys.huji.ac.il

Abstract. We study the competing order and chaos in a first-order quantum phase transition with a high barrier. The boson model Hamiltonian employed, interpolates between its U(5) (spherical) and SU(3) (deformed) limits. A classical analysis reveals regular (chaotic) dynamics at low (higher) energy in the spherical region, coexisting with a robustly regular dynamics in the deformed region. A quantum analysis discloses, amidst a complicated environment, persisting regular multiplets of states associated with partial U(5) and quasi SU(3) dynamical symmetries.

1. Introduction

Quantum phase transitions (QPTs) are qualitative changes in the ground state properties of a physical system induced by a variation of parameters λ in the quantum Hamiltonian $\hat{H}(\lambda)$ [1,2]. Such ground-state transformations have received considerable attention in recent years and have found a variety of applications in many areas of physics and chemistry [3]. The particular type of QPT is reflected in the topology of the underlying mean-field (Landau) potential $V(\lambda)$. Most studies have focused on second-order (continuous) QPTs [4], where $V(\lambda)$ has a single minimum which evolves continuously into another minimum. The situation is more complex for discontinuous (first-order) QPTs, where $V(\lambda)$ develops multiple minima that coexist in a range of λ values and cross at the critical point, $\lambda=\lambda_c$. The interest in such QPTs stems from their key role in phase-coexistence phenomena at zero temperature. Examples are offered by the metal-insulator Mott transition [5], heavy-fermion superconductors [6], quantum Hall bilayers [7] and shape-coexistence in mesoscopic systems, such as atomic nuclei [8].

The competing interactions in the Hamiltonian that drive these ground-state phase transitions can affect dramatically the nature of the dynamics and, in some cases, lead to the emergence of quantum chaos. This effect has been observed in quantum optics models of N two-level atoms interacting with a single-mode radiation field [9], where the onset of chaos is triggered by continuous QPTs. In the present contribution, we address the mixed regular and chaotic dynamics associated with a first order QPT [10–12]. For that purpose, we employ an interacting boson model which describes such QPTs between spherical and axially-deformed nuclei. The model is amenable to both classical and quantum treatments, has a rich group structure and inherent geometry, which makes it an ideal framework for studying the intricate interplay of order and chaos and the role of symmetries in such shape-phase transitions.

2. Symmetries, geometry and quantum phase transitions in the IBM

The interacting boson model (IBM) [13] describes low-lying quadrupole collective states in nuclei in terms of N interacting monopole (s) and quadrupole (d) bosons representing valence nucleon pairs. The bilinear combinations $\mathcal{G}_{ij} \equiv b_i^\dagger b_j = \{s^\dagger s,\, s^\dagger d_m,\, d_m^\dagger s,\, d_m^\dagger d_{m'}\}$ span a U(6) algebra, which serves as the spectrum generating algebra. The IBM Hamiltonian is expanded in terms of these generators, $\hat{H} = \sum_{ij}\epsilon_{ij}\,\mathcal{G}_{ij} + \sum_{ijk\ell} u_{ijk\ell}\,\mathcal{G}_{ij}\mathcal{G}_{k\ell}$, and consists of Hermitian, rotational-invariant interactions which conserve the total number of s- and d- bosons, $\hat{N} = \hat{n}_s + \hat{n}_d = s^\dagger s + \sum_m d_m^\dagger d_m$. A dynamical symmetry (DS) occurs if the Hamiltonian can be written in terms of the Casimir operators of a chain of nested sub-algebras of U(6). The Hamiltonian is then completely solvable in the basis associated with each chain. The three dynamical symmetries of the IBM and corresponding bases are

$$\begin{array}{llll} \mathrm{U}(6)\supset\mathrm{U}(5)\supset\mathrm{O}(5)\supset\mathrm{O}(3) & \vert N, n_d, \tau, n_\Delta, L\rangle & \text{spherical vibrator} & (1a)\\ \mathrm{U}(6)\supset\mathrm{SU}(3)\supset\mathrm{O}(3) & \vert N, (\lambda,\mu), K, L\rangle & \text{axially}-\text{deformed rotor} & (1b)\\ \mathrm{U}(6)\supset\mathrm{O}(6)\supset\mathrm{O}(5)\supset\mathrm{O}(3) & \vert N, \sigma, \tau, n_\Delta, L\rangle & \gamma-\text{unstable deformed rotor} & (1c) \end{array}$$

The associated analytic solutions resemble known limits of the geometric model of nuclei [14], as indicated above. The basis members are classified by the irreducible representations (irreps) of the corresponding algebras. Specifically, the quantum numbers $N, n_d, (\lambda,\mu), \sigma, \tau$ and L label the relevant irreps of U(6), U(5), SU(3), O(6), O(5) and O(3), respectively. n_Δ and K are multiplicity labels needed for complete classification in the reductions O(5) $\supset$ O(3) and SU(3) $\supset$ O(3), respectively. Each basis is complete and can be used for a numerical diagonalization of the Hamiltonian in the general case. A geometric visualization of the model is obtained by a potential surface, $V(\beta,\gamma) = \langle\beta,\gamma;N\vert\hat{H}\vert\beta,\gamma;N\rangle$, defined by the expectation value of the Hamiltonian in the intrinsic condensate state [15, 16]

$$\vert\beta,\gamma;N\rangle = (N!)^{-1/2}[\Gamma_c^\dagger(\beta,\gamma)]^N\vert 0\rangle\ , \tag{2a}$$

$$\Gamma_c^\dagger(\beta,\gamma) = \left[\beta\cos\gamma d_0^\dagger + \beta\sin\gamma(d_2^\dagger + d_{-2}^\dagger)/\sqrt{2} + \sqrt{2-\beta^2}s^\dagger\right]/\sqrt{2}\ . \tag{2b}$$

Here (β,γ) are quadrupole shape parameters analogous to the variables of the collective model of nuclei. Their values $(\beta_{\rm eq},\gamma_{\rm eq})$ at the global minimum of $V(\beta,\gamma)$ define the equilibrium shape for a given Hamiltonian. For one- and two-body interactions, the shape can be spherical $(\beta_{\rm eq}=0)$ or deformed $(\beta_{\rm eq}>0)$ with $\gamma_{\rm eq}=0$ (prolate), $\gamma_{\rm eq}=\pi/3$ (oblate), or γ-independent.

The dynamical symmetries of Eq. (1), correspond to phases of the system, and provide analytic benchmarks for the dynamics of stable nuclear shapes. Quantum phase transitions (QPTs) between such stable shapes have been studied extensively in the IBM framework [16, 17] and are manifested empirically in nuclei [8]. The Hamiltonians employed mix interaction terms from different DS chains, *e.g.*, $\hat{H}(\lambda) = \lambda\hat{H}_a + (1-\lambda)\hat{H}_b$. The coupling coefficient (λ) responsible for the mixing, serves as the control parameter and the surface, $V(\lambda) \equiv V(\lambda;\beta,\gamma)$, serves as the Landau potential. In general, under such circumstances, solvability is lost, there are no remaining non-trivial conserved quantum numbers and all eigenstates are expected to be mixed. However, for particular symmetry breaking, some intermediate symmetry structure can survive. The latter include partial dynamical symmetry (PDS) [18] and quasi-dynamical symmetry (QDS) [19]. In a PDS, the conditions of an exact dynamical symmetry (solvability of the complete spectrum and existence of exact quantum numbers for all eigenstates) are relaxed and apply to only part of the eigenstates. In a QDS, particular states continue to exhibit selected characteristic properties (*e.g.*, energy and B(E2) ratios) of the closest dynamical symmetry, in the face of strong-symmetry breaking interactions. This "apparent" symmetry is due to the coherent nature of the mixing. As discussed below, both PDS and QDS are relevant to quantum phase transitions [19, 20].

Symmetries in Science XVI IOP Publishing
Journal of Physics: Conference Series **538** (2014) 012012 doi:10.1088/1742-6596/538/1/012012

In view of the central role of the Landau potential, $V(\beta,\gamma)$, for QPTs, it is convenient to resolve the Hamiltonian into two parts, $\hat{H} = \hat{H}_{\rm int} + \hat{H}_{\rm col}$ [21]. The intrinsic part ($\hat{H}_{\rm int}$) determines the potential surface while the collective part ($\hat{H}_{\rm col}$) is composed of kinetic terms which do not affect the shape of $V(\beta,\gamma)$. For first-order QPTs, the resolution allows the construction of an intrinsic Hamiltonian with a high-barrier [22], and by treating it separately, one avoids the complication of rotation-vibration couplings that can obscure the simple pattern of mixed dynamics, reported below. Henceforth, we confine the discussion to the dynamics of the intrinsic Hamiltonian.

3. Intrinsic Hamiltonian for a first-order QPT

Focusing on first-order QPTs between stable spherical ($\beta_{\rm eq} = 0$) and prolate-deformed ($\beta_{\rm eq} > 0$, $\gamma_{\rm eq} = 0$) shapes, the intrinsic Hamiltonian reads

$$\hat{H}_1(\rho)/\bar{h}_2 = 2(1-2\rho^2)\hat{n}_d(\hat{n}_d-1) + 2R_2^\dagger(\rho)\cdot\tilde{R}_2(\rho)\ , \tag{3a}$$

$$\hat{H}_2(\xi)/\bar{h}_2 = \xi P_0^\dagger P_0 + P_2^\dagger\cdot\tilde{P}_2\ , \tag{3b}$$

where $\hat{n}_d$ is the d-boson number operator, $R^\dagger_{2\mu}(\rho) = \sqrt{2}s^\dagger d^\dagger_\mu + \rho\sqrt{7}(d^\dagger d^\dagger)^{(2)}_\mu$, $P^\dagger_0 = d^\dagger\cdot d^\dagger - 2(s^\dagger)^2$ and $P^\dagger_{2\mu} = 2s^\dagger d^\dagger_\mu + \sqrt{7}(d^\dagger d^\dagger)^{(2)}_\mu$. Here $\tilde{R}_{2\mu} = (-1)^\mu R_{2,-\mu}$, $\tilde{P}_{2\mu} = (-1)^\mu P_{2,-\mu}$ and the dot implies a scalar product. Scaling by $\bar{h}_2 \equiv h_2/N(N-1)$ is used throughout, to facilitate the comparison with the classical limit. The control parameters that drive the QPT are ρ and ξ, with $0\leq\rho\leq\frac{1}{\sqrt{2}}$ and $0\leq\xi\leq 1$.

The intrinsic Hamiltonian in the spherical phase, $\hat{H}_1(\rho)$, has by construction the intrinsic state of Eq. (2) with $\beta_{\rm eq} = 0$ as zero energy eigenstate. For large N, its normal modes involve quadrupole vibrations about the spherical global minimum of the potential surface, with frequency $\epsilon = 4\bar{h}_2 N$. For $\rho = 0$ it reduces to

$$\hat{H}_1(\rho=0)/\bar{h}_2 = 2\hat{n}_d(\hat{n}_d - 1) + 4(\hat{N} - \hat{n}_d)\hat{n}_d\ . \tag{4}$$

Since $\hat{n}_d$ is the linear Casimir operator of U(5), $\hat{H}_1(\rho = 0)$ has U(5) dynamical symmetry (DS). The spectrum is completely solvable $E_{\rm DS} = [2n_d(n_d-1) + 4(N-n_d)n_d]\bar{h}_2$, and the eigenstates, $|N, n_d, \tau, n_\Delta, L\rangle$, are those of the U(5) chain, Eq. (1a). The spectrum resembles that of an anharmonic spherical vibrator, describing quadrupole excitations of a spherical equilibrium shape. The lowest U(5) multiplets involve states with quantum numbers ($n_d = 0, \tau = 0, L = 0$), ($n_d = 1, \tau = 1, L = 2$), ($n_d = 2, \tau = 2, L = 2, 4; \tau = 0, L = 0$), ($n_d = 3, \tau = 3, L = 6, 4, 3, 0; \tau = 1, L = 2$).

For $\rho > 0$, $\hat{H}_1(\rho)$ has an additional $\rho[(d^\dagger d^\dagger)^{(2)}\cdot\tilde{d}s + s^\dagger d^\dagger\cdot(\tilde{d}\tilde{d})^{(2)}]$ term, which breaks the U(5) DS, and induces U(5) and O(5) mixing subject to $\Delta n_d = \pm 1$ and $\Delta\tau = \pm 1, \pm 3$. The explicit breaking of O(5) symmetry leads to non-integrability and, as will be shown in subsequent discussions, is the main cause for the onset of chaos in the spherical region. Although $\hat{H}_1(\rho > 0)$ is not diagonal in the U(5) chain, it retains the following selected solvable U(5) basis states

$$|N, n_d = \tau = L = 0\rangle \qquad E_{\rm PDS} = 0\ , \tag{5a}$$

$$|N, n_d = \tau = L = 3\rangle \qquad E_{\rm PDS} = 12\left(N - 2 + 3\rho^2\right)\bar{h}_2\ , \tag{5b}$$

while other eigenstates are mixed with respect to U(5). As such, it exhibits U(5) partial dynamical symmetry [U(5)-PDS].

The intrinsic Hamiltonian in the deformed phase, $\hat{H}_2(\xi)$, has by construction the intrinsic state of Eq. (2) $|\beta_{\rm eq} = \frac{2}{\sqrt{3}}, \gamma_{\rm eq} = 0; N\rangle$ as zero energy eigenstate. For large N, its normal modes involve both β and γ vibrations about the deformed global minimum of $V(\beta,\gamma)$, with frequencies $\epsilon_\beta = 4\bar{h}_2 N(2\xi + 1)$ and $\epsilon_\gamma = 12\bar{h}_2 N$. For $\xi = 1$, the Hamiltonian reduces to

$$\hat{H}_2(\xi = 1)/\bar{h}_2 = -\hat{C}_{\rm SU(3)} + 2\hat{N}(2\hat{N} + 3)\ , \tag{6}$$

Symmetries in Science XVI IOP Publishing
Journal of Physics: Conference Series **538** (2014) 012012 doi:10.1088/1742-6596/538/1/012012

It involves the quadratic Casimir of SU(3) and hence has SU(3) DS. The spectrum is completely solvable, $E_{\rm DS}/\bar{h}_2 = [-(\lambda^2+\mu^2+\lambda\mu+3\lambda+3\mu)+2N(2N+3)]\bar{h}_2$, and the eigenstates, $|N,(\lambda,\mu),K,L\rangle$, are those of the SU(3) chain , Eq. (1b). The spectrum resembles that of an axially-deformed rotor with degenerate K-bands arranged in SU(3) (λ,μ) multiplets, K being the angular momentum projection on the symmetry axis. The rotational states in each band have angular momenta $L=0,2,4\ldots$, for $K=0$ and $L=K,K+1,K+2,\ldots$, for $K>0$. The lowest SU(3) multiplets are $(2N,0)$ which contains the ground band $g(K=0)$, and $(2N-4,2)$ which contains the $\beta(K=0)$ and $\gamma(K=2)$ bands.

For $\xi<1$, $\hat{H}_2(\xi)$ has an additional term, $(\xi-1)P_0^\dagger P_0$, which breaks the SU(3) DS and most eigenstates are mixed with respect to SU(3). However, the following states

$$\begin{aligned}
&|N,(2N,0)K=0,L\rangle \quad L=0,2,4,\ldots,2N\\
&E_{\rm PDS}=0 \qquad\qquad (7a)\\
&|N,(2N-4k,2k)K=2k,L\rangle \quad L=K,K+1,\ldots,(2N-2k) \qquad k>0\,,\\
&E_{\rm PDS}=6k\,(2N-2k+1)\,\bar{h}_2 \qquad\qquad (7b)
\end{aligned}$$

remain solvable with good SU3) symmetry. As such, $\hat{H}_2(\xi<1)$ exhibits SU(3) partial dynamical symmetry [SU(3)-PDS]. The selected states of Eq. (7) span the ground band $g(K=0)$ and $\gamma^k(K=2k)$ bands.

The intrinsic Hamiltonians, $\hat{H}_1(\rho)$ and $\hat{H}_2(\xi)$ of Eq. (3), with $0\le\rho\le\frac{1}{\sqrt{2}}$ and $0\le\xi\le1$, interpolate between the U(5) (ρ=0) and SU(3) (ξ=1) DS limits. The two Hamiltonians coincide at the critical point $\rho_c=\frac{1}{\sqrt{2}}$ and ξ_c=0: $\hat{H}_1(\rho_c)=\hat{H}_2(\xi_c)$. Both Hamiltonians support subsets of solvable PDS states, Eqs. (5) and (7), whose analytic properties provide unique signatures for their identification in the quantum spectrum.

4. Classical limit and topology of the Landau potential

The classical limit of the IBM is obtained through the use of Glauber coherent states. This amounts to replacing $(s^\dagger,\,d_\mu^\dagger)$ by six c-numbers $(\alpha_s^*,\,\alpha_\mu^*)$ rescaled by $\sqrt{N}$ and taking $N\to\infty$, with $1/N$ playing the role of $\hbar$ [23]. Number conservation ensures that phase space is 10-dimensional and can be phrased in terms of two shape (deformation) variables, three orientation (Euler) angles and their conjugate momenta. The shape variables can be identified with the β,γ variables introduced through Eq. (2). Setting all momenta to zero, yields the classical potential which is identical to $V(\beta,\gamma)$ mentioned above. In the classical analysis presented below we consider, for simplicity, the dynamics of $L=0$ vibrations, for which only two degrees of freedom are active. The rotational dynamics with $L>0$ is examined in the subsequent quantum analysis.

For the intrinsic Hamiltonian of Eq. (3), constrained to $L=0$, the above procedure yields the following classical Hamiltonian

$$\begin{aligned}
\mathcal{H}_1(\rho)/h_2 &= \mathcal{H}_{d,0}^2+2(1-\mathcal{H}_{d,0})\mathcal{H}_{d,0}+2\rho^2p_\gamma^2\\
&\quad+\rho\sqrt{2(1-\mathcal{H}_{d,0})}\left[(p_\gamma^2/\beta-\beta p_\beta^2-\beta^3)\cos3\gamma+2p_\beta p_\gamma\sin3\gamma\right]\,, \qquad (8a)\\
\mathcal{H}_2(\xi)/h_2 &= \mathcal{H}_{d,0}^2+2(1-\mathcal{H}_{d,0})\mathcal{H}_{d,0}+p_\gamma^2\\
&\quad+\sqrt{1-\mathcal{H}_{d,0}}\left[(p_\gamma^2/\beta-\beta p_\beta^2-\beta^3)\cos3\gamma+2p_\beta p_\gamma\sin3\gamma\right]\\
&\quad+\xi\left[\beta^2p_\beta^2+\tfrac{1}{4}(\beta^2-T)^2-2(1-\mathcal{H}_{d,0})(\beta^2-T)+4(1-\mathcal{H}_{d,0})^2\right]\,. \qquad (8b)
\end{aligned}$$

Here the coordinates $\beta\in[0,\sqrt{2}]$, $\gamma\in[0,2\pi)$ and their canonically conjugate momenta $p_\beta\in[0,\sqrt{2}]$ and $p_\gamma\in[0,1]$ span a compact classical phase space. The term, $\mathcal{H}_{d,0}=(T+\beta^2)/2$ with $T=p_\beta^2+p_\gamma^2/\beta^2$, denotes the classical limit of $\hat{n}_d$ (restricted to $L=0$) and forms an isotropic harmonic oscillator Hamiltonian in the β and γ variables. Notice that the classical Hamiltonian of Eq. (8) contains complicated momentum-dependent terms originating from the two-body interactions in the Hamiltonian (3), not just the usual quadratic kinetic energy T. Setting $p_\beta = p_\gamma = 0$ in Eq. (8) leads to the following classical potential

$$V_1(\rho)/h_2 = 2\beta^2 - 2\rho\sqrt{2-\beta^2}\beta^3\cos 3\gamma - \tfrac{1}{2}\beta^4 \,, \quad (9a)$$
$$V_2(\xi)/h_2 = 2(1-3\xi)\beta^2 - \sqrt{2(2-\beta^2)}\beta^3\cos 3\gamma + \tfrac{1}{4}(9\xi-2)\beta^4 + 4\xi \,. \quad (9b)$$

The variables β and γ can be interpreted as polar coordinates in an abstract plane parametrized by Cartesian coordinates $x = \beta\cos\gamma$ and $y = \beta\sin\gamma$. Using these relations together with $p_x = p_\beta\cos\gamma - (p_\gamma/\beta)\sin\gamma$ and $p_y = (p_\gamma/\beta)\cos\gamma + p_\beta\sin\gamma$, one can cast the classical Hamiltonian, Eq. (8) and potential, Eq. (9), in Cartesian form. Thus, $\mathcal{H}_{d,0} = (p_x^2 + p_y^2 + x^2 + y^2)/2$ and the potentials $V(\beta,\gamma) = V(x,y)$ depend on the combinations $\beta^2 = x^2+y^2$, $\beta^4 = (x^2+y^2)^2$ and $\beta^3\cos 3\gamma = x^3 - 3xy^2$.

The control parameters ρ and ξ determine the landscape and extremal points of the potentials $V_1(\rho;\beta,\gamma)$ and $V_2(\xi;\beta,\gamma)$, Eq. (9). Important values of these parameters at which a pronounced change in structure is observed, are the spinodal point (ρ^*) where a second (deformed) minimum occurs, an anti-spinodal point (ξ^{**}) where the first (spherical) minimum disappears and a critical point (ρ_c, ξ_c) in-between, where the two minima are degenerate. For the potentials under discussion, the values of the control parameters at these points are

$$\rho^* = \tfrac{1}{2} \quad , \quad (\rho_c = \tfrac{1}{\sqrt{2}}, \xi_c = 0) \quad , \quad \xi^{**} = \tfrac{1}{3} \;. \quad (10)$$

The critical point separates the spherical and deformed phases. The spinodal and anti-spinodal points embrace it and mark the boundary of the phase coexistence region.

In general, the only γ-dependence in the potentials (9) is due to the $\sqrt{2-\beta^2}\beta^3\cos 3\gamma$ term. This induces a three-fold symmetry about the origin $\beta = 0$. As a consequence, the deformed extremal points are obtained for $\gamma = 0, \frac{2\pi}{3}, \frac{4\pi}{3}$ (prolate shapes), or $\gamma = \frac{\pi}{3}, \pi, \frac{5\pi}{3}$ (oblate shapes). It is therefore possible to restrict the analysis to $\gamma = 0$ and allow for both positive and negative values of β, corresponding to prolate and oblate deformations, respectively. The potentials $V(\beta,\gamma=0) = V(x, y=0)$ for several values of ξ, ρ, are shown at the bottom rows of Figs. 2-4.

The relevant potential in the spherical phase is $V_1(\rho;\beta,\gamma)$, Eq. (9a), with $0 \leq \rho \leq \rho_c$. In this case, $\beta_{\rm eq} = 0$ is a global minimum of the potential at an energy $V_{\rm sph} = 0$, representing the spherical equilibrium shape. The limiting value at the domain boundary is $V_{\rm lim} = V_1(\rho;\beta = \sqrt{2},\gamma) = 2h_2$. For $\rho = 0$, (the U(5) limit), the potential is independent of γ and has $\beta_{\rm eq} = 0$ as a single minimum. For $\rho > 0$, the potential depends on γ, and $\beta = 0$ remains a single minimum for $0 \leq \rho < \rho^*$. At the spinodal point ρ^*, $V_1(\rho)$ acquires an additional deformed local minimum at an energy $V_{\rm def} > 0$, and a barrier develops between the two minima. The spherical and deformed minima cross and become degenerate at the critical point (ρ_c, ξ_c).

The relevant potential in the deformed phase is $V_2(\xi;\beta,\gamma)$, Eq. (9b), with $\xi \geq \xi_c$. In this case, $[\beta_{\rm eq} = \frac{2}{\sqrt{3}}, \gamma_{\rm eq} = 0]$ is a global minimum of the potential at an energy $V_{\rm def} = 0$, representing the deformed equilibrium shape. The limiting value of the domain boundary is $V_{\rm lim} = V_2(\xi;\beta = \sqrt{2},\gamma) = (2+\xi)h_2$. The two potentials coincide at the critical point (ρ_c, ξ_c), $V_2(\xi_c;\beta,\gamma) = V_1(\rho_c;\beta,\gamma)$, and the barrier height is $V_{\rm bar} = \frac{1}{2}h_2(1-\sqrt{3})^2 = 0.268h_2$. The spherical minimum turns local in $V_2(\xi)$ for $\xi > \xi_c$ with energy $V_{\rm sph} = 4h_2\xi > 0$, and disappears at the anti-spinodal point ξ^{**}. For $\xi > \xi^{**}$, $\beta = 0$ turns into a maximum and the potential remains with a single deformed minimum, reaching the SU(3) limit at $\xi = 1$.

The indicated changes in the topology of the potential surfaces upon variation of the control parameters (ρ, ξ), identify three regions with distinct structure.

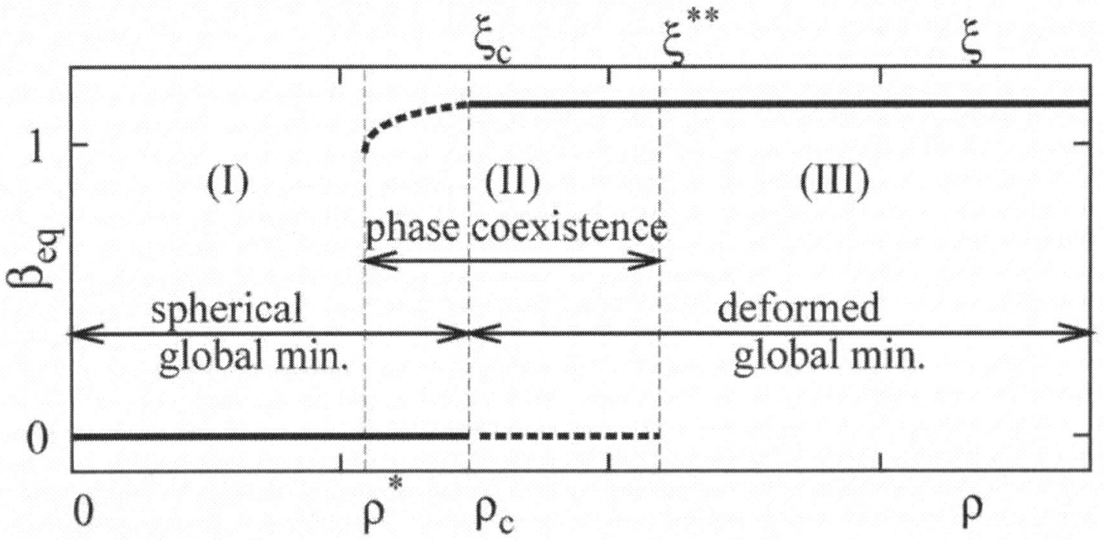

Figure 1. Behavior of the order parameter, $\beta_{\rm eq}$, as a function of the control parameters (ρ,ξ) of the intrinsic Hamiltonian, Eq. (3). Here ρ^{*}, $(\rho_c,\,\xi_c)$, ξ^{**}, are the spinodal, critical and anti-spinodal points, respectively, with values given in Eq. (10). The deformation at the global (local) minimum of the Landau potential (9) is marked by solid (dashed) lines. $\beta_{\rm eq}=0$ ($\beta_{\rm eq}=\frac{2}{\sqrt{3}}$) on the spherical (deformed) side of the QPT. Region I (III) involves a single spherical (deformed) shape, while region II involves shape-coexistence.

I. The region of a stable spherical phase, $\rho\in[0,\rho^{*}]$, where the potential has a single spherical minimum.

II. The region of phase coexistence, $\rho\in(\rho^{*},\rho_c]$ and $\xi\in[\xi_c,\xi^{**})$, where the potential has both spherical and deformed minima which cross and become degenerate at the critical point.

III. The region of a stable deformed phase, $\xi>\xi^{**}$, where the potential has a single deformed minimum.

The potential surface in each region serves as the Landau potential of the QPT, with the equilibrium deformations as order parameters. The latter evolve as a function of the control parameters (ρ,ξ) and exhibit a discontinuity typical of a first order transition. As depicted in Fig 1, the order parameter β_{eq} is a double-valued function in the coexistence region (in-between ρ^{*} and ξ^{**}) and a step-function outside it. In what follows, we examine the nature of the classical and quantum dynamics in each region.

5. Classical analysis

Chaotic properties of the IBM have been studied extensively [24], albeit, with a simplified Hamiltonian, giving rise to an extremely low barrier. A new element in the present study is the presence of a high barrier at the critical-point, $V_{\rm bar}/h_2=0.268$, compared to $V_{\rm bar}/h_2=0.0018$ in previous works. This allows the uncovering of a rich pattern of regularity and chaos across a generic first-order QPT in a wide coexistence region.

The classical dynamics constraint to $L=0$, can be depicted conveniently via Poincaré surfaces of sections in the plane $y=0$, plotting the values of x and the momentum p_x each time a trajectory intersects the plane [25]. Regular trajectories are bound to toroidal manifolds within the phase space and their intersections with the plane of section lie on 1D curves (ovals).

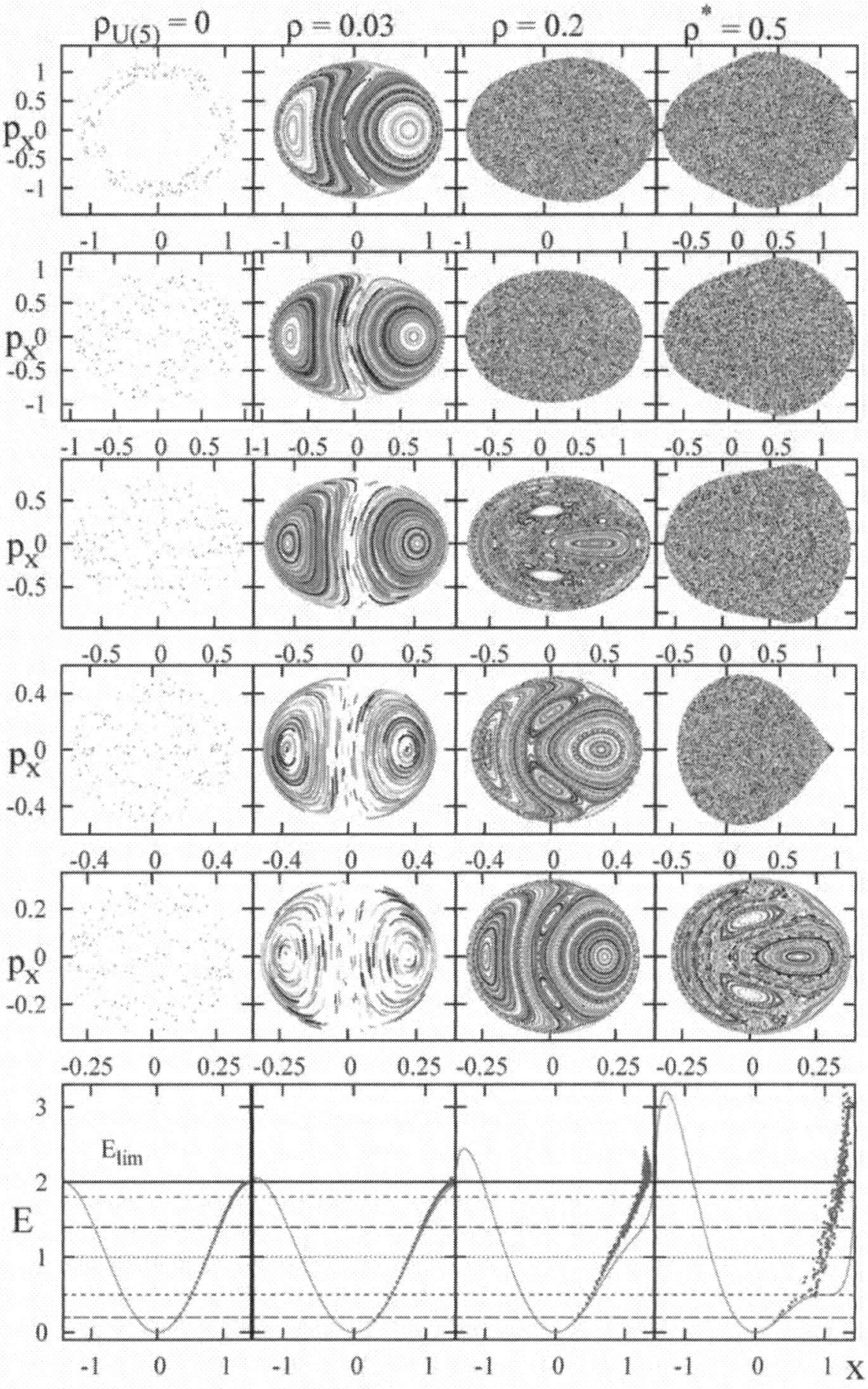

Figure 2. Poincaré sections in the stable spherical phase (region I). Upper five rows depict the classical dynamics of $\mathcal{H}_1(\rho)$ (8a) with $h_2 = 1$, for several values of $\rho \leq \rho^{**}$. The bottom row displays the Peres lattices $\{x_i, E_i\}$, portraying the quantum dynamics for $(N = 80, L = 0)$ eigenstates of $\hat{H}_1(\rho)$ (3a), overlayed on the classical potentials $V_1(\rho; x, y = 0)$ (9a). The five energies at which the sections were calculated consecutively, are indicated by horizontal lines.

*** Formatting Issue - Best Available Paper/Graphic ***

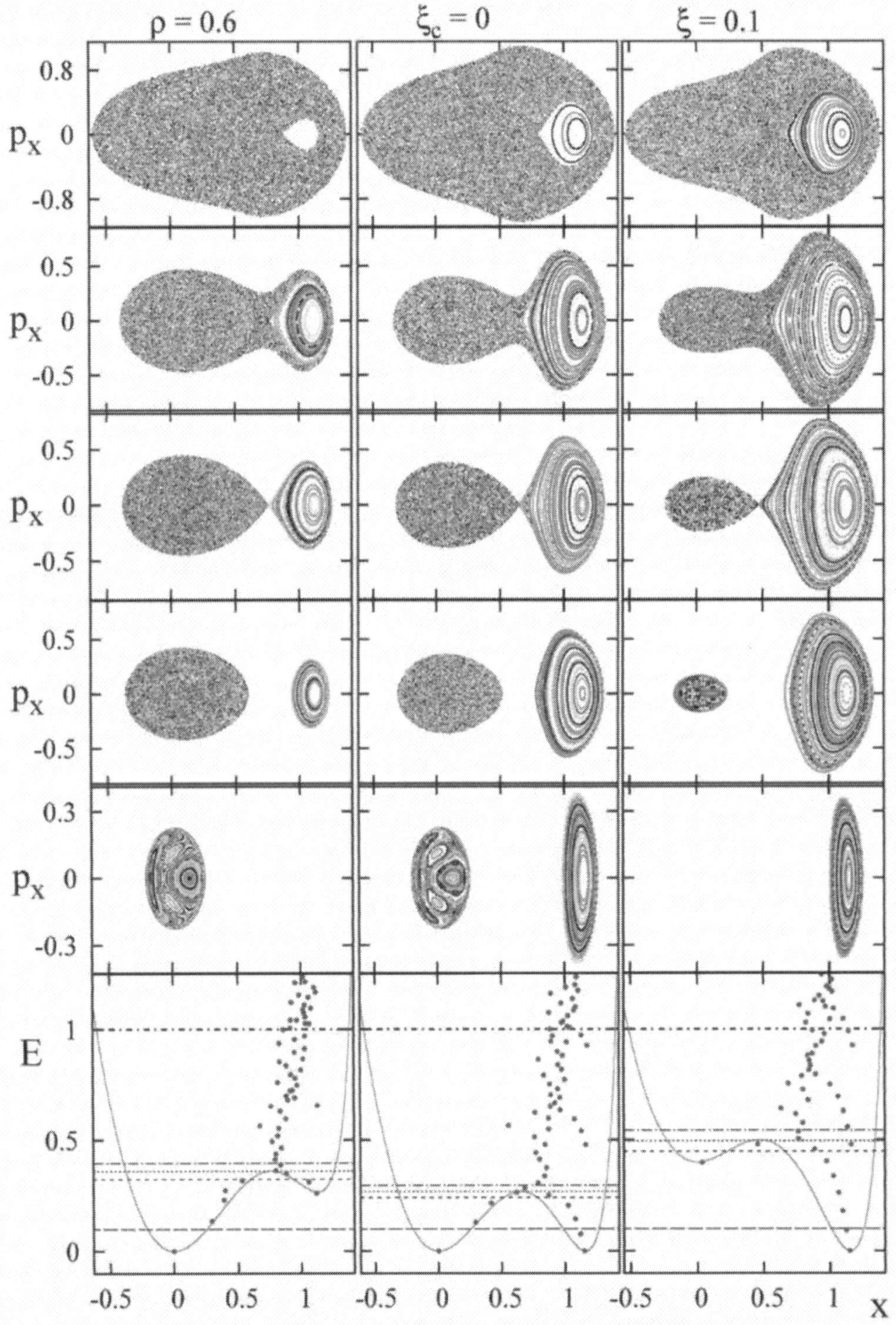

Figure 3. Poincaré sections in the region of phase-coexistence (region II). The panels are as in Fig. 2, but for the classical Hamiltonians $\mathcal{H}_1(\rho)$ (8a) with $\rho^{**} < \rho \leq \rho_c$, and $\mathcal{H}_2(\xi)$ (8b) with $\xi_c \leq \xi < \xi^{**}$. The classical potentials are $V_1(\rho; x, y = 0)$ (9a) and $V_2(\xi; x, y = 0)$ (9b). The Peres lattices involve the quantum Hamiltonians $\hat{H}_1(\rho)$ (3a) and $\hat{H}_2(\xi)$ (3b).

*** Formatting Issue - Best Available Paper/Graphic ***

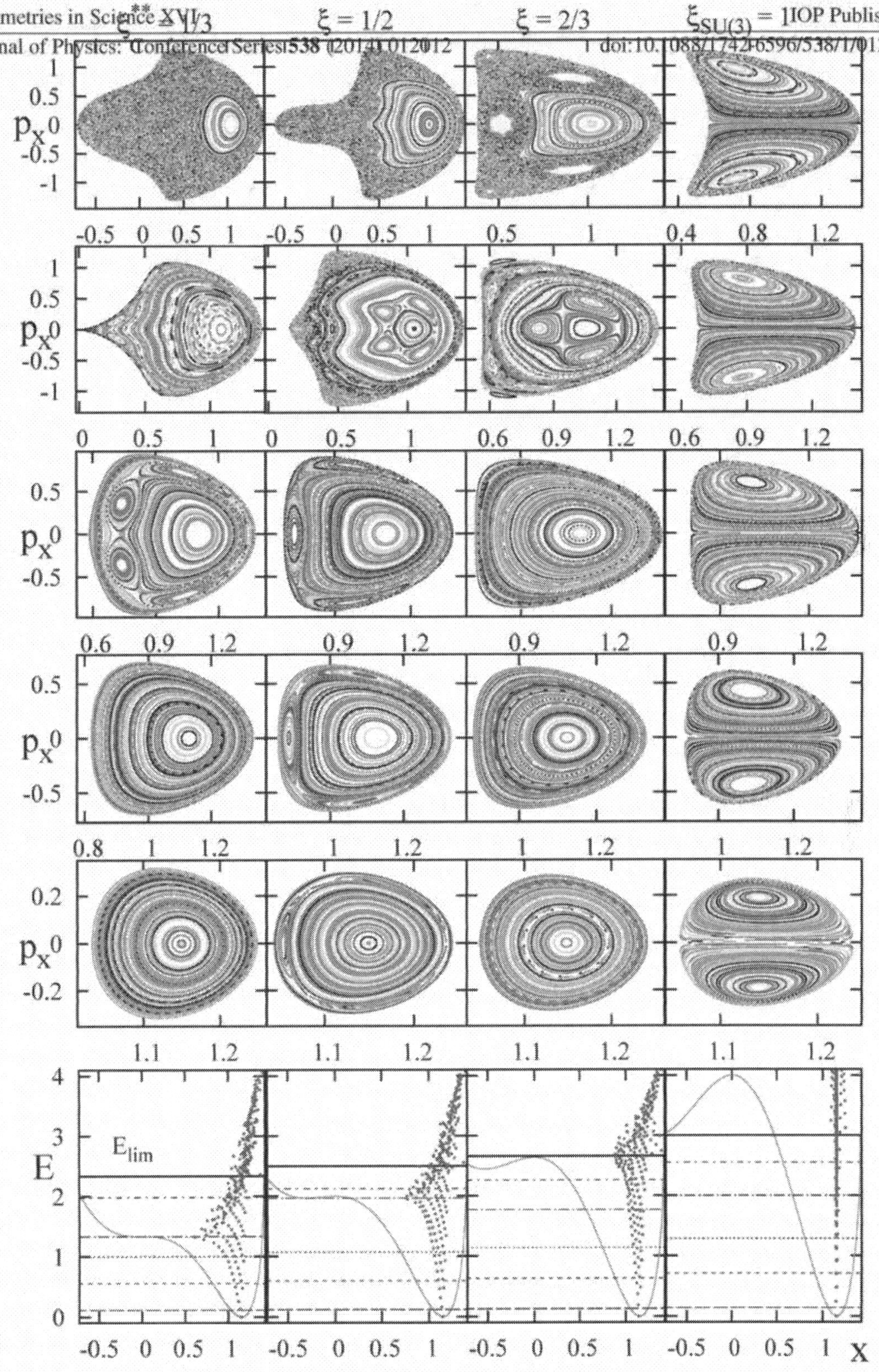

Figure 4. Poincaré sections in the stable deformed phase (region III). The panels are as in Fig. 2, but for the classical Hamiltonian $\mathcal{H}_2(\xi)$ (8b) and potential $V_2(\xi; x, y = 0)$ (9b), with $\xi \geq \xi^{**}$. The Peres lattices involve the quantum Hamiltonian $\hat{H}_2(\xi)$ (3b).

In contrast, chaotic trajectories randomly cover kinematically accessible areas of the section.

The Poincaré sections associated with the classical Hamiltonian of Eq. (8) are shown in Figs. 2-3-4 for representative energies, below the domain boundary, and control parameters (ρ, ξ) in regions I-II-III, respectively. The bottom row in each figure displays the corresponding classical potential $V(\beta, \gamma = 0) = V(x, y = 0)$, Eq. (9). In region I ($0 \le \rho \le \rho^*$), for $\rho = 0$, $\mathcal{H}_1(\rho = 0) = \mathcal{H}_{d,0}(2 - \mathcal{H}_{d,0})$, involves the 2D harmonic oscillator Hamiltonian and $V_1(\rho = 0) \propto 2\beta^2 - \beta^4/2$. The system is in the U(5) DS limit and is completely integrable. The orbits are periodic and, as shown in Fig. 2, appear in the surface of section as a finite collection of points. The sections for $\rho = 0.03$ in Fig. 2, show the phase space portrait typical of an anharmonic (quartic) oscillator with two major regular islands, weakly perturbed by the small $\rho \cos 3\gamma$ term. The orbits are quasi-periodic and appear as smooth one-dimensional invariant curves. For small β, $V_1(\rho) \approx \beta^2 - \rho\sqrt{2}\beta^3 \cos 3\gamma$ and resembles the well-known Hénon-Heiles potential (HH) [26]. Correspondingly, as shown for $\rho = 0.2$ in Fig. 2, at low energy, the dynamics remains regular and two additional islands show up. The four major islands surround stable fixed points and unstable (hyperbolic) fixed points occur in-between. At higher energy, one observes a marked onset of chaos and an ergodic domain. The chaotic component of the dynamics increases with ρ and maximizes at the spinodal point $\rho^* = 0.5$. The chaotic orbits densely fill two-dimensional regions of the surface of section.

The dynamics changes profoundly in region II of phase coexistence ($\rho^* < \rho \le \rho_c$ and $\xi_c \le \xi < \xi^{**}$). The Poincaré sections before at and after the critical point, ($\rho = 0.6$, $\xi_c = 0$, $\xi = 0.1$) are shown in Fig. 3. In general, the motion is predominantly regular at low energies and gradually turning chaotic as the energy increases. However, the classical dynamics evolves differently in the vicinity of the two wells. As the local deformed minimum develops, robustly regular dynamics attached to it appears. The trajectories form a single island and remain regular even at energies far exceeding the barrier height $V_{\rm bar}$. This behavior is in marked contrast to the HH-type of dynamics in the vicinity of the spherical minimum, where a change with energy from regularity to chaos is observed, until complete chaoticity is reached near the barrier top. The clear separation between regular and chaotic dynamics, associated with the two minima, persists all the way to the barrier energy, $E = V_{\rm bar}$, where the two regions just touch. At $E > V_{\rm bar}$, the chaotic trajectories from the spherical region can penetrate into the deformed region and a layer of chaos develops, and gradually dominates the surviving regular island for $E \gg V_{\rm bar}$. As ξ increases, the spherical minimum becomes shallower, and the HH-like dynamics diminishes.

Fig. 4 displays the classical dynamics in region III ($\xi^{**} \le \xi \le 1$). The local spherical minimum and the associated HH-like dynamics disappear at the anti-spinodal point $\xi^{**} = 1/3$. The regular motion, associated with the single deformed minimum, prevails for $\xi \ge \xi^{**}$. Here a single stable fixed point, surrounded by a family of elliptic orbits, continues to dominate the Poincaré section. In certain regions of the control parameter ξ and energy, the section landscape changes from a single to several regular islands, reflecting the sensitivity of the dynamics to local degeneracies of normal modes. A notable exception to such variation is the SU(3) DS limit ($\xi = 1$), for which the system is integrable and the phase space portrait is the same for any energy.

6. Quantum analysis

Quantum manifestations of classical chaos are often detected by statistical analyses of energy spectra [25]. In a quantum system with mixed regular and irregular states, the statistical properties of the spectrum are usually intermediate between the Poisson and the Gaussian orthogonal ensemble (GOE) statistics. Such global measures of quantum chaos are, however, insufficient to reflect the rich dynamics of an inhomogeneous phase space structure encountered in Figs. 2-4, with mixed but well-separated regular and chaotic domains. To do so, one needs to distinguish between regular and irregular subsets of eigenstates in the same energy intervals. For that purpose, we employ the spectral lattice method of Peres [27], which provides additional properties of individual energy eigenstates. The Peres lattices are constructed by plotting the expectation values $O_i = \langle i|\hat{O}|i\rangle$ of an arbitrary operator, $[\hat{O}, \hat{H}] \neq 0$, versus the energy $E_i = \langle i|\hat{H}|i\rangle$ of the Hamiltonian eigenstates $|i\rangle$. The lattices $\{O_i, E_i\}$ corresponding to regular dynamics can be shown to display an ordered pattern, while chaotic dynamics leads to disordered meshes of points [27, 28].

In the present analysis we choose the Peres operator to be $\hat{n}_d$. The lattices correspond to the set of points $\{x_i, E_i\}$, with $x_i \equiv \sqrt{2\langle i|\hat{n}_d|i\rangle/N}$ and $|i\rangle$ being the eigenstates of the quantum Hamiltonian (3). The expectation value of $\hat{n}_d$ in the condensate of Eq. (2) is related to the deformation β and the coordinate x in the classical potential (9). Accordingly, the particular choice of lattices $\{x_i, E_i\}$ can distinguish regular from irregular states and associate them with a given region in phase space, through the classical-quantum correspondence $\beta = x \leftrightarrow x_i$.

The Peres lattices for $L = 0$ eigenstates of the intrinsic Hamiltonian (3) with $N = 80$, are shown on the bottom rows of Figs. 2-4, overlayed on the classical potentials $V(x, y = 0)$ of Eq (9). For $\rho = 0$, the Hamiltonian (4) has U(5) dynamical symmetry with a solvable spectrum, a function of n_d. For large N and replacing x_i by β, the Peres lattice coincides coincides with $V_1(\rho = 0)$, a trend seen exactly for $\rho = 0$ and approximately at $\rho = 0.03$ in Fig. 2. For $\rho = 0.2$, at low energy a few lattice points still follow the potential curve $V_1(\rho)$, but at higher energies one observes sizeable deviations and disordered meshes of lattice points, in accord with the onset of chaos in the classical Hénon-Heiles system. The disorder in the Peres lattice enhances at the spinodal point $\rho^* = 0.5$, where the chaotic component of the classical dynamics maximizes. As seen in Figs. 3-4, whenever a deformed minimum occurs in the potential, the Peres lattices exhibit regular sequences of states localized within and above the deformed well. They form several chains of lattice points close in energy, with the lowest chain originating at the deformed ground state. A close inspection reveals that the x_i-values of these regular states, lie in the intervals of x-values occupied by the regular tori in the Poincaré sections. Similarly to the classical tori, these regular sequences persist to energies well above the barrier $V_{\rm bar}$. The lowest sequence consists of $L = 0$ bandhead states of the ground $g(K = 0)$ and $\beta^n(K = 0)$ bands. Regular sequences at higher energy correspond to $\beta^n\gamma^2(K = 0)$, $\beta^n\gamma^4(K = 0)$ bands, etc. In contrast, the remaining states, including those residing in the spherical minimum, do not show any obvious patterns and lead to disordered (chaotic) meshes of points at high energy. For $\xi > \xi^{**}$, a larger number and longer sequences of regular $K = 0$ bandhead states are observed in the vicinity of the single deformed minimum ($x \approx 1$), as its depth increases.

Peres lattices can also be used to visualize the dynamics of quantum states with non-zero angular momenta. Examples for $N = 50$, $L = 0, 2, 3, 4$, eigenstates of the critical-point Hamiltonian, $\hat{H}_1(\rho_c) = \hat{H}_2(\xi_c)$, are shown in Fig. 5. The right column in the figure combines the separate-L lattices and overlays them on the relevant classical potential. Rotational states with $L = 0, 2, 4, \ldots$, comprise the regular $K = 0$ bands mentioned above, and are accompanied by sequences $L = 2, 3, 4, \ldots$, forming $K = 2$ bands. Additional K-bands (not shown in Fig. 5), corresponding to multiple β and γ vibrations about the deformed shape, can also be identified. The states in each regular band share a common intrinsic structure, as indicated by their nearly equal values of $\langle \hat{n}_d \rangle$ and a similar coherent decomposition of their wave functions in the SU(3) basis, to be discussed in Section 7. These ordered band structures show up in the vicinity of the deformed well and are not present in the disordered (chaotic) portions of the Peres lattice. Their occurrence and persistence in the spectrum throughout the coexistence region, including the critical-point, is somewhat unexpected, in view of the strong mixing and abrupt structural changes taking place.

7. Symmetry aspects

Away from the U(5) and SU(3) limits ($\rho > 0$ and $\xi < 1$), both dynamical symmetries are broken in the intrinsic Hamiltonian (3). The competition between terms with different symmetry

Symmetries in Science XVI IOP Publishing
Journal of Physics: Conference Series **538** (2014) 012012 doi:10.1088/1742-6596/538/1/012012

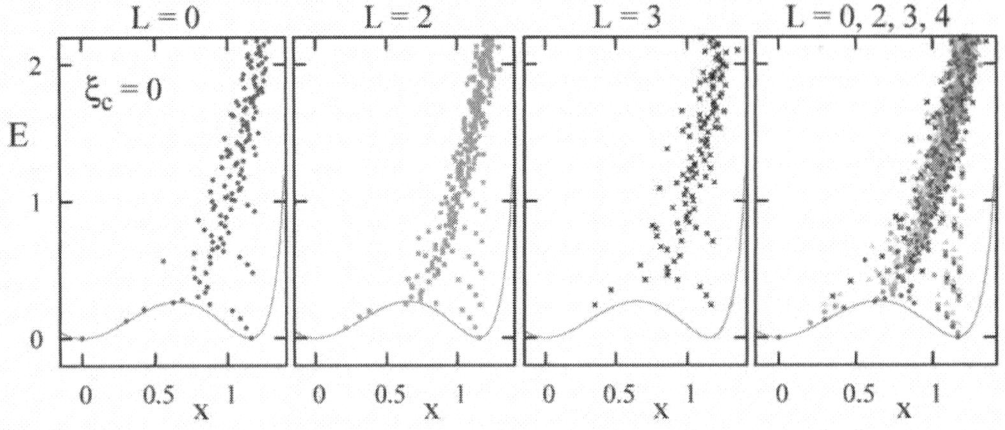

Figure 5. Peres lattices $\{x_i, E_i\}$, for $L = 0, 2, 3, 4$, eigenstates of the critical-point Hamiltonian $\hat{H}_1(\rho_c) = \hat{H}_2(\xi_c)$ with $h_2 = 1$ and $N = 50$. The right column combines the separate-L lattices and overlays them on the corresponding classical potential.

character, drives the system through a first-order QPT with a characteristic pattern of mixed dynamics. It is of great interest to study the symmetry properties of the Hamiltonian eigenstates across the QPT and, in particular, seek for a symmetry-based explanation for the persistence of regular subsets of states amidst a complicated environment.

Consider an eigenfunction of the Hamiltonian, $|L_i\rangle$, with angular momentum L and ordinal number i (enumerating the occurrences of states with the same L, with increasing energy). Its expansion in the U(5) DS basis, $|N, n_d, \tau, n_\Delta, L\rangle$ of Eq. (1a), and in the SU(3) DS basis, $|N, (\lambda, \mu), K, L\rangle$ of Eq. (1b), reads

$$\begin{aligned} |L_i\rangle &= \sum_{n_d,\tau,n_\Delta} C^{(L_i)}_{n_d,\tau,n_\Delta} |N, n_d, \tau, n_\Delta, L_i\rangle \, , \\ &= \sum_{(\lambda,\mu),K} C^{(L_i)}_{(\lambda,\mu),K} |N, (\lambda,\mu), K, L_i\rangle \, . \end{aligned} \tag{11}$$

The U(5) (n_d) probability distribution, $P^{(L_i)}_{n_d}$, and the SU(3) $[(\lambda, \mu)]$ probability distribution, $P^{(L_i)}_{(\lambda,\mu)}$, are calculated as

$$P^{(L_i)}_{n_d} = \sum_{\tau,n_\Delta} |C^{(L_i)}_{n_d,\tau,n_\Delta}|^2 \, , \tag{12a}$$

$$P^{(L_i)}_{(\lambda,\mu)} = \sum_{K} |C^{(L_i)}_{(\lambda,\mu),K}|^2 \, . \tag{12b}$$

The purity of eigenstates with respect to a DS basis can be evaluated by means of the U(5) and SU(3) Shannon entropies defined as

$$S_{\rm U5}(L_i) = -\frac{1}{\ln D_5} \sum_{n_d} P^{(L_i)}_{n_d} \ln P^{(L_i)}_{n_d} \, , \tag{13a}$$

$$S_{\rm SU3}(L_i) = -\frac{1}{\ln D_3} \sum_{(\lambda,\mu)} P^{(L_i)}_{(\lambda,\mu)} \ln P^{(L_i)}_{(\lambda,\mu)} \, . \tag{13b}$$

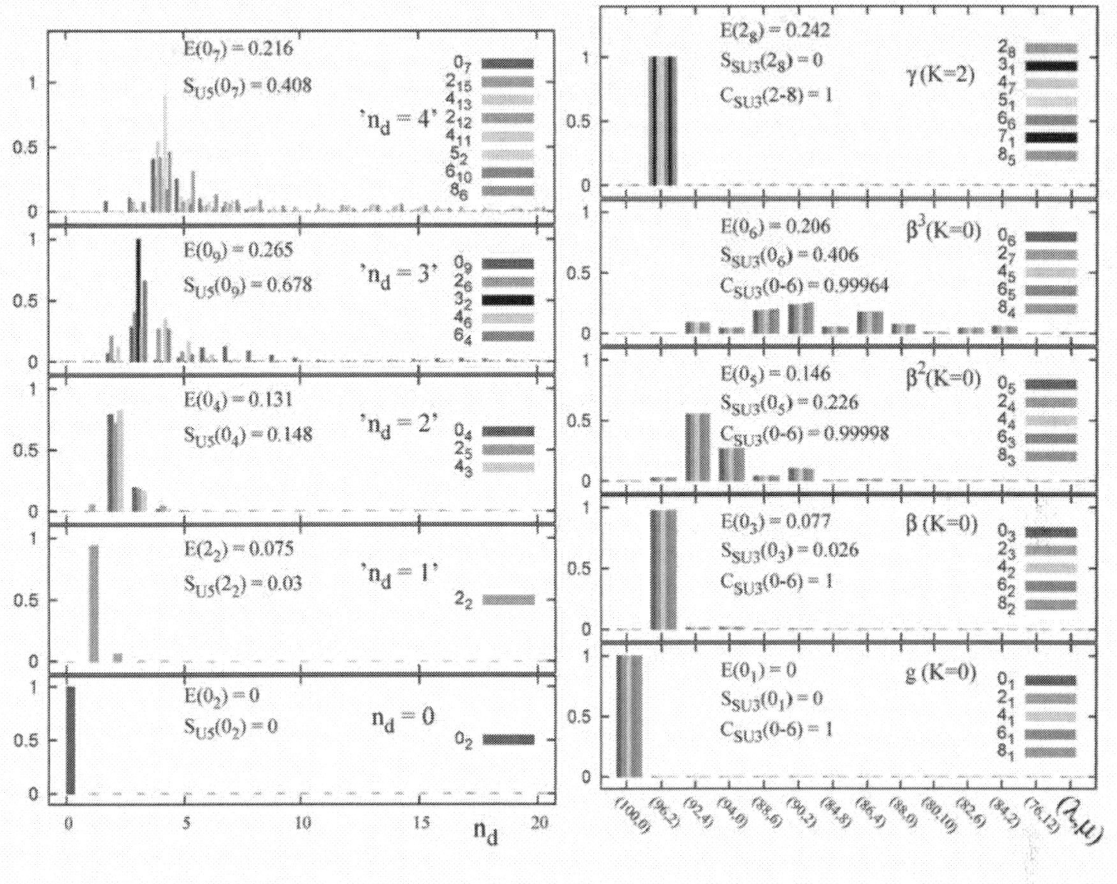

Figure 6. U(5) n_d-probability distribution, $P_{n_d}^{(L_i)}$ (12a) [left column], and SU(3) (λ,μ)-probability distribution, $P_{(\lambda,\mu)}^{(L_i)}$ (12b) [right column], for selected eigenstates of the critical-point (ρ_c, ξ_c) Hamiltonian (3), with $N = 50$. The U(5) Shannon entropy, $S_{\rm U5}(L_i)$ (13a), and SU(3) correlator, $C_{\rm SU3}(0{-}6)$ (14), are indicated for spherical and deformed type of states, respectively.

The normalization D_5 (D_3) counts the number of possible n_d $[(\lambda,\mu)]$ values for a given L. A Shannon entropy vanishes when the considered state is pure with good G-symmetry $[S_{\rm G}(L_i){=}0]$, and is positive for a mixed state. The maximal value $[S_{\rm G}(L_i){=}1]$ is obtained when $|L_i\rangle$ is uniformly spread among the irreps of G, *i.e.* for $P_G^{(L_i)} = 1/D_G$. Intermediate values, $0 < S_{\rm G}(L_i) < 1$, indicate partial fragmentation of the state $|L_i\rangle$ in the respective DS basis.

Focusing on the critical-point Hamiltonian, the states shown on the left column of Fig. 6 were selected on the basis of having the largest components with $n_d = 0, 1, 2, 3, 4$, within the given L spectra. States with different L values are arranged into panels labeled by 'n_d' to conform with the structure of the n_d-multiplets of the U(5) DS limit, Eq. (4). Each panel depicts the U(5) n_d-probabilities, $P_{n_d}^{(L_i)}$ (12a), for states in the multiplet and lists the U(5) Shannon entropy $S_{\rm U5}(L_i)$ of a representative eigenstate. In particular, the zero-energy $L{=}0_2^+$ state is seen to be a pure $n_d{=}0$ state, with $S_{\rm U5}{=}0$, which is the solvable U(5)-PDS eigenstate of Eq. (5a). The state 2_2^+ has a pronounced $n_d{=}1$ component (96%) and the states ($L = 0_4^+$, 2_5^+, 4_3^+) in the third panel, have a pronounced $n_d{=}2$ component and a low value of $S_{\rm U5} < 0.15$. All the above states with '$n_d \leq 2$' have a dominant single n_d component, and hence qualify as 'spherical' type of states. These multiplets comprise the lowest left-most states shown in the combined Peres lattices of

Fig. 5. In contrast, the states in the panels '$n_d = 3$' and '$n_d = 4$' of Fig. 6, are significantly fragmented. Notable exceptions are the $L = 3_2^+$ state, which is the solvable U(5)-PDS state of Eq. (5b) with $n_d = 3$, and the $L = 5_2^+$ state with a dominant $n_d = 4$ component. The existence in the spectrum of specific spherical-type of states with either $P^{(L)}_{n_d} = 1$ [$S_{\rm U5}(L) = 0$] or $P^{(L)}_{n_d} \approx 1$ [$S_{\rm U5}(L) \approx 0$], exemplifies the presence of an exact or approximate U(5) PDS at the critical-point.

The states shown on the right column of Fig. 6 have a different character. They belong to the five lowest regular sequences seen in the combined Peres lattices of Fig. 5. They have a broad n_d-distribution, hence are qualified as 'deformed'-type of states, forming rotational bands: $g(K=0)$, $\beta(K=0)$, $\beta^2(K=0)$, $\beta^3(K=0)$ and $\gamma(K=2)$. Each panel depicts the SU(3) (λ,μ)-distribution, $P^{(L_i)}_{(\lambda,\mu)}$ (12b) for the rotational states in each band. The ground $g(K = 0)$ and the $\gamma(K = 2)$ bands are pure [$S_{\rm SU3} = 0$] with $(\lambda,\mu) = (2N, 0)$ and $(2N - 4, 2)$ SU3) character, respectively. These are the solvable bands of Eq. (7) with SU(3) PDS. The non-solvable K-bands are mixed with respect to SU(3), but the mixing is similar for the different L-states in the same band. Such strong but coherent (L-independent) mixing is the hallmark of SU(3) quasi-dynamical symmetry (QDS). It results from the existence of a single intrinsic state for each such band and imprints an adiabatic motion and increased regularity [29].

The coherent decomposition characterizing SU(3) QDS, implies strong correlations between the SU(3) components of different L-states in the same band. This can be used as a criteria for the identification of rotational bands. We focus here on the $L = 0, 2, 4, 6$, members of $K = 0$ bands. Given a $L = 0_i^+$ state, among the ensemble of possible states, we associate with it those $L_j > 0$ states which show the maximum correlation, $\max_j\{\pi(0_i, L_j)\}$. Here $\pi(0_i, L_j)$ is a Pearson coefficient whose values lie in the range $[-1, 1]$. Specifically, $\pi(0_i, L_j) = 1, -1, 0$, indicate a perfect correlation, a perfect anti-correlation, and no linear correlation, respectively, among the SU(3) components of the 0_i and L_j states. To quantify the amount of coherence (hence of SU(3)-QDS) in the chosen set of states, we employ the following product of the maximum correlation coefficients [30]

$$C_{\rm SU3}(0_i{-}6) \equiv \max_j\{\pi(0_i, 2_j)\}\; \max_k\{\pi(0_i, 4_k)\}\; \max_\ell\{\pi(0_i, 6_\ell)\}\ . \tag{14}$$

We consider the set of states $\{0_i,\, 2_j,\, 4_k,\, 6_\ell\}$ as comprising a $K = 0$ band with SU(3)-QDS, if $C_{\rm SU3}(0_i{-}6) \approx 1$. As expected, we find the values $C_{\rm SU3}(0_i{-}6) \approx 1$ for all the 'deformed' K-bands, shown in the right column of Fig. 6. It should be noted that the coherence property of a band of states, as measured by $C_{\rm SU3}(0_i{-}6)$, is independent of its purity, as measured by $S_{\rm SU3}(L_i)$. Thus, in Fig. 6, the pure $g(K = 0)$ and $\gamma(K = 2)$ bands with SU(3) PDS have $C_{\rm SU3}(0_i{-}6) = 1$ and $S_{\rm SU3} = 0$, while the mixed $\beta^3(K = 0)$ band has $C_{\rm SU3}(0_i{-}6) = 0.9996$ and $S_{\rm SU3} = 0.406$.

The top panel of Fig. 7 displays the values of the SU(3) Shannon entropy (13b) for $L = 0$ eigenstates of the Hamiltonian (3), with $N = 50$. The vertical axis lists the energy E of the states, while the horizontal axis lists 35 values of the control parameters (ρ, ξ). Vertical dashed lines which embrace each control parameter, correspond to the value $S_{SU3}(L{=}0) = 0$ (left) and $S_{SU3}(L{=}0) = 1$ (right). Thus, states which are pure with respect to SU(3) are represented by points on the vertical dashed line to the left of the given control parameter. Departures from this vertical line, $0 < S_{SU3}(L{=}0) \leq 1$, indicate the amount of SU(3) mixing. The bottom panel of Fig. 7, displays the values of the SU(3) correlation coefficient $C_{\rm SU3}(0{-}6)$, Eq. (14), correlating sequences of $L = 0, 2, 4, 6$ states, throughout the entire spectrum. The energy E, listed on the vertical axis, corresponds to the energy of the $L = 0$ eigenstate in each sequence. The vertical dashed lines correspond now to the value $C_{\rm SU3}(0{-}6) = 1$ (right) and $C_{\rm SU3}(0{-}6) = 0$ (left). Thus, a highly-correlated sequence of $L = 0, 2, 4, 6$ states, comprising a $K = 0$ band and manifesting SU(3)-QDS, are represented by points lying on or very close to the vertical dashed line to the right of the given control parameter, corresponding to $C_{\rm SU3}(0{-}6) \approx 1$. Slight departures from this vertical line, $C_{\rm SU3}(0{-}6) < 1$, indicate a reduction of SU(3) coherence.

*** Formatting Issue - Best Available Paper/Graphic ***

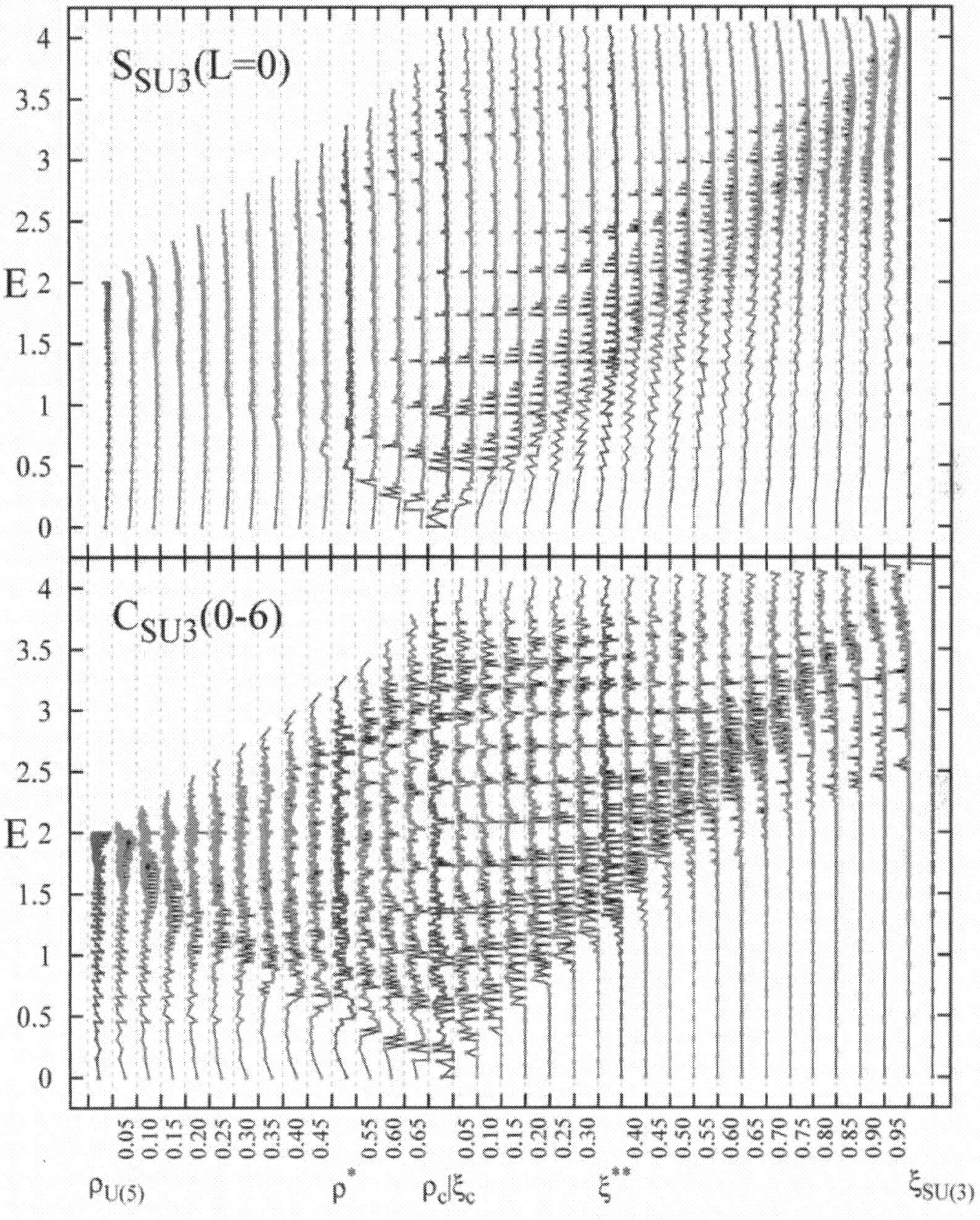

Figure 7. SU(3) Shannon entropy $S_{\rm SU3}(L=0)$ (13b) [top panel], and SU(3) correlator, $C_{\rm SU3}(0{-}6)$ (14) [bottom panel] for energy eigenstates of the Hamiltonian (3) with $N=50$, as a function of the control parameters (ρ, ξ). The vertical dashed lines are explained in the text.

At the SU(3) DS limit ($\xi_{\rm SU(3)} = 1$), the SU(3) entropy, $S_{\rm SU3}(L) = 0$, vanishes for all states. In this case, the L-states in a given K-band belong to a single SU(3) irrep, hence necessarily $C_{\rm SU3}(0{-}6) = 1$. For $\xi < 1$, $S_{\rm SU3}(L=0) > 0$ acquires positive values, reflecting an SU(3) mixing. The SU(3) breaking becomes stronger at higher energies and as ξ approaches $\xi_c = 0$ from above, resulting in higher values of $S_{\rm SU3}(L=0)$. A notable exception to this behavior is

Symmetries in Science XVI IOP Publishing
Journal of Physics: Conference Series **538** (2014) 012012 doi:10.1088/1742-6596/538/1/012012

the deformed ground state ($L = 0_1$) of $\hat{H}_2(\xi)$, which maintains $S_{\rm SU3}(L=0_1) = 0$ throughout region III ($\xi^{**} \leq \xi \leq 1$) and in part of region II ($\xi_c \leq \xi < \xi^{**}$), in accord with its SU(3)-PDS property, Eq. (7a). In contrast to the lack of SU(3)-purity in all excited $L = 0$ states, the SU(3) correlation function maintains a value close to unity, $C_{\rm SU3}(0\!-\!6) \approx 1$. This indicates that the SU(3) mixing is coherent and that these $L = 0$ states serve as bandhead states of $K = 0$ bands with a pronounced SU(3) QDS. This band-structure is observed throughout region III in extended energy domains, consistent with the classical analysis, which revealed a robustly regular dynamics in this region.

In region I ($0 \leq \rho \leq \rho^*$), all states show high values of $S_{\rm SU3}(L=0) \approx 1$ and $C_{\rm SU3}(0\!-\!6) < 1$, indicating considerable SU(3) mixing and lack of SU(3) coherence. This is in line with the presence of spherical-states, at low energy, and of more complex-type of states at higher energy, and the absence of rotational bands in this region. In region II of phase coexistence ($\rho^* < \rho \leq \rho_c$ and $\xi_c \leq \xi < \xi^{**}$), one encounters both points with $C_{\rm SU3}(0\!-\!6) \approx 1$, and points with $C_{\rm SU3}(0\!-\!6) < 1$. This reflects the presence of deformed states arranged into regular bands, exemplifying SU(3) QDS, and at the same time, the presence of spherical states and other states of a different nature. These results highlight the relevance of U(5)-PDS (partial purity) and SU(3)-QDS (coherence) for clarifying the survival of regular subsets of states in the presence of more complicated type of states, a situation encountered in QPTs of non-integrable systems.

Acknowledgments
This work is supported by the Israel Science Foundation. M.M. acknowledges the Golda Meir Fellowship Fund and the Czech Science Foundation (P203-13-07117S).

[1] Hertz J A 1976 *Phys. Rev. B* **14** (1976) 1165
[2] Gilmore R and Feng D H 1978 *Nucl. Phys. A* **301** 189; Gilmore R 1979 *J. Math. Phys.* **20** 891
[3] Carr L ed 2010 *Understanding Quantum Phase Transitions* (CRC press)
[4] S. Sachdev 1999 *Quantum Phase Transitions* (Cambridge: Cambridge Univ. Press)
[5] Marianetti C A , Kotliar G and Ceder G 2004 *Nature Mater.* **3** 627
[6] Pfleiderer C 2009 *Rev. Mod. Phys.* **81** 1551
[7] Karmakar B, Pellegrini V, Pinczuk A, Pfeiffer L N and West K W 2009 *Phys. Rev. Lett.* **102** 036802
[8] Cejnar P, Jolie J and Casten R F 2010 *Rev. Mod. Phys.* **82** 2155
[9] Emary C and Brandes T 2003 *Phys. Rev. Lett.* **90** 044101; *Phys. Rev. E* **67** 066203
[10] Macek M and Leviatan A 2011 *Phys. Rev. C* **84** 041302(R)
[11] Leviatan A and Macek M 2012 *Phys. Lett. B* **714** 110
[12] Macek M and Leviatan A 2013 in preparation
[13] Iachello F and Arima A 1987 *The Interacting Boson Model* (Cambridge: Cambridge Univ. Press)
[14] Bohr A and Mottelson B R 1998 *Nuclear Structure Vol. II* (Singapore: World Scientific)
[15] Ginocchio J N and Kirson M W 1980 *Phys. Rev. Lett.* **44** 1744
[16] Dieperink A E L, Scholten O and Iachello F 1980 *Phys. Rev. Lett.* **44** 1747
[17] Cejnar P and Jolie J 2009 Prog. Part. Nucl. Phys. **62** 210
[18] Leviatan A 1996 *Phys. Rev. Lett.* **77** 818; Leviatan A 2011 *Prog. Part. Nucl. Phys.* **66** 93
[19] Rowe D J 2004 *Phys. Rev. Lett.* **93** 122502; Rosensteel G and Rowe D J 2005 *Nucl. Phys. A* **759** 92
[20] Leviatan A 2007 *Phys. Rev. Lett.* **98** 242502
[21] Leviatan A *Ann. Phys. (NY)* **179** 201
[22] Leviatan A 2006 *Phys. Rev. C* **74** 051301(R)
[23] Hatch R L and Levit S 1982 *Phys. Rev. C* **25** 614
[24] Whelan N and Alhassid Y 1993 *Nucl. Phys. A* **556** 42
[25] Reichl L E 1992 *The Transition to Chaos in Conservative Classical Systems: Quantum Manifestations*, (New York: Springer-Verlag).
[26] Hénon M H and Heiles C 1964 *Astron. J.* **69** 73
[27] A. Peres A 1984 *Phys. Rev. Lett.* **53** 1711
[28] Stránský P, Hruška P and Cejnar P 2009 *Phys. Rev. E* **79** 066201
[29] Macek M, Dobeš J , Stránský P and Cejnar P 2010 *Phys. Rev. Lett.* **105** 072503
[30] Macek M, Dobeš J and Cejnar P 2010 *Phys. Rev. C* **82** 014308

Journal of Physics: Conference Series **538** (2014) 012013 doi:10.1088/1742-6596/538/1/012013

Atomistic-level non-equilibrium model for chemically reactive systems based on steepest-entropy-ascent quantum thermodynamics

Guanchen Li, Omar Al-Abbasi, and Michael R von Spakovsky

Center for Energy Systems Research, Mechanical Engineering Department
Virginia Tech, Blacksburg, VA 24061

E-mail: Michael.von.Spakovsky@vt.edu, guanchen@vt.edu

Abstract. This paper outlines an atomistic-level framework for modeling the non-equilibrium behavior of chemically reactive systems. The framework called steepest-entropy-ascent quantum thermodynamics (SEA-QT) is based on the paradigm of intrinsic quantum thermodynamic (IQT), which is a theory that unifies quantum mechanics and thermodynamics into a single discipline with wide applications to the study of non-equilibrium phenomena at the atomistic level. SEA-QT is a novel approach for describing the state of chemically reactive systems as well as the kinetic and dynamic features of the reaction process without any assumptions of near-equilibrium states or weak-interactions with a reservoir or bath. Entropy generation is the basis of the dissipation which takes place internal to the system and is, thus, the driving force of the chemical reaction(s). The SEA-QT non-equilibrium model is able to provide detailed information during the reaction process, providing a picture of the changes occurring in key thermodynamic properties (e.g., the instantaneous species concentrations, entropy and entropy generation, reaction coordinate, chemical affinities, reaction rate, etc). As an illustration, the SEA-QT framework is applied to an atomistic-level chemically reactive system governed by the reaction mechanism $F + H_2 \Leftrightarrow FH + H$.

1. Introduction

Studying the kinetics and dynamics of an atomistic-level system is a central topic of interest in physics and thermodynamics. Classical mechanics and quantum mechanics, which are built on Newton's and Schrödinger's equations of motion, respectively, provide an almost perfect description of the zero-entropy phenomena occurring in nature. However, both lack the 2nd law of thermodynamics and are, thus, unable to provide a description of states at stable equilibrium with non-zero values of the entropy. To do so, both statistical thermodynamics (ST) and intrinsic quantum thermodynamics (IQT) use principles derived from the 2nd law, namely, the maximum entropy and minimum energy principles, to arrive in constrained optimizations at a description of such states. However, this still leaves a gap not described between the zero-entropy and the maximum-entropy worlds, a gap which is important for a complete understanding of system behavior since it contains information about the kinetics and dynamics of non-equilibrium, non-zero-entropy, system-state evolutions. To fill this gap, IQT [1-29] has emerged over the last several decades as a theory which unifies quantum mechanics and thermodynamics and provides a complete picture of the kinetics and dynamics of system behavior in this region [25].

Of course, underlying any such description is the 2nd law of thermodynamics and the dissipative phenomena which occur in a realm where the beautiful symmetry of the time reversibility of classical mechanics and quantum mechanics and that of the quasi-equilibrium processes of classical thermodynamics is broken. A traditional way of studying the state evolutions in this region is that of quantum thermodynamics (QT) [30,31], which discards the isolation of the system, forming a so-called "open-system" model that includes "weak-interactions" with the environment. This approach places the system into a larger composite of system plus environment so that the dynamics of the dissipative phenomena present can be modeled as linear in the near-equilibrium realm of the composite. Thus, the QT master equations of the Lindblad type [32-35] retain the linear and reversible characteristics of the Schrödinger equation of motion, viewing dissipative phenomena as effects external to the system. This limits, among other things, this theory to the near-equilibrium and weak-interaction regimes and eschews entropy and entropy generation as intrinsic to the system.

In contrast, IQT regards the entropy and entropy generation as intrinsic and directly incorporates the irreversible and non-linear features of the dissipative process into the theory and its equation of motion. In this way, IQT avoids the necessity of introducing an external environment and does not limit itself to the near-equilibrium realm of weak system-environment interactions. Furthermore, in the limit of time reversibility, the IQT equation of motion reduces to the linear Schrödinger-von Neumann equation of quantum mechanics, which describes the linear dynamics of zero-entropy-state evolutions.

Now, of all the non-equilibrium phenomena of interest, those of chemically reactive systems are some of the most frequently studied in science and engineering. Starting with the Van't Hoff-Arrhenius law [36,37], the study of the kinetics and dynamics of chemical reactions remains an important topic of theoretical chemistry even today [38,39]. The result is a set of theories which attempt to explain the mechanisms of chemical reactions and to predict their properties using classical mechanics or quantum mechanics or a mix of both.

Conventional theory describes the kinetic features of a reactive system using configuration space (or phase space), while potential energy surfaces (PESs) serve as constraints in configuration space. In this space, the chemical reaction is regarded as a motion from the reactant region to the product region, and the kinetics are represented by a path found, for example, via the Minimum Energy Principle [40] or by a group of possible trajectories [41,42] on the PES. The associated dynamics of each reaction mechanism is then viewed as resulting from inelastic collisions [43] and the motions of single particles for which temperature plays an important role [44]. In contrast, the kinetics of each reaction mechanism provides the 'environment' for particle motion and contributes to the important concepts of potential energy barrier [45] and activated complex (e.g, in Transition State Theory) [46,47], which underlie the exponential term in the Van't Hoff-Arrhenius law. With the microscopic velocities and a reaction path or set of trajectories known, a reaction rate calculation can be made via the kinetics in configuration space. Taking into account the uncertainty in the position or momentum of each particle, an ensemble of reactant particles resulting in a Maxwellian velocity distribution can be used to arrive at reasonable expectation values for reactant concentrations, which in turn provides a link to the statistical thermodynamic value of the entropy. Of course, quantum theory can be used to modify and improve, for example, the classical calculations [45] of the PES in order to include important quantum effects such as tunneling and the discrete nature of the energy spectrum of each particle. These so-called quasi-classical calculations can then be further improved by replacing them with a fully quantum calculation using quantum scattering theory based on either the time-dependent or time-independent Schrödinger equation [48,49].

These conventional theories, whether classical, quasi-classical, or quantum, suffer from two disadvantages. First, both temperature and the Maxwellian velocity distribution play an important role in the theories but are valid only at stable equilibrium. The problem with this is that stable equilibrium only exists at the end of the reaction process whereas all prior states are in non-equilibrium, possibly very far from equilibrium. Second, the chemical reaction process in these theories is described by motion in configuration or phase space and the evolution of particle position and momentum is the principal driving force for the chemical reaction. This evolution is determined using an equation of motion which classically is given by Hamilton's equations or quantum mechanically by Schrödinger's equation. Either is reversible and, thus, there is the inevitable contradiction of using a reversible equation of motion to describe what inherently is an irreversible process. This compromise is made since the theories lack an equation of motion at a fundamental level of description for dissipative processes. Of course, turning at a classical level to the Boltzmann equation of motion, which only provides a phenomenological description, or at the quantum level to QT and its master equations of the Lindblad type, which do provide a fundamental description, only begs the question since both suffer from the so-called Loschmidt paradox [50] of how an underlying dynamics, which for both is reversible or unitary, can lead to an irreversible result.

These two essential weaknesses are addressed in IQT and its mathematical framework steepest-entropy-ascent quantum thermodynamics (SEA-QT). To begin with, the SEA-QT framework links the zero-entropy and maximum entropy worlds by providing a comprehensive description of the non-equilibrium states of the system. Entropy, which is universally defined and intrinsic to the system, is used instead of temperature to provide information for the microscopic description. All the necessary information for system state evolution is derived from the energy eigenstructure and entropy of the system via the non-linear SEA-QT equation of motion. Furthermore, the irreversibility of the chemical reaction is no longer regarded as a phenomenon resulting from particle collisions and the movement of single microscopic particles although both can be present and taken into account but instead as a dissipative process derived from the principle of steepest entropy ascent (SEA) or local maximal entropy generation (LMEG), which is a macroscopic or group effect at a fundamental level of description. Second, the SEA-QT framework describes the system in a state space spanned by eigenstates of the reactants and products and not in a configuration or phase space. Recalling that quantum effects are very important at the atomistic level, the use of state space has the advantage of directly representing the uncertain position and momentum of each energy eigenstate. Furthermore, the energy eigenstructure, which is a fundamental feature of the system, is described by eigenlevels of the system Hamiltonian and does not require the very time consuming calculation of a PES and of particle or wave packet trajectories on the PES (also very time consuming).

With these advantages in mind, the SEA-QT framework is applied here to the well-studied chemical reaction mechanism $F + H_2 \Leftrightarrow FH + H$. A system using a finite number of energy eigenlevels as its energy eigenstructure is used and compared with one based on an infinite number of energy eignlevels. In Section 2, the mathematics for expressing system states in state space is described and the energy eignstructure of the system is built from single particle eigenstates. The SEA-QT equation of motion is then presented and briefly discussed. In Section 3, results are presented which demonstrate the crucial role that entropy plays in dissipation phenomena. Results for the chemical reaction path and reaction rate are also presented and linked to experimental data. The chemical kinetic results for the finite-level system are then compared to those for the infinite-level system. Section 4 provides some further discussion and conclusions.

2. Method

2.1 State description

The description of state of a chemically reactive system is at the core of the SEQ-QT framework [24,28,29]. The SEA-QT description is based on an algebraic representation, which is quite different from the conventional description in configuration space (e.g., as in Transition State Theory) or the wave packet description using multi-body eigenfunctions (e.g., as in the Quantum Scattering Method). This avoids having to express particle position and momentum under conditions of uncertainty and brings to the fore the energy and entropy as being of primary significance. Thus, the kinetics and dynamics in the SEA-QT framework depend principally on the critical concepts of energy and entropy for defining coordinates in state space and not on momentum and position as in phase space or configuration space. This provides a clearer vantage point from which to see the driving role, which entropy plays in the non-equilibrium evolutions of state of chemically reactive systems. The behavior in phase space or configuration space, which is not fundamental but which is extremely complex, is simply another view of system state evolution.

The state space of the chemically reactive system is spanned by the eigenstates of the reactants and those of the products. Intermediate states between reactants and products are represented by a mixture of reactant states and product states. This state space can be represented by the Hilbert space $\mathscr{H}$ as follows:

$$\mathscr{H} = \mathscr{H}_1 \oplus \mathscr{H}_2 \tag{1}$$

where $\mathscr{H}_1$ is the subspace of reactants and $\mathscr{H}_2$ is the subspace of products. Each subspace of the state is generated from single molecular state spaces by the cross product. In this paper, the reaction

$$F + H_2 \Leftrightarrow FH + H \tag{2}$$

is modeled so that each subspace is given by

$$\mathscr{H}_1 = \mathscr{H}_{H_2} \oplus \mathscr{H}_F \tag{3}$$

$$\mathscr{H}_2 = \mathscr{H}_{FH} \oplus \mathscr{H}_H \tag{4}$$

The eigenstate of each molecule is generated from the one-particle eigenstates available to each molecule. For example, the state space of H_2 is expanded to include degrees of freedom for translation, rotation, vibration such that

$$\mathscr{H}_{H_2} = span\left\{ \left| t_x t_y t_z \right\rangle_{H_2} \otimes \left| lm \right\rangle_{H_2} \otimes \left| v \right\rangle_{H_2} \right\} \tag{5}$$

where $\left| t_x t_y t_z \right\rangle$ represents the translational eigenstates, $\left| lm \right\rangle$ the rotational ones, and $\left| v \right\rangle$ the vibrational ones. Inside each *ket* are the quantum numbers for the one-particle eigenstate.

Starting from the one-particle eigenstates, one can represent the system-level eigenstate $\left| \vec{n} \right\rangle$ of the reactants or products by a quantum number group composed of translational, rotational, and vibrational one-particle eigenstates, i.e.,

$$\left| \vec{n} \right\rangle = \left| 1(t_x t_y t_z lmv)_{H_2} (t_x t_y t_z)_F \right\rangle \quad \text{for the reactants} \tag{6}$$

$$\left| \vec{n} \right\rangle = \left| 2(t_x t_y t_z lmv)_{FH} (t_x t_y t_z)_H \right\rangle \quad \text{for the products} \tag{7}$$

In Eqs. (6) and (7), each eigenstate for the reaction mechanism of Eq. (2) of the chemically reactive system is represented by a total of 10 indices. The first refers to the subspace (1 or 2), 6 to the quantum numbers for the diatomic molecules H_2 and FH, and 3 to the quantum numbers for the monatomic molecules H and F. The thermodynamic state of the system can then be written in a superposition of system eigenstates such that

$$\left| \Psi \right\rangle = \sum_{\vec{n}} \gamma(\vec{n}) \left| \vec{n} \right\rangle \tag{8}$$

where $\gamma(\vec{n})=\sqrt{p(\vec{n})}$ is the square root of the probability $p(\vec{n})$ of system energy eigenlevel $|\vec{n}\rangle$. Given the energy eigenstructure of the system, the system's thermodynamic state can also be represented by the probability distribution $p(\vec{k})$ where the $\vec{k}$ are all of the possible quantum number groups of $|\vec{n}\rangle$.

2.2 Energy eigenstructure

As developed in [24] for chemically reactive systems, the Hamiltonian and identity operators are the generators of motion in the kinetics and dynamics of the SEA-QT framework. In addition, the state of the system is represented by a probability distribution of available energy eigenlevels. The evolution of system state results in a different distribution for each state at every instant of time. Thus, generating the system-level energy eigenstructure of available levels for the chemically reactive system is an important step in the development of the SEA-QT model for the system. This system-level structure is constructed from sets of one-particle eigenlevels, which for the reactive systems based on Eq. (1) considered here consist of the translational energies of the center of mass (ε_t), rotational energies (ε_r), and vibrational energies (ε_v). For *F* and *H*, which lack an internal structure, only translational energies need be considered. For translation,

$$\varepsilon_t(t_x,t_y,t_z)=\frac{\hbar^2}{8m}\left(\left(\frac{t_x}{L_x}\right)^2+\left(\frac{t_y}{L_y}\right)^2+\left(\frac{t_z}{L_z}\right)^2\right) \tag{9}$$

where ε_t is the one-particle translational energy eigenvalue; $\hbar$ Plank's modified constant; m the center of mass of the particle; (t_x,t_y,t_z) the quantum numbers in the x, y, and z directions; and (L_x,L_y,L_z) the dimensions of system volume in the x, y, and z directions.

For the internal structure of the diatomic molecules H_2 and FH, the one-particle rotational energy eigenvalue is given by

$$\varepsilon_r(l,m)=\frac{J(J+1)\,\hbar^2}{2\mu r^2} \tag{10}$$

where μ the reduced mass of the diatomic molecule, r the distance between the two atoms of the diatomic molecule, l the azimuthal quantum number, and m the magnetic quantum number. For vibration, the one-particle energy eigenvalue from the oscillator potential is written as

$$\varepsilon_v(v)=\left(v+\tfrac{1}{2}\right)\hbar\omega \tag{11}$$

where ω is the angular frequency of the potential and v the quantum number for vibration.

The system-level energy eigenvalues for each subspace can now be found from

$$\begin{aligned} E_{\vec{n}} &= E\left(\left|1(t_xt_yt_zlmv)_{H_2}(t_xt_yt_z)_F\right\rangle\right) \\ &= \varepsilon_{t,H_2}(t_x,t_y,t_z)+\varepsilon_{t,F}(t_x,t_y,t_z)+\varepsilon_{r,H_2}(l,m)+\varepsilon_{v,H_2}(v) \end{aligned} \tag{12}$$

for subspace 1 and from

$$\begin{aligned} E_{\vec{n}} &= E\left(\left|2(t_xt_yt_zlmv)_{H_2}(t_xt_yt_z)_F\right\rangle\right) \\ &= \varepsilon_{t,FH}(t_x,t_y,t_z)+\varepsilon_{t,H}(t_x,t_y,t_z)+\varepsilon_{r,FH}(l,m)+\varepsilon_{v,FH}(v) \end{aligned} \tag{13}$$

for subspace 2. Combining these values from the two subspaces results in the system energy eigenstructure.

2.3 Property Operators

In quantum mechanics, an observable is represented by a single Hermitian operator. In a like manner, the SEA-QT framework is written in terms of operators. To begin with, the density operator of the system is given by

$$\rho = \sum_{\vec{n}} p(\vec{n}) P_{\vec{n}} \tag{14}$$

where $P_{\vec{n}}$ is the projector of a given eigenstate $|\vec{n}\rangle$ of the system expressed as

$$P_{\vec{n}} = |\vec{n}\rangle\langle\vec{n}| \tag{15}$$

Since $P_{\vec{n}}$ is the basis in operator space, all other operators can be written as a superposition of the projector. Thus, the particle number operators are given by

$$N_{H_2} = N_F = \sum_{\substack{\vec{n}\text{ of}\\ \text{reactant}}} |\vec{n}\rangle\langle\vec{n}| \tag{16}$$

$$N_{FH} = N_H = \sum_{\substack{\vec{n}\text{ of}\\ \text{product}}} |\vec{n}\rangle\langle\vec{n}| \tag{17}$$

while the Hamiltonian operator is expressed as

$$H = \sum_{\vec{n}} E_{\vec{n}} |\vec{n}\rangle\langle\vec{n}| \tag{18}$$

As to the reaction coordinate and reaction rate operators for the system considered here, they are written as

$$\varepsilon = N_{FH} \tag{19}$$

$$r = \frac{dN_{FH}}{dt} \tag{20}$$

Of course, the SEA-QT framework as presented above provides only a very brief outline of the overall framework. A more general and comprehensive treatment can be found in Beretta and von Spakovsky [24] as well as in Al-Abbasi [29].

Finally, for any observable A, its expectation value can be calculated at any given instant of time by

$$\langle A \rangle = Tr(\rho A) \tag{21}$$

since $\rho = \rho(t)$ is known via the SEA-QT equation of motion. Here $Tr(\cdot)$ is the trace of the operator. Eq. (21) is equally applicable for the observables of finite- and infinite-level systems with the exception of the entropy for which the expectation value for a finite-level system in a state of non-equilibrium is determined from

$$\langle S \rangle = -k_B Tr(\rho \ln \rho) \tag{22}$$

Here k_B is the Boltzmann constant. For an infinite-level system in a non-equilibrium state, however, Eq. (22) must be modified as explained in the following section.

2.4 Density of State Method

To deal with an infinite-level system in the non-equilibrium realm, the expression for the expectation value of the entropy is modified using the density of states method developed by the first and third authors. The details of this method will be reported in a future paper, which is now in preparation. The same approach can be used for a continuous energy eigenspectrum system as well as a degenerate system. Using this method, the energy eigenstructure can be approximated by a degenerate group of sampled energy eigenlevels where the degeneracy is proportional to the density of state of each sample energy value. As will be shown in a future paper, a set of energy eigenlevels in a small interval can be approximated by one energy value with degeneracy to an accuracy on the order of the length of the interval. With this modification, each degenerate energy value is equivalent to the set of energy eigenlevels for a given interval and contributes to the SEA-QT equation of motion and the modified system entropy, i.e.,

$$\langle S\rangle = -k_B Tr\left(\rho \, ln \frac{\rho}{N(\vec{n})}\right) \tag{23}$$

with the same kinetics and dynamics that would be present if all the infinite number of energy eigenlevels were explicitly taken into account. Note that $N(\vec{n})$ is the degeneracy for energy level $|\vec{n}\rangle$.

2.5 SEA-QT equation of motion

The SEA-QT equation of motion is presented directly here without derivation. For details of the derivation, the reader is referred to [18,24,29]. The equation of motion for a reactive system is written here in operator form and is expressed as

$$\frac{d\rho}{dt} = -\frac{i}{\hbar}[H,\rho] + \frac{1}{2k_B\tau}\{\Delta M,\rho\} \tag{24}$$

where the first term on the right captures the linear Hamiltonian dynamics and the second the non-linear dissipative dynamics. When the system is in a zero-entropy state, the second term vanishes, and the SEA-QT equation of motion reduces to the time-dependent Schrödinger-von Neumann equation. In the dissipative term, τ is a relaxation time, which is the characteristic time of the dissipation process and can either be a constant or a function of ρ. $\{\}$ is the anti-commutator operator, while $\Delta M = M - \langle M\rangle$ is the deviation from the mean of the non-equilibrium Massieu operator defined as $M = S - H/\theta_H$ where $\theta_H = \langle \Delta H \Delta H\rangle / \langle \Delta S \Delta H\rangle$ is a constant-energy, non-equilibrium temperature and S and H are the entropy and Hamiltonian operators, respectively.

2.6 Initial state generation

To generate an initial density or state operator $\rho^{in}(\vec{n})$ for the SEA-QT equation of motion, the following procedure is used. It is assumed that the reactants are a Gibbs-Dalton mixture of ideal gases in an isolated container. Initially, the mixture is in a state of stable equilibrium of a non-reactive system at some temperature T. The mixture is then ignited to move the state of the system into a state of non-equilibrium at which point the system becomes chemically reactive. The physical meaning of such a system can be that of a reactive system that requires ignition or a system whose reaction speed is much smaller than its diffusion speed. The initial canonical or stable equilibrium distribution $\rho^{se}(\vec{n})$ in reactant subspace at temperature T can be represented by

$$\rho^{se}(\vec{n}) = \frac{e^{-\beta E_{\vec{n}}}}{\sum e^{-\beta E_{\vec{k}}}} \quad \text{for } \vec{n} \text{ and } \vec{k} \text{ in the reactant subspace} \tag{25}$$

where $\beta = 1/k_B T$. Initially, this distribution is that of the system as a whole as well since only reactants are present. The ignition process is described by a perturbation of this canonical distribution $\rho^{se}(\vec{n})$ in both the reactant and product subspaces such that

$$\rho^{in}(\vec{n}) = \frac{(1-\delta)e^{-\beta E_{\vec{n}}}}{\sum e^{-\beta E_{\vec{k}}}} \quad \text{for } \vec{n} \text{ and } \vec{k} \text{ in the reactant subspace} \tag{26}$$

$$\rho^{in}(\vec{n}) = \frac{\delta e^{-\beta E_{\vec{n}}}}{\sum e^{-\beta E_{\vec{k}}}} \quad \text{for } \vec{n} \text{ and } \vec{k} \text{ in the product subspace} \tag{27}$$

where δ is a very small number.

2.7 Numerical approach

Since the number of system energy eigenlevels is extremely large, a finite number from all possible $|\vec{n}\rangle$ is chosen for the finite-level system model. This finite number is sufficient to describe the entropy-driven phenomena of the reactive system and link the predicted results with published experimental data. The energy eigenlevels chosen are listed in Table 1. For the infinite-level system, the density of state method briefly described above in Section 2.4 is used. The kinetics for both systems is the same. Differences in results manifest themselves in the dynamics with the latter system, of course, more closely representing the actual reactive system.

As to the equation of motion, it is solved here using a fourth order Runge-Kutta explicit scheme [51-53] with the relative tolerance error set to 1e-5.

Table 1. Quantum numbers considered for each of the molecules and atoms in the SEA-QT model [28,29].

Species	Translational quantum nos.[a]	Vibrational quantum nos.	Rotational quantum nos.
F	t=1,...,5500		
H_2	t=1,...,1000	v =0	J =0,1,2,3
FH	t=1,...,1000	v =0,1,2,3,4	J =0,1,...,10
H	t=1,...,5000		

[a] Although the translational principal quantum number t *(defined as the square root of* $t_x^2+t_y^2+t_z^2$) varies here in the range shown in the second column of the table for each species, only a sampling (10) of these quantum numbers across this range is used for each species in the SEA-QT model.

3. Results

3.1 Entropy and concentration evolution

The SEA-QT chemical reaction model is based on the idea that the entropy generation via the principle of steepest entropy ascent is the driving force behind the chemical reaction. The first reactive system considered here initially consists of 1 mol of H_2 and 1 mol of F contained in a cubic tank, 10 cm on a side. For this system, only the infinite-level model based on our density of state method is used.

Figure 1 shows the evolution in time of the expectation values of the entropy and the particle number. It is evident that the expectation values of the species particle numbers evolve in a fashion similar to that of the entropy and that this is consistent with the idea that the entropy generation is the driving force of the chemical reaction. The entropy is, thus, not simply a phenomenological property calculated to check whether or not a system behaves consistently with the 2nd law of thermodynamics but instead a means for applying the 2nd law to the description and prediction of the non-equilibrium kinetics and dynamics of chemically reactive systems. In addition, note that Figure 1b verifies that the mass of the system is conserved at every instant of time throughout the evolution of state of the system. Although not shown here, similar graphs for the expectation values of the energies of the species show that energy is also conserved at every instant of time throughout the evolution.

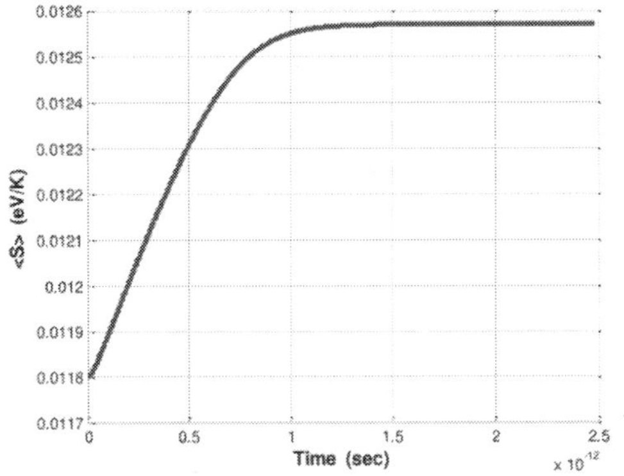

(a)

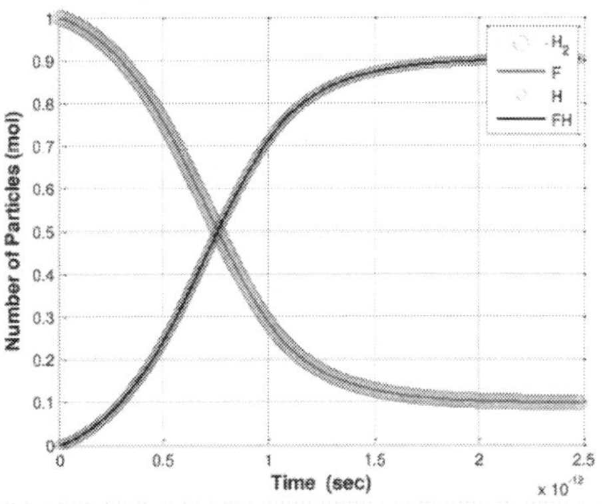

(b)

Figure 1. Evolution of the expectation values of system (a) entropy and (b) species particle number for the infinite energy eigenlevel system initially at a temperature of 298 K.

3.2 Chemical reaction rate

Solution of SEA-QT equation of motion provides system information at every instant of time, which can be used to calculate the chemical reaction rate. The traditional reaction rate constant $k(T)$ or thermal rate constant reported in the literature is a function of the initial stable equilibrium temperature of the reactive system and is a single value for each temperature, i.e., it does not vary in time during the reaction since it is a value gleamed from the initial reaction rate. In contrast, the SEA-QT framework is able to predict the reaction rate constant at every instant of time independent of temperature, referencing this temperature only in a post-processing procedure which allows the SEA-QT results to be compared to those in the literature. The SEA-QT framework is, thus, consistent with the fact that the chemical reaction process is a non-equilibrium process for which no temperature is defined. Moreover, the state of a chemically reactive system varies a lot not only with respect to the species concentrations but as well with respect to the entropy generation. The SEA-QT framework, therefore, provides a more complete picture of all the intermediate states of the system and the instantaneous chemical reaction rates in the non-equilibrium region.

Now, for the reaction mechanism considered here, i.e., $F + H_2 \Leftrightarrow FH + H$, the reaction rate is given by

$$r(t) = r_f(t) - r_b(t) \tag{27}$$

$$r_f(t) = k_f[F(t)][H_2(t)] \tag{28}$$

$$r_b(t) = k_b[FH(t)][H(t)] \tag{29}$$

where r is the net reaction rate, r_f the forward reaction rate, and r_b the backward reaction rate. k_f and k_b are the forward and backward reaction rate "constants", while $[A(t)]$ is the concentration of species A, which can be found from its particle number expectation value and the geometric dimensions of the system. The reaction order for each species in the reaction coincides here with the stoichiometric coefficient of each species in the reaction mechanism. However, this is not generally the case [54].

At stable equilibrium, Eqs. (27) to (29) provide a zero rate condition. With this condition and the assumption that the detailed balance condition also holds for the time-dependent rate constant, another relation is found, i.e.,

$$\frac{k_f(t)}{k_b(t)} = \frac{k_f(t^{se})}{k_b(t^{se})} = \frac{[F(t^{se})][H_2(t^{se})]}{[FH(t^{se})][H(t^{se})]} \tag{30}$$

which can be used with Eqs. (27) to (29) to calculate the $k_f(t)$ and $k_b(t)$ at every instant of time. In this last equation, t^{se} is the time at stable equilibrium.

The reaction rate constant values provided in the literature are those for the forward reaction at the beginning of the reaction process when the backward reaction is negligible. To make a comparison with the experimental data found in Heidner *et al.* [55], initial values for $k_f(t=t_0)=k_f(T^{in})$ for different initial equilibrium temperatures are calculated with the SEA-QT equation of motion by adjusting the relaxation time τ in the equation of motion to match the experimental data. The results for the infinite-level system are presented in Table 2, which lists the relaxation time calculated from the experimental data and the T^{eq}. The latter are the final stable equilibrium temperatures of the system corresponding to each initial stable equilibrium temperature. Of course, it is important to point out that the kinetic path of the chemical reaction process predicted by the SEA-QT framework is independent of the speed of system state evolution, i.e., of the relaxation time. No matter which relaxation time we use, the non-equilibrium path in state space remains the same. This result can be shown from the equation of motion where the relaxation time does not influence the relative value of each probability. We fit the relaxation time with the experimental data in order to show our state-space path result in the proper time scale.

Table 2. Equilibrium temperatures and relaxation times of the SEA-QT infinite-level model matched to the experimental values of the reaction rate constants found in Heidner *et al.* [55].

T^{in}(K)	T^{eq}(K)	$k_f(T^{in})/10^{-11}$ (cm^3/molecule-sec) [55]	$\tau/10^{-14}$ (sec)
298	3906.84	2.93	15.3045
400	3981.67	4.88	7.4923
500	4054.37	6.57	4.7362
600	4126.64	8.01	3.4038
700	4198.57	9.23	2.6453

Figure 2 shows a typical temperature evolution for the chemically reactive system. As mentioned above, intermediate states are non-equilibrium states for which the temperature is not defined. Nonetheless, using the surrogate system approach outlined in [29], a stable equilibrium temperature can be associated with each of these system states by using the entropy of each intermediate state at a given system energy and the fundamental relation for the entropy of a Gibbs-Dalton mixture of ideal gases to solve for the corresponding equilibrium temperature. For the infinite-level system, this temperature can also be gleamed from the slope of the curve of stable equilibrium states in the energy-entropy plane corresponding to the surrogate system [29] associated with a given intermediate (i.e., non-equilibrium) state at given values of the energy and entropy.

In Figure 3a, the characteristics of the reaction rates in the non-equilibrium realm for the infinite-level system are presented. The rates show a behaviour similar to that for the entropy generation rate, which puts in evidence once more the idea that the chemical reaction is driven by the entropy generation. The evolutions of the forward and backward reaction rates are consistent with what would be expected for a chemical reaction, namely, that the forward reaction rate dominates initially and then begins to diminish its dominance over the backward rate as the reaction proceeds until both cancel each other at stable equilibrium consistent with the initial detailed balance assumption.

Finally, Figure 3b shows an extension of the conventional reaction rate "constant" to one that has a different value at every instant of time. This more complete description of the reaction process has a clear physical meaning, namely, that the evolving state of the system in a dissipative process

Journal of Physics: Conference Series **538** (2014) 012013 doi:10.1088/1742-6596/538/1/012013

results in varying reaction speeds. This is consistent with the temperature evolution shown in Figure 2, i.e., that the rapid rate of increase in temperature seen initially coincides with the rapid increase in the net reaction rate after which it slows as the net reaction rate decreases.

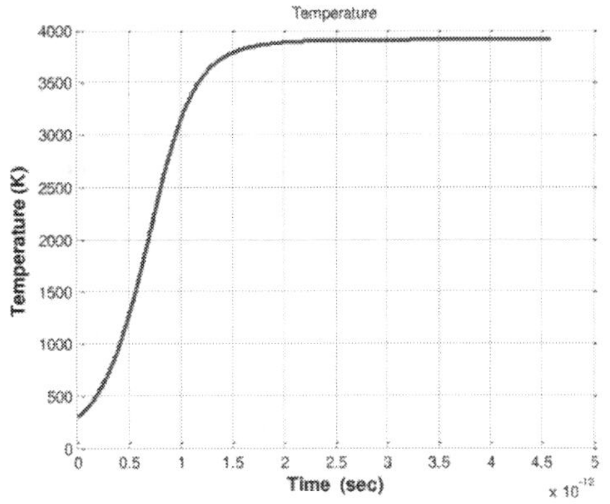

Figure 2. Evolution of temperature for the infinite-level system initially at a temperature of 298 K.

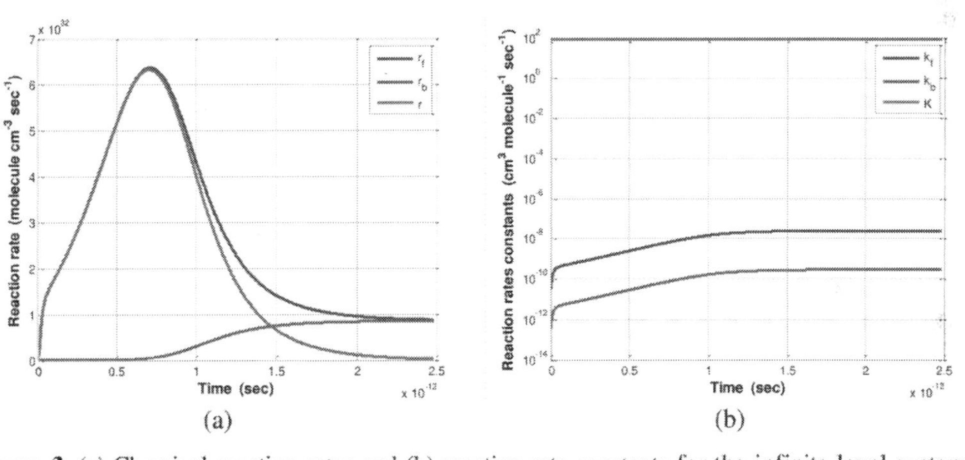

Figure 3. (a) Chemical reaction rates and (b) reaction rate constants for the infinite-level system initially at a temperature of 298 K.

3.3 Comparison of infinite-level model and finite-level model

Use of the density of states approach that we have developed has allowed us to extend the finite-level system approach reported in [24,28,29] from the microscale to the macroscale. However, to demonstrate the improvement of the infinite-level model results over that for the finite-level model at the microscale, the number of particles considered and the size of the system are reduced to one particle of H_2 and one of F in a tank with dimensions 4 nm by 2 nm by 2 nm. Anything larger results in computational convergence problems with the finite-level system model.

Figures 4 and 5 show the evolution in time of the expectation values of the entropy and entropy generation rate for the finite- and infinite-level systems. A number of differences can be seen. The first is that the maximum entropy value and the entropy generation peak value for the finite-level system are, respectively, a little less than one and a little greater than one order of magnitude smaller than that for the infinite-level system. In contrast, the evolution time to stable equilibrium for the infinite-level system is about half an order of magnitude smaller than that for the finite-level system. In addition, although the shapes of the entropy generation rate curves are similar for the two systems, that for the latter is much more asymmetric with a wider peak than that for the former suggesting that the irreversible phenomena present differ much more in intensity throughout the evolution in state of the latter system.

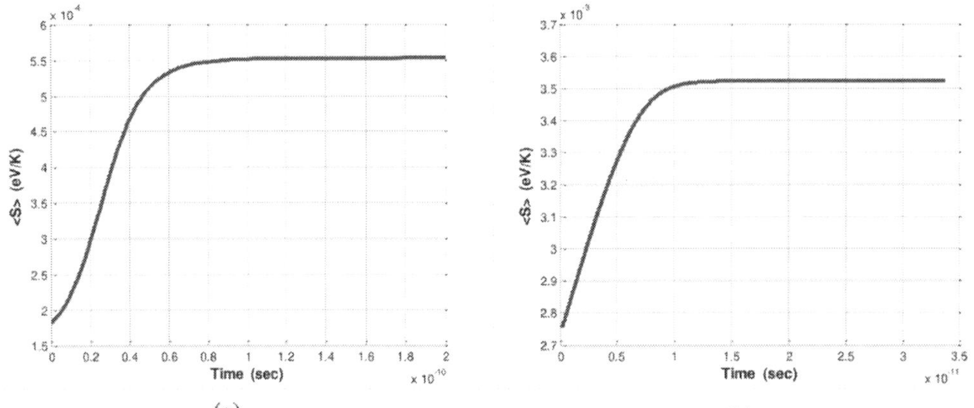

(a) (b)

Figure 4. Evolution of the expectation values of system entropy for (a) the finite energy eigenlevel system and (b) the infinite energy eigenlevel system initially at a temperature of 298 K.

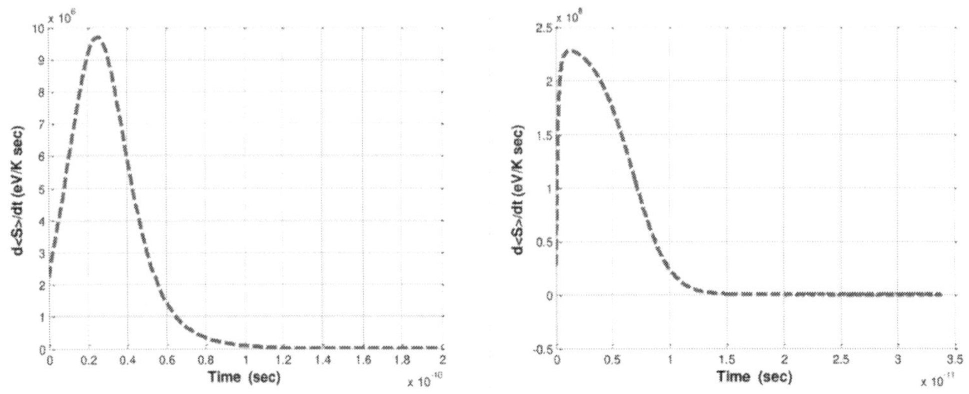

(a) (b)

Figure 5. Entropy generation for (a) the finite energy eigenlevel system and (b) the infinite energy eigenlevel system initially at a temperature of 298 K.

Clearly, for a fixed system energy, the more energy eigenlevels available over which the energy can be distributed, the higher the system entropy. In fact, using the infinite-level model, one can validate the SEA-QT model prediction of the expectation value of the entropy at stable equilibrium by comparing it with that found from standard calculations using the maximum entropy principle of ST or IQT. Indeed good agreement is seen. Similarly, the larger the number of energy eigenlevels available, the larger the entropy generation is expected to be and the faster the dynamics. Again, this is indeed what is observed.

Finally, Figures 6 and 7 show the reaction rates and reaction rate constants for the finite- and infinite-level systems, respectively. Clearly, as seen in Figure 6, the peak reaction rates for the latter system are about half an order of magnitude large than those for the former, which explains the order of magnitude reduction in the evolution time for the latter seen here and in previous figures. As expected, the trends are comparable to those seen for the entropy generation rates in Figure 5. The differences in the reaction rate constants k_f and k_b seen in Figure 7 are, on the other hand, more pronounced, showing a difference of two orders of magnitude between the k_f and k_b for each system model as well as differences of as much as one order of magnitude between the finite- and infinite-level systems even though the initial forward reaction rate constant k_f for both is set to be the same.

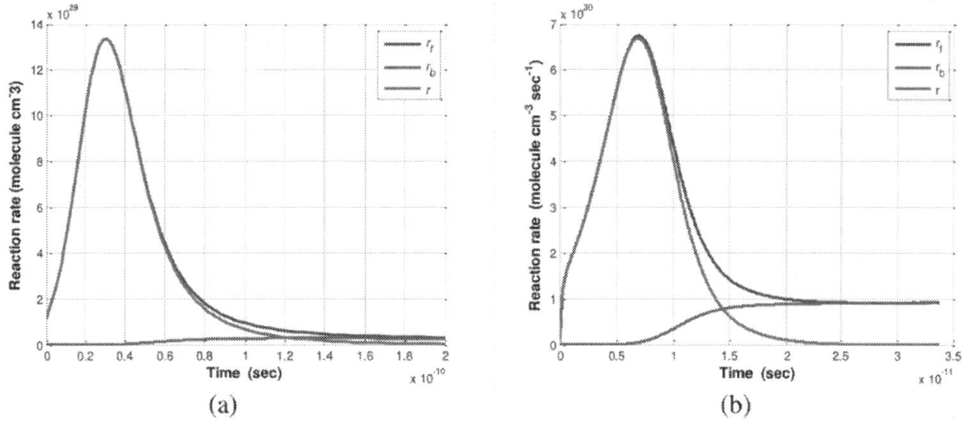

(a) (b)

Figure 6. Chemical reaction rates for (a) the finite energy eigenlevel system and (b) the infinite energy eigenlevel system initially at a temperature of 298 K.

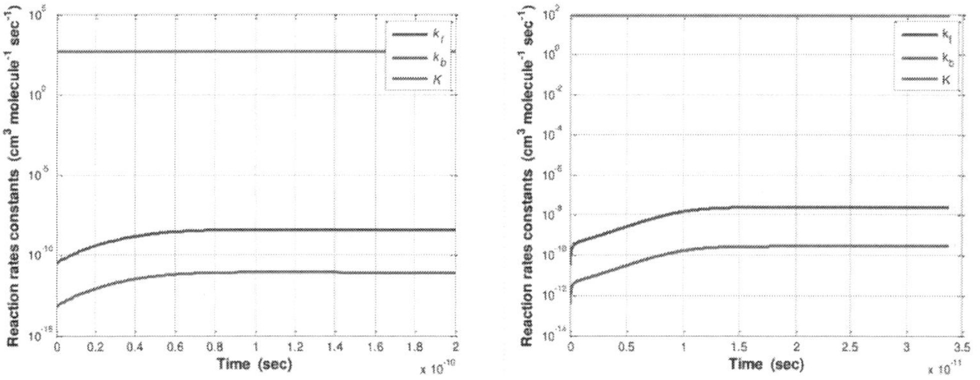

(a) (b)

Figure 7. Reaction rate constant for (a) the finite energy eigenlevel system and (b) the infinite energy eigenlevel system initially at a temperature of 298 K.

4. Conclusions

The SEA-IQT chemical reaction framework provides a new fundamental approach to studying chemically reactive systems, especially their non-equilibrium behaviour. A major benefit of this framework is that it accounts for both the quantum mechanical and thermodynamic features of the system and the process it undergoes. It is different from conventional state-to-state chemical reaction models in that it uses the system energy eigenstructure and a non-linear equation of motion to quickly and straightforwardly establish the reaction characteristics of a system in the non-equilibrium realm, providing access to detailed information about the reaction process, which would otherwise not be available. In addition, the evolution of key system properties provides a clearer physical picture of non-equilibrium behaviour and leads to a deeper understanding of the role played by the 2nd law of thermodynamics in the dissipative process.

A final point is that the computational burden involved in these calculations is minimal when compared to comparable methods in the literature. To date, model predictions have been made on a PC workstation and complete in a matter of seconds to a few hours depending on the complexity of the chemical reaction pathways that are considered. For those produced for this paper, the computational times are at the lower end. For a multi-phase, multi-component, nano-scale system with heat conduction, external applied field, diffusion, and multiple coupled electrochemical reaction pathways, the computational times are at the upper end. This is a huge advantage compared to the computational cost required by, for example, scattering calculations. Furthermore, since the solution is obtained via solving a set of first order ordinary differential equations, the memory requirement for solving the system models formulated using this framework is minimal, which contrasts with methods that depend on 3D grids of configurational space where the dimensionality grows exponentially. For this reason and the others demonstrated above, the SEA-QT framework has a great potential to contribute to the field of reaction kinetics for all phases of matter.

Acknowledgements

Funding for this research was provided by the U.S. Office of Naval Research under ONR grant N00014-11-1-0266.

Reference

[1] Hatsopoulos G N and Gyftopoulos E P 1976 *Found. Phys.* **6** 15-31, 127-41, 439-55 and 561-70
[2] Park J L 1968 *Am. J. Phys.* **36** 211
[3] Simmons Jr R F and Park J L 1981 *Found. Phys.* **11** 297
[4] Beretta G P 1981 Dr. Sc. thesis, MIT, Cambridge, MA (arXiv:quant-ph/0509116)
[5] Park J L and Simmons Jr R F 1983 The knots of thermodynamics: Old and New Questions in Physics *Cosmology, Philosophy, and Theoretical Biology* ed A van der Merwe (New York: Plenum).
[6] Beretta G P, Gyftopoulos E P, Park J L and Hatsopoulos G N 1984 *Il Nuovo Cimento B* **82** 169
[7] Beretta G P, Gyftopoulos E P and Park J L 1985 *Nuovo Cimento B* **87** 77
[8] Beretta G P 1985 *Int. J. Theor. Phys* **24(12)** 1233
[9] Beretta G P 1986 *Frontiers of Nonequilibrium Statistical Physics: Proc. of the NATO Advanced Study Institute (Santa Fe, 1984) Series B: Physics* **135** ed G T Moore and M O Scully (New York: Plenum) p 193 and p 205.
[10] Beretta G P 1987 *Found. Phys.* **17** 365
[11] Çubukçu E 1993 Dr. Sc. thesis, MIT, Cambridge, MA
[12] Gyftopoulos E P and Cubukçu E 1997 *Phys. Rev. E* **55** 3851-8.
[13] Beretta G P 1986 *Lecture Notes in Physics* **278** 441
[14] Gyftopoulos E P and von Spakovsky M R 2003 *J. of En. Res. Tech. ASME transactions* **125** 1, 1-8
[15] Beretta G P and Gyftopoulos E P 2004 *J. Chem. Phys.* **121** 2718
[16] Beretta G P 2005 *Mod. Phys. Letters A.* **20** 977
[17] Beretta G P 2006 *Phys. Rev. E.* **73** 026113
[18] Beretta G P 2008 *Entropy.* **10** 160
[19] Beretta G P 2009 *Reports on Math. Phys.* **64** 139
[20] Beretta G P 2010 *J. Phys.: Conf. Ser.* **237** 012004
[21] von Spakovsky M R 2008 *Am. Inst. Phys. CP Series* **1033** 1, 302
[22] Smith C E, Sciacovelli A, von Spakovsky M R and Verda V 2010 *J. of Phys.: Conf. Series.* **237** 012022
[23] Smith C E and von Spakovsky M R 2012 *J. of Phys.: Conf. Series* **380** 012015

[24] Beretta G P and von Spakovsky M R 2014 Steepest-entropy-ascent quantum thermodynamic framework for describing the non-equilibrium behavior of a chemically reactive system at an atomistic level (http://arXiv:physics.chem-ph/0899107) Appendix by Al-Abbassi O
[25] von Spakovsky M R 2013 Intrinsic Quantum Thermodynamic: What It Is and What Can Be Done with It *12th Joint European Thermodynamics Conference (Brescia, Italy, 1-5 July 2013)*
[26] Cano-Andrade S, Beretta G P and von Spakovsky M R 2013 Steepest-Entropy-Ascent Quantum Thermodynamic Non-equilibrium Modeling of Decoherence of a Composite System of Two Interacting Spin-½ Systems *ASME IMECE'2013* Paper No. IMECE2013–63596, N.Y., N.Y.
[27] Cano-Andrade S, Beretta G P and von Spakovsky M R 2013 Non-Equilibrium Thermodynamic Modeling of an Atom-Field State Evolution with Comparisons to Published Experimental Data *12th Joint European Thermodynamics Conference (Brescia, Italy, 1-5 July 2013)*
[28] Al-Abbasi O, von Spakovsky M R and Beretta G P 2013 Intrinsic Quantum Thermodynamic Prediction of the Non-equilibrium Atomistic-Level Behavior of Chemically Reactive Systems *12th Joint European Thermodynamics Conference (Brescia, Italy, 1-5 July 2013)*
[29] Al-Abbasi O 2013 Ph.D. dissertation, Virginia Tech, Blacksburg, VA (http://vtechworks.lib.vt.edu/handle/10919/24069)
[30] Weiss U 2008 *Quantum Dissipative Systems* (series in Modern Condensed Matter Physics, Singapore-London: World Scientific) **13**
[31] Blum K 1996 *Density Matrix Theory and Applications* 2nd ed (Physics of Atoms and Molecules Series, New York: Plenum Press)
[32] Kossakowski A 1972 *Bull. Acad. Sci. and Math.* **20** 1021
[33] Kossakowski A 1972 *Rep. on Math. Phys.* **3** 247
[34] Ingarden R S and Kossakowski A 1975 *Annals of Phys.* **89** 451
[35] Lindblad G 1976 *Comm. in Math. Phys.* **48** 2, 119
[36] Arrhenius S 1889 *Z. Phys. Chem.* (Leipzig) **4** 226
[37] Van't Hoff J H 1984 *in Etudes de Dynamiques Chimiques* ed F. Muller and Co Amsterdam p 144; translated by T. Ewan as *Studies in Chemical Dynamics* (London, 1896).
[38] Pollak E and Talkner P 2005 Reaction rate theory: What it was, where is it today, and where is it going? *Chaos: An Interdisciplinary Journal of Nonlinear Science* **15** 026116-026116-11.
[39] Hänggi P, Talkner P and Borkovec M 1990 Reaction-rate theory: fifty years after Kramers *Rev. Mod. Phys.***62** 251
[40] Quapp W and Heidrich D 1984 Analysis of the concept of minimum energy path on the potential energy surface of chemically reacting systems, *Theoretical Chemistry Accounts: Theory, Computation, and Modeling (Theoretica Chimica Acta)* **66** 245-60.
[41] Aoiz F J, Bañares L, Herrero V J, Stark K and Werner H J 1996 Reaction cross sections and rate constants for the F+H_2 (D_2) → HF(DF)+H(D) reactions from quasiclassical trajectory calculations on a potential energy surface *Chem. Phys. Lett.* **254** 341-8
[42] Wilkins R L 1972 Monte Carlo Calculations of Reaction Rates and Energy Distributions Among Reaction Products. I. F + H_2--> HF + H *J. Chem. Phys.* **57** 912-7.
[43] (2012, June 14). *Collision theory*. Wikipeida (http://en.wikipedia.org/w/index.php?title=Collision_theory&oldid=497586960)
[44] Henriksen N E and Hansen F Y 2008 *Theories of molecular reaction dynamics : the microscopic foundation of chemical kinetics* (Oxford: Oxford University Press)
[45] Karplus M, Porter R N and Sharma R D 1965 Exchange Reactions with Activation Energy. I. Simple Barrier Potential for (H, H_2) *J. Chem. Phys.* **43** 3259-87

[46] Laidler K J and King M C 1983 Development of transition-state theory *J. Chem. Phys.* **87** 2657-64
[47] Truhlar D G, Garrett B C and Klippenstein S J 1996 Current Status of Transition-State Theory *J. Chem. Phys.* **100** 12771-800
[48] Schatz G C 1996 Scattering Theory and Dynamics: Time-Dependent and Time-Independent Methods *J. Chem. Phys.* **100** 12839-47
[49] Althorpe, S. C. and Clary, D. C., 2003 Quantum scattering calculations on chemical reactions *Annu. Rev. Phys. Chem.* **54** 493-529
[50] Loschmidt J 1876 *Sitzungsber. Kais. Akad. Wiss. Wien, Math. Naturwiss.* **73** 128–42
[51] Fehlberg E 1969 Low-order classical Runge-Kutta formulas with step size control and their application to some heat transfer problems *NASA Technical Report* 315.
[52] Fehlberg E 1970 *Computing* (*Arch. Elektron. Rechnen*) **6** 61–71.
[53] Hairer E, Nørsett S and Wanner G 1993 *Solving Ordinary Differential Equations I: Nonstiff Problems*, 2nd ed. (Berlin: Springer-Verlag).
[54] Carr R W 2007 *Modeling of Chemical Reactions* (Elsevier Series Comprehensive Chemical Kinetics, Amsterdam: Elsevier)
[55] Heidner R, Bott J, Gardner C and Melzer J 1980 Absolute rate coefficients for $F+H_2$ and $F+D_2$ at T=295--765 K *J. Chem. Phys.*72 4815-21

The KP hierarchy with self-consistent sources: construction, Wronskian solutions and bilinear identities

Runliang Lin[1], **Xiaojun Liu**[2] **and Yunbo Zeng**[1]

[1] Department of Mathematical Sciences, Tsinghua University, Beijing 100084, P.R. China
[2] Department of Applied Mathematics, China Agricultural University, Beijing 100083, P.R. China

E-mail: rlin@math.tsinghua.edu.cn, xjliu@cau.edu.cn, yzeng@math.tsinghua.edu.cn

Abstract. In this paper, we will present some of our results on the soliton hierarchy with self-consistent sources (SHSCSs). The Kadomtsev–Petviashvili (KP) hierarchy will be used as an illustrative example to show the method to construct the SHSCSs. Some properties of the KP hierarchy with self-consistent sources will also be given, such as the dressing approach, the Wronskian solutions (including soliton solutions), its bilinear identities and the tau function.

1. Introduction

The soliton equation with self-consistent sources (SESCSs) was proposed by Mel'nikov [16] and has important applications in hydrodynamics, plasma physics and solid state physics. Later on, some SESCSs were studied by inverse scattering method (*without* explicit Lax pair) [16, 17], Matrix theory [18] and D-bar method [4].

Since the constrained flows of soliton equations can be viewed as the stationary case of the SESCSs, one can derive the auxiliary linear problems for the SESCSs on the basis of the Lax pair for the constrained flows (the latter can be obtained systematically by the adjoint representation) [24]. With the help of the constrained flows, some SESCSs were studied by inverse scattering method (with *explicit* Lax pair), and some soliton solutions were obtained [8, 23]. It was shown that the velocities of the solitons can be changed by the choice of sources [8, 23]. A similar observation can be found in [5, 19].

The KdV equation with self-consistent sources reads [16, 8]

$$u_t = -\frac{1}{4}(6uu_x + u_{xxx}) - \frac{1}{2}\partial_x \sum_{j=1}^{N} \phi_j^2, \tag{1a}$$

$$\phi_{j,xx} + (\lambda_j + u)\phi_j = 0, \qquad j = 1, \cdots, N, \tag{1b}$$

where $u = u(x,t)$, $\phi_j = \phi_j(x,t)$, N is a natural number, λ_j's are parameters.

One can get the soliton solutions for KdV equation with sources (1) by inverse scattering method [8], e.g., the following is its two-soliton solution

$$u(x,t) = \frac{12\{3 + 4\cosh[2x - 2t - 2\int_0^t \beta_2(z)\,dz] + \cosh[4x - 16t - 2\int_0^t \beta_1(z)\,dz]\}}{\{\cosh[3x - 9t - \int_0^t(\beta_1(z) + \beta_2(z))\,dz] + 3\cosh[x - 7t - \int_0^t(\beta_1(z) - \beta_2(z))\,dz]\}^2}, \tag{2a}$$

$$\phi_1(x,t) = \frac{4\sqrt{6\beta_1(t)}\cosh[x - t - \int_0^t \beta_2(z)\,dz]}{\cosh[3x - 9t - \int_0^t(\beta_1(z) + \beta_2(z))\,dz] + 3\cosh[x - 7t - \int_0^t(\beta_1(z) - \beta_2(z))\,dz]}, \tag{2b}$$

$$\phi_2(x,t) = \frac{4\sqrt{3\beta_2(t)}\sinh[2x - 8t - \int_0^t \beta_1(z)\,dz]}{\cosh[3x - 9t - \int_0^t(\beta_1(z) + \beta_2(z))\,dz] + 3\cosh[x - 7t - \int_0^t(\beta_1(z) - \beta_2(z))\,dz]}, \tag{2c}$$

where $\beta_1(t)$ and $\beta_2(t)$ are arbitrary continuous functions of t. If we choose special $\beta_1(t)$ and $\beta_2(t)$, the soliton with smaller amplitude may propagate faster than that with bigger amplitude (e.g., as plotted in Figure 1 with $\beta_1(t) = 1$ and $\beta_2(t) = 9$). This phenomenon is *completely different* from that of solitons to the original KdV hierarchy (*without sources*). Some other choices of $\beta_i(t)$ can give a great variety of dynamics of soliton solutions. Similar cases were also studied in [2-7,16-19,25].

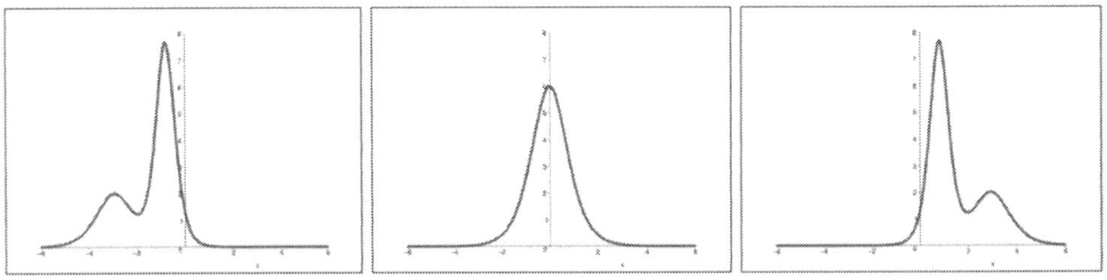

(a) (b) (c)

Figure 1. The plot of two-soliton solution $u(x,t)$ in Eq. (2) with $\beta_1(t) = 1$ and $\beta_2(t) = 9$ (for $t = -0.06$ (a), $t = 0$ (b) and $t = 0.06$ (c)).

Later, some explicit solutions (such as solitons, positons, negatons) of some SESCSs were obtained by Darboux transformation (see the references in [9]) and Hirota method [25]. A source generalization method was proposed by Hu and coworkers [6].

In the study on SESCSs, it is found that two types of self-consistent sources can be added to a soliton equation [7, 16]. Let's take the Kadomtsev–Petviashvili (KP) equation as an example. The first type of KP equation with self-consistent sources (KPwS-I) is [7, 16]

$$(4u_t - 12uu_x - u_{xxx})_x - 3u_{yy} + 4\sum_{i=1}^{N}(q_i r_i)_{xx} = 0, \tag{3a}$$

$$q_{i,y} = q_{i,xx} + 2uq_i, \quad i = 1, \ldots, N, \tag{3b}$$

$$r_{i,y} = -r_{i,xx} - 2ur_i, \tag{3c}$$

where N is a natural number. The second type of KP equation with self-consistent sources

(KPwS-II) is [7, 16]

$$4u_t - 12uu_x - u_{xxx} - 3\partial^{-1}u_{yy} = 3\sum_{i=1}^{N}\Big[q_{i,xx}r_i - q_i r_{i,xx} + (q_i r_i)_y\Big], \tag{4a}$$

$$q_{i,t} = q_{i,xxx} + 3uq_{i,x} + \frac{3}{2}q_i\partial^{-1}u_y + \frac{3}{2}q_i\sum_{j=1}^{N}q_j r_j + \frac{3}{2}u_x q_i, \tag{4b}$$

$$r_{i,t} = r_{i,xxx} + 3ur_{i,x} - \frac{3}{2}r_i\partial^{-1}u_y - \frac{3}{2}r_i\sum_{j=1}^{N}q_j r_j + \frac{3}{2}u_x r_i, \tag{4c}$$

where ∂^{-1} stands for the inverse of $\partial \equiv \partial_x$.

In the beginning, the two types of SESCSs (3) and (4) were studied individually. In order to find a unified framework to study the two types of SESCSs, a systematical method was proposed by the authors on the basis of Sato's theory [13]. This is a systematic method to generate the SESCSs, and can be used to study the case of BKP, CKP [22], q-deformed KP [10, 11], and some other cases. A generalized dressing method was also derived for these soliton hierarchy with sources, and their Wronskian solutions (including soliton solutions) were obtained [14]. Recently, a bilinear identity for the KP hierarchy with self-consistent sources (KPHwS) and their Hirota's bilinear equations (*in a simpler form*) were obtained [12].

This paper presents some of our results on the SESCSs. In Section 2, the KP hierarchy is used as an illustrative example to show the method to construct a soliton hierarchy with self-consistent sources on the basis of Sato's theory. In Section 3, a dressing approach is given for the KP hierarchy with self-consistent sources (KPHwS) and its Wronskian solutions (including soliton solutions) are shown. In Section 4, the bilinear identities and tau-function for the KPHwS are given, Hirota's bilinear form for the KP equation with sources is obtained, which is in a simpler form compared with the existing result. In Section 5, we will give a conclusion, remarks and some problems for further exploration.

2. Construction of KP hierarchy with self-consistent sources

In this section, we will use the KP hierarchy as an example to show how to construct a soliton hierarchy with self-consistent sources [13] on the basis of Sato's theory.

It is known that there is a squared eigenfunction symmetry (or "ghost flow") [1, 20]

$$\partial_z L = \Big[\sum_{i=1}^{N} q_i\partial^{-1}r_i, L\Big], \tag{5a}$$

$$\partial_{t_n} q_i = L_+^n(q_i), \tag{5b}$$

$$\partial_{t_n} r_i = -(L_+^n)^*(r_i), \qquad i = 1, \dots, N, \tag{5c}$$

to the original KP hierarchy

$$\partial_{t_n} L = [B_n, L], \qquad B_n \equiv L_+^n, \tag{5d}$$

where $L = \partial + \sum_{i=1}^{\infty} u_i\partial^{-i}$, and for an pseudo-differential operator $P = \sum_{i=-\infty}^{n} p_i\partial^i$, we define $P_+ = \sum_{i=0}^{n} p_i\partial^i$, $P^* = \sum_{i=-\infty}^{n} (-\partial)^i p_i$. The idea of generating the KP hierarchy with self-consistent sources (KPHwS) is to modify a specific flow (say t_k-flow) by the squared eigenfunction symmetry to be a new flow (denoted by $\bar{t}_k$-flow) as

$$\partial_{\bar{t}_k} L = [L_+^k + \sum_{i=1}^{N} q_i \partial^{-1} r_i, L], \tag{6a}$$

$$\partial_{t_n} L = [L_+^n, L], \quad (n = k) \tag{6b}$$

$$\partial_{t_n} q_i = L_+^n(q_i), \tag{6c}$$

$$\partial_{t_n} r_i = -(L_+^n)^*(r_i), \qquad i = 1, \ldots, N. \tag{6d}$$

It can be shown that the $\bar{t}_k$-flow commutes with the other t_n-flow ($n = k$) which gives the KP hierarchy with self-consistent sources (KPHwS) [13].

Proposition 1. *(see [13]) The community of the $\bar{t}_k$-flow and the t_n-flow ($n = k$) in (6) give rise to the following KP hierarchy with self-consistent sources (KPHwS)*

$$B_{n,\bar{t}_k} - (B_k + \sum_{i=1}^{N} q_i \partial^{-1} r_i)_{t_n} + [B_n, B_k + \sum_{i=1}^{N} q_i \partial^{-1} r_i] = 0 \tag{7a}$$

$$q_{i,t_n} = B_n(q_i), \tag{7b}$$

$$r_{i,t_n} = -B_n^*(r_i), \quad i = 1, \cdots, N. \tag{7c}$$

For example (see [13]), the system (7) with $n = 2$ and $k = 3$ gives the KPwS-I (3) with $y \equiv t_2$, $t \equiv \bar{t}_k$, $u \equiv u_1$, and the system (7) with $n = 3$ and $k = 2$ gives the KPwS-II (4) with $y \equiv \bar{t}_k$, $t \equiv t_3$, $u \equiv u_1$.

In fact, the two types of reductions on the KPHwS (7) give two types of $(1+1)$-dimensional soliton hierarchy with self-consistent sources (see [13]), i.e., the k-constrained KP hierarchy and the Gelfand–Dickey hierarchy with sources, which include two types of KdV equations with sources and two types of Boussinesq equations with sources.

3. Dressing approach and Wronskian solutions for the KP hierarchy with sources (7)

Here we denote the Wronskian determinant as

$$\mathrm{Wr}(h_1, \cdots, h_N) = \begin{vmatrix} h_1 & h_2 & \cdots & h_N \\ h_1' & h_2' & \cdots & h_N' \\ \vdots & \vdots & \ddots & \vdots \\ h_1^{(N-1)} & h_2^{(N-1)} & \cdots & h_N^{(N-1)} \end{vmatrix}.$$

Then by modifying the dressing approach for the original KP hierarchy, we can get a dressing approach for the KP hierarchy with self-consistent sources as the following [14]:

Proposition 2. *(see [14]) Let W, q_i and r_i be defined as*

$$W = \frac{\mathrm{Wr}(h_1, \cdots, h_N, \partial)}{\mathrm{Wr}(h_1, \cdots, h_N)} \equiv \frac{1}{\mathrm{Wr}(h_1, \cdots, h_N)} \begin{vmatrix} h_1 & h_2 & \cdots & h_N & 1 \\ h_1' & h_2' & \cdots & h_N' & \partial \\ \vdots & \vdots & \ddots & \vdots & \vdots \\ h_1^{(N)} & h_2^{(N)} & \cdots & h_N^{(N)} & \partial^N \end{vmatrix}, \tag{8a}$$

$$q_i = -\alpha_{i,\bar{t}_k} W(g_i) \quad r_i = (-1)^{N-i}\frac{\mathrm{Wr}(h_1,\cdots,h_{i-1},h_{i+1},\cdots,h_N)}{\mathrm{Wr}(h_1,\cdots,h_N)}, i=1,\ldots,N \tag{8b}$$

$$h_i = f_i + \alpha_i(\bar{t}_k) g_i, \quad i=1,\ldots,N, \tag{8c}$$

$$\partial_{t_n} f_i = \partial^n(f_i), \quad \partial_{\bar{t}_k} f_i = \partial^k(f_i), \quad i=1,\ldots,N, \tag{8d}$$

$$\partial_{t_n} g_i = \partial^n(g_i), \quad \partial_{\bar{t}_k} g_i = \partial^k(g_i), \tag{8e}$$

where the α_i's are arbitrary differentiable function of $\bar{t}_k$, $\mathrm{Wr}(h_1,\cdots,h_N,\partial)$ is understood as an expansion with respect to its last column, in which all sub-determinants are collected on the left of the differential symbols, then $L = W\partial W^{-1}$, q_i and r_i satisfy the KP hierarchy with self-consistent sources (7).

For example, the one-soliton solution for the KP equation with a second type of self-consistent sources (4) with $N=1$ can be obtained by the dressing approach (Proposition 2)

$$u = \frac{(\lambda_1-\mu_1)^2}{4}\mathrm{sech}^2(\Omega_1), \tag{9a}$$

$$q_1 = \sqrt{\alpha_{1,y}}(\lambda_1-\mu_1)\,\mathrm{e}^{\frac{\xi_1+\eta_1}{2}}\mathrm{sech}(\Omega_1), \tag{9b}$$

$$r_1 = \frac{1}{2\sqrt{\alpha_1}}\mathrm{e}^{-\frac{\xi_1+\eta_1}{2}}\mathrm{sech}(\Omega_1), \tag{9c}$$

where $\Omega_i = \frac{\xi_i-\eta_i}{2} - \frac{1}{2}\ln(\alpha_i)$, $\exp(\lambda_i x + \lambda_i^2 y + \lambda_i^3 t) \equiv \mathrm{e}^{\xi_i}$, $\exp(\mu_i x + \mu_i^2 y + \mu_i^3 t) \equiv \mathrm{e}^{\eta_i}$, λ_i and μ_i are different parameters (see [14]).

In [14], a gauge transformation between the KP hierarchy with sources and the mKP hierarchy with sources is constructed, and a Wronskian solution (including soliton solutions) for the mKP hierarchy with sources is also obtained with the help of dressing approach (Proposition 2) for the KP hierarchy with sources.

4. The bilinear identities and Hirota's bilinear form for the KP hierarchy with sources (7)

The Sato theory is of fundamental importance in the study of integrable systems (see [3] and references therein). It reveals the infinite dimensional Grassmannian structure of the space of τ-functions, where the τ-functions are solutions of Hirota's bilinear form of the KP hierarchy. The key point to this important discovery is a bilinear residue identity for wave functions called *bilinear identity*. Bilinear identity plays an important role in the proof of existence for τ-functions. It also serves as the generating function of Hirota's bilinear equations for the KP hierarchy [2, 15, 21].

With the help of an auxiliary parameter "z" corresponding to the "ghost flow", we can find the bilinear identities for the KP hierarchy with self-consistent source (7).

Proposition 3. *(see [12]) The bilinear identity for the KP hierarchy with self-consistent sources (7) with $N=1$ is given by the following sets of residue identities with auxiliary variable "z":*

$$\mathrm{res}_\lambda\, w(z-\bar{t}_k,\mathbf{t},\lambda)\cdot w^*(z-\bar{t}'_k,\mathbf{t}',\lambda) = 0, \tag{10a}$$

$$\mathrm{res}_\lambda\, w_z(z-\bar{t}_k,\mathbf{t},\lambda)\cdot w^*(z-\bar{t}'_k,\mathbf{t}',\lambda) = q(z-\bar{t}_k,\mathbf{t})r(z-\bar{t}'_k,\mathbf{t}'), \tag{10b}$$

$$\mathrm{res}_\lambda\, w(z-\bar{t}_k,\mathbf{t},\lambda)\cdot\partial^{-1}\left(q(z-\bar{t}'_k,\mathbf{t}')w^*(z-\bar{t}'_k,\mathbf{t}',\lambda)\right) = -q(z-\bar{t}_k,\mathbf{t}), \tag{10c}$$

$$\mathrm{res}_\lambda\, \partial^{-1}\left(r(z-\bar{t}_k,\mathbf{t})w(z-\bar{t}_k,\mathbf{t},\lambda)\right)\cdot w^*(z-\bar{t}'_k,\mathbf{t}',\lambda) = r(z-\bar{t}'_k,\mathbf{t}'), \tag{10d}$$

where the residue with respect to λ can be simply considered as the coefficient of λ^{-1} in the Laurent expansion, the inverse of ∂ is understood as pseudo-differential operator acting on an exponential function, e.g., $\partial^{-1}(rw) = (\partial^{-1}rW)(e^{\xi})$, $W = \sum_{i\geq 0} w_i\partial^{-i}$ $(w_0 = 1)$, $\xi(\mathbf{t},\lambda) = \bar{t}_k\lambda^k + \sum_{i\neq k} t_i\lambda^i$, $\mathbf{t} = (t_1, t_2, \cdots, t_{k-1}, \bar{t}_k, t_{k+1}, \cdots)$, $\mathbf{t}' = (t'_1, t'_2, \cdots, t'_{k-1}, \bar{t}'_k, t'_{k+1}, \cdots)$, $w(z,\mathbf{t},\lambda) = We^{\xi(\mathbf{t},\lambda)}$, $w^*(z,\mathbf{t},\lambda) = (W^*)^{-1}e^{-\xi(\mathbf{t},\lambda)}$.

Remark 4.1. *In this section, we will only show the formulae on the KP hierarchy with self-consistent sources (7) with $N = 1$, $q \equiv q_1$ and $r \equiv r_1$. The same idea can be used to study the case of $N > 1$.*

The existence of the τ-function for the original bilinear identity of the KP hierarchy is proved in [3]. In our case, the wave functions $w(z - \bar{t}_k, \mathbf{t}, \lambda)$ and $w^*(z - \bar{t}_k, \mathbf{t}, \lambda)$ satisfy exactly the same bilinear identity (10a) as the original KP case if one considers z as an additional parameter. So it is reasonable to assume the existence of a τ-function and make the following ansatz:

$$w(z - \bar{t}_k, \mathbf{t}, \lambda) = \frac{\tau(z - \bar{t}_k + \frac{1}{k\lambda^k}, \mathbf{t} - [\lambda])}{\tau(z - \bar{t}_k, \mathbf{t})} \cdot \exp\xi(\mathbf{t},\lambda), \tag{11a}$$

$$w^*(z - \bar{t}_k, \mathbf{t}, \lambda) = \frac{\tau(z - \bar{t}_k - \frac{1}{k\lambda^k}, \mathbf{t} + [\lambda])}{\tau(z - \bar{t}_k, \mathbf{t})} \cdot \exp(-\xi(\mathbf{t},\lambda)), \tag{11b}$$

where $[\lambda] = (\frac{1}{\lambda}, \frac{1}{2\lambda^2}, \frac{1}{3\lambda^3}, \cdots)$. According to [2], we should make further assumptions:

$$q(z,\mathbf{t}) = \frac{\sigma(z,\mathbf{t})}{\tau(z,\mathbf{t})}, \qquad r(z,\mathbf{t}) = \frac{\rho(z,\mathbf{t})}{\tau(z,\mathbf{t})}. \tag{11c}$$

Then, similar to [2], we have the following results:

$$\partial^{-1}(r(z - \bar{t}_k, \mathbf{t})w(z - \bar{t}_k, \mathbf{t}, \lambda)) = \frac{\rho(z - \bar{t}_k + \frac{1}{k\lambda^k}, \mathbf{t} - [\lambda])}{\lambda\tau(z - \bar{t}_k, \mathbf{t})} e^{\xi(\mathbf{t},\lambda)}, \tag{12a}$$

$$\partial^{-1}(q(z - \bar{t}_k, \mathbf{t})w^*(z - \bar{t}_k, \mathbf{t}, \lambda)) = \frac{-\sigma(z - \bar{t}_k - \frac{1}{k\lambda^k}, \mathbf{t} + [\lambda])}{\lambda\tau(z - \bar{t}_k, \mathbf{t})} e^{-\xi(\mathbf{t},\lambda)}. \tag{12b}$$

After substituting (11) into (10) and some calculation, we get the following systems with Hirota bilinear derivatives $\tilde{D}$ and D_i's:

$$\sum_{i\geq 0} p_i(2\mathbf{y})p_{i+1}(-\tilde{D}) \exp\Big(\sum_{j\geq 1} y_j D_j\Big)\bar{\tau}(z,\mathbf{t}) \cdot \bar{\tau}(z,\mathbf{t}) = 0, \tag{13a}$$

$$\begin{aligned}&\sum_{i\geq 0} p_i(2\mathbf{y})p_{i+1}(-\tilde{D}) \exp\Big(\sum_{j\geq 1} y_j D_j\Big)\bar{\tau}_z(z,\mathbf{t}) \cdot \bar{\tau}(z,\mathbf{t})\\ &\quad - \sum_{i\geq 0} p_i(2\mathbf{y})\Big(\partial_z \log\bar{\tau}(z,\mathbf{t}+\mathbf{y})\Big)p_{i+1}(-\tilde{D})\bar{\tau}(z,\mathbf{t}+\mathbf{y}) \cdot \bar{\tau}(z,\mathbf{t}-\mathbf{y})\\ &\qquad = \exp\Big(\sum_{j\geq 1} y_j D_j\Big)\bar{\sigma}(z,\mathbf{t}) \cdot \bar{\rho}(z,\mathbf{t})\end{aligned} \tag{13b}$$

$$\sum_{i\geq 0} p_i(2\mathbf{y})p_i(-\tilde{D}) \exp\Big(\sum_{j\geq 1} y_j D_j\Big)\bar{\tau}(z,\mathbf{t}) \cdot \bar{\sigma}(z,\mathbf{t}) = \exp\Big(\sum_{j\geq 1} y_j D_j\Big)\bar{\sigma}(z,\mathbf{t}) \cdot \bar{\tau}(z,\mathbf{t}), \tag{13c}$$

$$\sum_{i\geq 0} p_i(2\mathbf{y})p_i(-\tilde{D}) \exp\Big(\sum_{j\geq 1} y_j D_j\Big)\bar{\rho}(z,\mathbf{t}) \cdot \bar{\tau}(z,\mathbf{t}) = \exp\Big(\sum_{j\geq 1} y_j D_j\Big)\bar{\tau}(z,\mathbf{t}) \cdot \bar{\rho}(z,\mathbf{t}), \tag{13d}$$

where the bar $^{-}$ over a function $f(z,\mathbf{t})$ is defined as $\bar{f}(z,\mathbf{t}) \equiv f(z-\bar{t}_k,\mathbf{t})$, e.g., $\bar{\tau}(z,\mathbf{t}-[\lambda]) \equiv \tau(z-(\bar{t}_k-\frac{1}{k\lambda^k}),\mathbf{t}-[\lambda])$, $\bar{\tau}(z,\mathbf{t}'+[\lambda]) \equiv \tau(z-(\bar{t}'_k+\frac{1}{k\lambda^k}),\mathbf{t}'+[\lambda])$, $\mathbf{y}=(y_1,y_2,\cdots)$, $\tilde{D}=(D_1,\frac{1}{2}D_2,\frac{1}{3}D_3,\cdots)$, D_i is the well-known Hirota bilinear derivative $D_i f\cdot g = f_{t_i}g - fg_{t_i}$, and $p_i(\mathbf{t})$ is the i-th *Schur polynomial*, whose generating function is given by

$$\exp\sum_{i=1}^{\infty} y_i\lambda^i = \sum_{i=0}^{\infty} p_i(\mathbf{y})\lambda^i.$$

Example 1 (First type of KP equation with a source (KPwS-I) [7, 13, 16, 18], i.e., the KPHwS (7) for $n=2$ and $k=3$). *The Hirota equations for the KPwS-I (3) can be obtained as*

$$D_x\tau_z\cdot\tau+\sigma\rho=0, \qquad \text{by (13b) with } y_j=0, \tag{14a}$$
$$(D_x^4+3D_{t_2}^2-4D_x(D_{\bar{t}_3}-D_z))\tau\cdot\tau=0, \qquad \text{by (13a) in } y_3, \tag{14b}$$
$$(D_{t_2}+D_x^2)\tau\cdot\sigma=0, \qquad \text{by (13c) in } y_2, \tag{14c}$$
$$(D_{t_2}+D_x^2)\rho\cdot\tau=0, \qquad \text{by (13d) in } y_2, \tag{14d}$$

where D_z is Hirota's derivative, i.e., $D_z f(z)\cdot g(z) = f_z g - f g_z$. Note that from the definition of $\bar{\tau}$, we know that $D_{\bar{t}_3}\bar{\tau}\cdot\bar{\tau} = (D_{\bar{t}_3}-D_z)\tau\cdot\tau$, which interprets the appearance of this term in the second equation.

Example 2 (Second type of KP with a source (KPwS-II) [7, 13, 16, 18], i.e., the KPHwS (7) for $n=3$ and $k=2$). *The Hirota equations for the KPwS-II (4) can be obtained as*

$$D_x\tau_z\cdot\tau+\sigma\rho=0, \qquad \text{by (13b) with } y_j=0, \tag{15a}$$
$$(D_x^4+3(D_{\bar{t}_2}-D_z)^2-4D_xD_{t_3})\tau\cdot\tau=0, \qquad \text{by (13a) in } y_3, \tag{15b}$$
$$((D_{\bar{t}_2}-D_z)+D_x^2)\tau\cdot\sigma=0, \qquad \text{by (13c) in } y_2, \tag{15c}$$
$$((D_{\bar{t}_2}-D_z)+D_x^2)\rho\cdot\tau=0. \qquad \text{by (13d) in } y_2, \tag{15d}$$
$$(4D_{t_3}-D_x^3+3D_x(D_{\bar{t}_2}-D_z))\tau\cdot\sigma=0, \qquad \text{by (13c) in } y_3, \tag{15e}$$
$$(4D_{t_3}-D_x^3+3D_x(D_{\bar{t}_2}-D_z))\rho\cdot\tau=0, \qquad \text{by (13d) in } y_3. \tag{15f}$$

It seems that the Hirota bilinear equations (15) obtained here for KPwS-II are simpler than the results by Hu and Wang [7].

5. Conclusion and discussions

In this paper, we presented some results on the soliton hierarchy with self-consistent sources, especially on the method based on the symmetry of the soliton hierarchy. The Kadomtsev–Petviashvili (KP) hierarchy is used as an illustrative example to show the method to construct the soliton hierarchy with self-consistent sources (SHSCSs) and some properties of the SHSCSs, such as dressing approach, Wronskian solutions (including soliton solutions), bilinear identities and tau function.

There are some important applications of the bilinear identities for the KP hierarchy with self-consistent sources (KPHwS). As we know, the quasi-periodic solutions for the KP hierarchy can be constructed by using a method in algebraic geometry, where the construction of wave functions (or Baker–Akhiezer functions as in quasi-periodic cases) are intimately related to the bilinear identities, Riemann surfaces and divisors on it. It is very interesting to consider the quasi-periodic solutions for the KPHwS (7) when bilinear identities have been obtained in this paper. Another interesting problem is to consider the bilinear identities for other soliton hierarchies with sources, such as BKP, 2D Toda and discrete KP, etc. We will investigate these problems in the future.

Acknowledgments
The authors are grateful for the referee's valuable suggestions to our manuscript. This work is supported by the National Natural Science Foundation of China (11171175, 11201477). XL is also supported by the Chinese Universities Scientific Fund (2011JS041).

References
[1] Aratyn H, Nissimov E and Pacheva S 1998 Method of squared eigenfunction potentials in integrable hierarchies of KP type *Commun. Math. Phys.* **193** 493–525 (arXiv:solv-int/9701017)
[2] Cheng Y and Zhang Y J 1994 Bilinear equations for the constrained KP hierarchy *Inverse Probl.* **10** L11–7
[3] Date E, Kashiwara M, Jimbo M and Miwa T 1983 Transformation groups for soliton equations *Nonlinear Integrable Systems – Classical Theory and Quantum Theory* ed M Jimbo, T Miwa and M Sato (Singapore: World Scientific) pp 39–119
[4] Doktorov E V and Shchesnovich V S 1995 Nonlinear evolutions with singular dispersion laws associated with a quadratic bundle *Phys. Lett. A* **207** 153–8
[5] Grinevich P G and Taimanov I A 2008 Spectral conservation laws for periodic nonlinear equations of the Melnikov type *Amer. Math. Soc. Transl.* **224(2)** 125–38 (arXiv:0801.4143)
[6] Hu X B and Wang H Y 2006 Construction of dKP and BKP equations with self-consistent sources *Inverse Probl.* **22** 1903–20
[7] Hu X B and Wang H Y 2007 New type of Kadomtsev–Petviashvili equation with self-consistent sources and its bilinear Bäcklund transformation *Inverse Probl.* **23** 1433–44
[8] Lin R L, Zeng Y B and Ma W X 2001 Solving the KdV hierarchy with self-consistent sources by inverse scattering method *Physica A* **291** 287–98
[9] Lin R L, Yao H S and Zeng Y B 2006 Restricted flows and the soliton equation with self-consistent sources *SIGMA* **2** 096 (arXiv:nlin/0701003)
[10] Lin R L, Liu X J and Zeng Y B 2008 A new extended q-deformed KP hierarchy *J. Nonlinear Math. Phys.* **15** 333–47
[11] Lin R L, Peng H and Mañas M 2010 The q-deformed mKP hierarchy with self-consistent sources, Wronskian solutions and solitons *J. Phys. A: Math. Theor.* **43** 434022 (15pp)
[12] Lin R L, Liu X J and Zeng Y B 2013 Bilinear identities and Hirota's bilinear forms for Kadomtsev–Petviashvili hierarchy *J. Nonlinear Math. Phys.* **20** 214–28
[13] Liu X J, Zeng Y B and Lin R L 2008 A new extended KP hierarchy *Phys. Lett. A* **372** 3819–23 (arXiv:0710.4015)
[14] Liu X J, Lin R L, Jin B and Zeng Y B 2009 A generalized dressing approach for solving the extended KP and the extended mKP hierarchy *J. Math. Phys.* **50** 053506 (arXiv:0905.1402)
[15] Loris I and Willox R 1997 Bilinear form and solutions of the k-constrained Kadomtsev–Petviashvili hierarchy *Inverse Probl.* **13** 411–20
[16] Mel'nikov V K 1983 On Equations for wave interactions *Lett. Math. Phys.* **7** 129–36
[17] Mel'nikov V K 1984 Some new nonlinear evolution equations integrable by the inverse problem method *Math. USSR SB+* **49** 461–89
[18] Mel'nikov V K 1987 A direct method for deriving a multi-soliton solution for the problem of interaction of waves on the x-y plane *Commun. Math. Phys.* **112** 639–52
[19] Mel'nikov V K 1989 Capture and confinement of solitons in nonlinear integrable systems *Commun. Math. Phys.* **120** 451–68
[20] Oevel W and Schief W 1994 Squared eigenfunctions of the (modified) KP hierarchy and scattering problems of Loewner type *Rev. Math. Phys.* **6** 1301–38
[21] Shen H F and Tu M H 2011 On the constrained B-type Kadomtsev–Petviashvili hierarchy: Hirota bilinear equations and Virasoro symmetry *J. Math. Phys.* **52** 032704
[22] Wu H X, Zeng Y B and Fan T Y 2008 A new multicomponent CKP hierarchy and solutions *J. Math. Phys.* **49** 093510
[23] Zeng Y B, Ma W X and Lin R L 2000 Integration of the soliton hierarchy with self-consistent sources *J. Math. Phys.* **41** 5453–89
[24] Zeng Y B and Li Y S 1993 The deduction of the Lax representation for constrained flows from the adjoint representation *J. Phys. A: Math. Gen.* **26** L273–8
[25] Zhang D J and Chen D Y 2003 The N-soliton solutions of the sine-Gordon equation with self-consistent sources *Physica A* **321** 467–81

Symmetries in Science XVI IOP Publishing
Journal of Physics: Conference Series **538** (2014) 012015 doi:10.1088/1742-6596/538/1/012015

Generalizations of the Ermakov system through the Quantum Arnold Transformation

Francisco F López-Ruiz[1] **and Julio Guerrero**[2]

[1] Departamento de Física Aplicada, Universidad de Cádiz, Campus de Puerto Real, 11510 Puerto Real, Cádiz, Spain.

[2] Departamento de Matemática Aplicada, Universidad de Murcia, Campus de Espinardo, 30100 Murcia, Spain.

E-mail: paco.lopezruiz@uca.es, juguerre@um.es

Abstract. An Ermakov system consists of a pair of coupled non-linear differential equations which share a joint constant of motion named Ermakov invariant. One of those equations, non-linear, is frequently referred to as the Ermakov-Pinney equation; the other equation may be thought of as describing a dynamical system: a harmonic oscillator with time-dependent frequency. In this paper, we revise the Quantum Arnold Transformation, a unitary operator mapping the solutions of the Schrödinger equation for time-dependent (even damped) harmonic oscillators, described by the Generalized Caldirola-Kanai equation, into solutions for the free particle. With this tool, we elucidate the existence of Ermakov-type invariants in classically linear systems at the classical and quantum levels. We also provide more general Ermakov-type systems and the corresponding invariants, together with a physical interpretation.

1. Introduction

Since 2011, the authors have participated in a series of publications in which the Quantum Arnold Transformation (QAT) was presented [1], developed [2, 3] and applied [4, 5, 6] to several purposes. In particular, in [3], an explicit relation with the Ermakov-Pinney equation and the corresponding Ermakov system was provided. The Ermakov-Pinney equation is related to the Ermakov invariant [7, 8, 9] and appears in many branches of physics, such as Cosmology [10], BEC [11], etc. The Ermakov system appears in BEC [12, 13] and what is known as Kepler-Ermakov systems [14].

The QAT is a unitary map that relates the Hilbert space of solutions of the time-dependent Schrödinger equation for a Generalized Caldirola-Kanai oscillator (a quantum version of a classical system whose equation of motion is a Linear Second-Order Differential Equation, LSODE) into the corresponding Schrödinger equation for the free particle.

The feeling of the authors is that the QAT provides a framework in which many known relations, in particular for the Ermakov System, are better understood and some new ones are found. It is the purpose of this paper to deepen in the relation between the QAT and the Ermakov system, and to present examples that may help the reader to get a better grasp of the approach to the Ermakov system through the QAT. Also, new generalizations of the Ermakov system are possible within the QAT framework in a quite easy way.

The paper is organized as follows: in Section 2 we describe the transformation due to Arnold [15] and set some of the notation. The corresponding quantum version, the QAT, is revised in Section 3. We use the QAT in Section 4 to construct the quantum Arnold-Ermakov-Pinney transformation, which maps solutions of the time-dependent Schrödinger equation for a LSODE-system into solutions of the time-dependent Schrödinger equation for any other LSODE-system; a generalization of the Ermakov-Pinney equation is also provided. Finally, Section 5 is devoted to provide several examples in order to clarify the ideas previously exposed.

2. The Classical Arnold Transformation

The context of the classical Arnold transformation [15] is that of Lie point symmetries of ordinary differential equations. A Lie point symmetry of an ordinary differential equation (ODE) is a coordinate transformation that sends solutions into solutions. The problem of determining the Lie point symmetries of an ODE is rather old, and S. Lie gave the main results at the end of the nineteenth century [16]. One of these results was that a second-order differential equation (SODE) $y'' = F(x,y,y')$ has the maximal number of Lie point symmetries ($sl(3,\mathbb{R})$) if it can be transformed into the free equation by a point transformation:

$$y'' = F(x,y,y') \overset{\substack{\tilde{x}=\tilde{x}(x,y)\\ \tilde{y}=\tilde{y}(x,y)}}{\Longrightarrow} \tilde{y}'' = 0\,. \tag{1}$$

This linearization is possible if the ODE is of the form:

$$y'' = E_3(x,y)(y')^3 + E_2(x,y)(y')^2 + E_1(x,y)y' + E_0(x,y)\,, \tag{2}$$

with $E_i(x,y)$ satisfying some *integrability conditions* (see, for instance, [16, 17, 18]).

There is a nice geometric interpretation of this condition in terms of projective geometry. The non-linear SODE (2) is obtained by projection from the geodesic equations in a two-dimensional Riemannian manifold. The coefficients $E_i(x,y)$ are in one-to-one correspondence with Thomas projective parameters Π, and the integrability conditions that they satisfy are the conditions for the Riemann tensor to be zero (see [17, 18]).

V.I. Arnold named this process *rectification* or *straightening* of the trajectories, and studied the case of linear SODE (LSODE), giving explicitly the point transformation for this case [15]. Specifically, given a general Linear Second-Order Differential Equation (LSODE):

$$\ddot{x} + \dot{f}\dot{x} + \omega^2 x = \Lambda\,, \tag{3}$$

where f, ω and Λ are functions of t, the Classical Arnold Transformation (CAT) is a point transformation that is a local (in time) diffeomorphism:

$$\begin{array}{rccc} A: & \mathbb{R}\times T & \to & \mathbb{R}\times\mathcal{T} \\ & (x,t) & \mapsto & (\kappa,\tau) \end{array} : \left\{ \begin{array}{ll} \tau & = \frac{u_1(t)}{u_2(t)} = \int_{t_0}^{t} \frac{W(t')}{u_2(t')^2}dt' \\ \kappa & = \frac{x-u_p(t)}{u_2(t)} \end{array} \right. , \tag{4}$$

where T and $\mathcal{T}$ are, in general, open intervals containing t_0 and 0, respectively, u_1 and u_2 are independent solutions of the homogeneous LSODE satisfying the canonicity conditions:

$$u_1(t_0) = 0 = u_2'(t_0)\,, \qquad u_1'(t_0) = 1 = u_2(t_0)\,, \tag{5}$$

u_p is a particular solution of the inhomogeneous LSODE satisfying $u_p(t_0) = u_p'(t_0) = 0$, and $W(t) = \dot{u}_1 u_2 - u_1\dot{u}_2 = e^{-f}$ is the Wronskian of the two solutions. Here t_0 is an arbitrary time, conveniently chosen to be $t_0 = 0$ (see [1] for details).

The CAT transforms the original LSODE (3) into that of the free particle, up to a factor:

$$\ddot{x} + \dot{f}\dot{x} + \omega^2 x = \Lambda \quad \xrightarrow{A} \quad \frac{W}{u_2^3}\,\ddot{\kappa} = 0\,. \tag{6}$$

The presence of this factor implies that patches of trajectories of (3) are transformed into patches of straight (free) trajectories. In fact, an arbitrary trajectory solution of (3) can be written as $x(t) = Au_1(t) + Bu_2(t) + u_p(t)$, and the CAT sends it to $\kappa(\tau) = A\tau + B$. While t varies in the interval T defined by two consecutive zeros of $u_2(t)$ (containing t_0), τ varies in the range of the map defined by $\frac{u_1(t)}{u_2(t)}$. In the case in which $u_2(t)$ has one zero, T is (left- or right-) unbounded, and, if it has no zeros, T is $\mathbb{R}$.

Even though the CAT is a local (in time) diffeomorphism, it can be defined for an arbitrary time t_0. Thus different CATs can be defined for different times t_0 and cover in this way a complete trajectory of (3). We shall show with the example of the harmonic oscillator how this can be done.

2.1. The example of the harmonic oscillator

For this case, and considering $\Lambda = 0$, the two solutions are:

$$u_1(t) = \frac{1}{\omega}\sin(\omega t)\,, \qquad u_2(t) = \cos(\omega t)\,. \tag{7}$$

The open interval T defined by two consecutive zeros of $u_2(t)$, and containing $t_0 = 0$, is $(-\frac{\pi}{2\omega}, \frac{\pi}{2\omega})$, and the CAT A and its inverse A^{-1} are then written as:

$$A: \quad \kappa = \frac{x}{u_2(t)} = \frac{x}{\cos(\omega t)}\,, \quad \tau = \frac{u_1(t)}{u_2(t)} = \frac{1}{\omega}\tan(\omega t)\,, \tag{8}$$

$$A^{-1}: \quad \begin{aligned} x &= \cos(\arctan(\omega\tau))\kappa\,, \qquad t = \tfrac{1}{\omega}\arctan(\omega\tau)\,. \\ &= \tfrac{\kappa}{\sqrt{1+\omega^2\tau^2}} \end{aligned} \tag{9}$$

In this case $\tau \in \mathbb{R}$. Pictorially, the CAT for the HO can be represented as in Figure 1, where velocities have also been included in the graphic for clarity. Here A maps the solid part of the helix (half a period of a harmonic oscillator trajectory) into the whole line (a free particle trajectory). The horizontal plane represents the space of all possible initial conditions at $t = 0 = \tau$. Note that both trajectories are tangent when projected onto this plane, due to the conditions (5). See [4] for more details in this case. For the CAT to map other patches of the HO trajectories into the free particle trajectories, different branches of the arctan function in the inverse CAT (9) should be used (and a different $t_0 \neq 0$ for the CAT). For each integer k, let us take $T_k = ((k-\frac{1}{2})\frac{\pi}{\omega}, (k+\frac{1}{2})\frac{\pi}{\omega})$ and $t_k = k\frac{\pi}{\omega}$. The solutions verifying conditions (5) at t_k are $u_i^{(k)}(t) = (-1)^k u_i(t) = u_i(t - t_k)\,,\ i = 1, 2$. Define a pair of CAT and inverse CAT from $\mathbb{R} \times T_k$ into $\mathbb{R}^2$ of the form: $A_{(k)}(x,t) = (\frac{x}{u_2^{(k)}(t)}, \frac{u_1^{(k)}(t)}{u_2^{(k)}(t)}) = (\kappa, \tau)$ and $A_{(k)}^{-1}(\kappa,\tau) = (x,t)$, where the k-th branch of the arctan function has been used in $A_{(k)}^{-1}$. An *unfolded* version of the CAT, $\tilde{A}$, can be built by joining all the patches $A_{(k)}$, defining an application that maps a complete trajectory $x(t)$ of the harmonic oscillator into a trajectory $\kappa(\tau)$ of the free particle. $\tilde{A}$ is periodic on t with period $\frac{\pi}{\omega}$, although discontinuous.

Other simple examples where this construction can be done are the damped particle and the damped harmonic oscillator, see [2] for details.

It should be stressed that the CAT can be used to find invariant expressions for the harmonic oscillator performing the transformation A on invariant expressions for the free particle (the reversed procedure can be realized using A^{-1}). We call this simple strategy *"importing"* conserved quantities from the free particle to the harmonic oscillator, a procedure that can be employed for different pairs of systems.

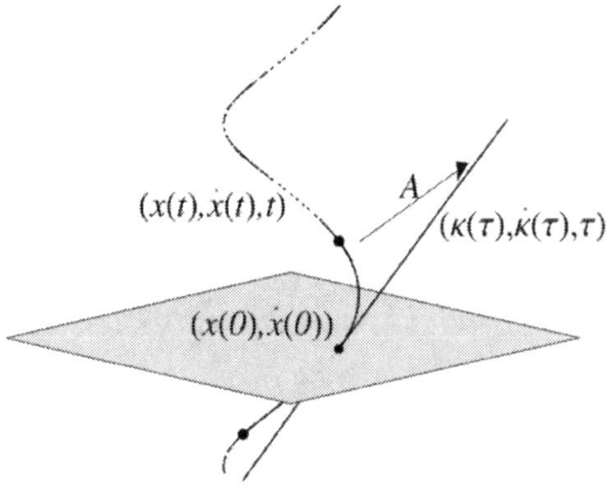

Figure 1. Depiction of the CAT for the harmonic oscillator (adapted from [4]).

3. The Quantum Arnold Transformation

An arbitrary LSODE system (3) can be derived from the Lagrangian (we take $\Lambda = 0$ for simplicity but the whole formalism can be developed with $\Lambda \neq 0$, see [1]):

$$L = \frac{1}{2} m e^{f} \left(\dot{x}^2 - \omega^2 x^2\right) , \tag{10}$$

and from this the Hamiltonian

$$H = \frac{p^2}{2m} e^{-f} + \frac{1}{2} m \omega^2 x^2 e^{f} \tag{11}$$

is derived, which is known as the Generalized Caldirola-Kanai (GCK) Hamiltonian for a damped oscillator (see [1] and references therein). The case in which $\dot{f} = \gamma$ and ω are constants corresponds to the original Caldirola-Kanai Hamiltonian for a damped harmonic oscillator [19, 20], and whose corresponding Lagrangian was given for the first time by Bateman [21]. Canonical quantization of the GCK Hamiltonian leads to the time-dependent Schrödinger equation:

$$i\hbar \frac{\partial \phi}{\partial t} = \hat{H} \phi = -\frac{\hbar^2}{2m} e^{-f} \frac{\partial^2 \phi}{\partial x^2} + \frac{1}{2} m \omega^2 x^2 e^{f} \phi . \tag{12}$$

The CAT A is a local (in time) diffeomorphism between the space of solutions of the LSODE system (3) and the space of solutions of the free particle. It is possible to extend it to a unitary transformation $\hat{A}$, the Quantum Arnold Transformation (QAT), between the Hilbert space of solutions $\phi(x,t)$ of the time-dependent Schrödinger equation for the GCK oscillator (12) at time t, $\mathcal{H}_t$, into the Hilbert space of solutions $\varphi(\kappa, \tau)$ of the time-dependent Schrödinger equation for the Galilean free particle

$$i\hbar \frac{\partial \varphi}{\partial \tau} = -\frac{\hbar^2}{2m} \frac{\partial^2 \varphi}{\partial \kappa^2} , \tag{13}$$

at time τ, $\mathcal{H}^G_\tau$. The desired extension is given by:

$$\begin{aligned} \hat{A} : \ \mathcal{H}_t &\longrightarrow \mathcal{H}^G_\tau \\ \phi(x,t) &\longmapsto \varphi(\kappa,\tau) = \hat{A}\left(\phi(x,t)\right) \\ &= A^* \left(\sqrt{u_2(t)}\, e^{-\frac{i}{2}\frac{m}{\hbar}\frac{1}{W(t)}\frac{\dot{u}_2(t)}{u_2(t)} x^2} \phi(x,t) \right) . \end{aligned} \tag{14}$$

Here A^* is the pullback of the CAT A, acting on functions (i.e. $A^*(f(x,t)) = f(A^{-1}(\kappa,\tau))$). The QAT can be diagrammatically represented as:

$$\begin{array}{ccc} \mathcal{H}^G_\tau & \xleftarrow{\hat{A}} & \mathcal{H}_t \\ {\scriptstyle \hat{U}_G(\tau)}\Big\uparrow & & \Big\uparrow{\scriptstyle \hat{U}(t)} \\ \mathcal{H}^G_0 \equiv \mathcal{H} & \xrightarrow[\hat{1}]{} & \mathcal{H} \equiv \mathcal{H}_0 \end{array} \tag{15}$$

where $\mathcal{H}_0 \equiv \mathcal{H}^G_0 \equiv \mathcal{H}$ is the common Hilbert space of solutions of the Schrödinger equation for both systems at $t = \tau = 0$ (we take, for simplicity, $t_0 = 0$, as before), $U(t)$ is the unitary time-evolution operator for the GCK oscillator and $\hat{U}_G(\tau)$ is the corresponding one for the Galilean free particle. The map at the bottom of the diagram is the identity due to conditions (5); otherwise a non-trivial unitary transformation appears (see [1]).

From the commutative diagram, it is clear that $\hat{A}$ is unitary given the unitarity of the evolution operators. However, it can also be checked explicitly that the scalar product of two states in $\mathcal{H}^G_\tau$ at a given time τ is the same than that of the transformed states by $\hat{A}$ in $\mathcal{H}_t$ at the corresponding time t:

$$\begin{aligned} \langle\varphi_1, \varphi_2\rangle_{\mathcal{H}^G_\tau} &= \int_{-\infty}^{+\infty} \mathrm{d}\kappa\, \varphi_1(\kappa,\tau)^* \varphi_2(\kappa,\tau) \\ &= \int_{-\infty}^{+\infty} \frac{\mathrm{d}x}{u_2(t)} \left(\sqrt{u_2(t)}\, e^{\frac{i}{2}\frac{m}{\hbar}\frac{1}{W(t)}\frac{\dot{u}_2(t)}{u_2(t)}x^2} \phi_1(x,t)^* \right) \\ &\qquad\qquad \times \left(\sqrt{u_2(t)}\, e^{-\frac{i}{2}\frac{m}{\hbar}\frac{1}{W(t)}\frac{\dot{u}_2(t)}{u_2(t)}x^2} \phi_2(x,t) \right) \\ &= \int_{-\infty}^{+\infty} \mathrm{d}x\, \phi_1(x,t)^* \phi_2(x,t) = \langle\phi_1, \phi_2\rangle_{\mathcal{H}_t}\,, \end{aligned} \tag{16}$$

where τ, κ, $\mathrm{d}\kappa$ and the integration limits have been transformed according to the CAT.

The QAT inherits from the CAT the local character in time, in the sense that it is valid only for $t \in T$ and $\tau \in \mathcal{T}$, although it can be defined for an arbitrary initial time t_0. To extend the QAT beyond T, we can proceed as in the classical case for the harmonic oscillator, considering the different branches of the inverse function of $\tau(t)$, defining an *unfolded* QAT, $\hat{\hat{A}}$.

It should be stressed that, if in the different branches of the *unfolded* CAT proper solutions verifying (5) are not used, changes in signs of the solutions can appear, which result in changes in phases in the different branches of the *unfolded* QAT. This phenomenon is related to the Maslov correction (see for instance [22]). In fact, it can be checked that, for the case of the harmonic oscillator previously considered, $\hat{A}_{(k)}(\phi(x,t)) = e^{ik\frac{\pi}{2}}\hat{A}(\phi((-1)^k x, t))$.

From the commutative diagram (15) and from (16) it is clear that the QAT is a unitary operator, and this has interesting and far-reaching consequences. Mimicking the process of "*importing*" that we mentioned for the CAT between the free particle and the harmonic oscillator at the end of Subsection 2.1, the QAT can be used to find invariant expressions for operators as well as symmetry generators from one system to the other, importing wave functions, scalar product, computing the time evolution operator, etc. One might say that the quantum free particle is somehow "linked" to any LSODE-type quantum system through the QAT. This is just a consequence of the fact that all those classically linear systems share a common symmetry group at the quantum level: the so called Schŕodinger group [23].

4. The Arnold-Ermakov-Pinney transformation

Having in mind the diagram (15), one might wonder what happens when two different LSODE-systems are related by QATs with the free-particle system as a middleman, that is, when a QAT and an inverse QAT are composed, as follows:

$$\begin{array}{ccccc} \mathcal{H}^{(1)}_{t_1} & \xrightarrow{\hat{A}_1} & \mathcal{H}^G_\tau & \xleftarrow{\hat{A}_2} & \mathcal{H}^{(2)}_{t_2} \\ {\scriptstyle \hat{U}^{(1)}(t_1)}\uparrow & & {\scriptstyle \hat{U}_G(\tau)}\uparrow & & \uparrow{\scriptstyle \hat{U}^{(2)}(t_2)} \\ \mathcal{H}\equiv\mathcal{H}^{(1)}_0 & \xrightarrow[\hat{1}]{} & \mathcal{H}^G_0\equiv\mathcal{H} & \xrightarrow[\hat{1}]{} & \mathcal{H}\equiv\mathcal{H}^{(2)}_0 \end{array} \tag{17}$$

That was shown in [3]: for the underlying classical construction, let A_1 and A_2 denote the CATs relating the LSODE-system 1 and LSODE-system 2 to the free particle, respectively, then $E = A_1^{-1}A_2$ relates LSODE-system 2 to LSODE-system 1. E can be written as:

$$\begin{aligned} E : \mathbb{R}\times T_2 &\to \mathbb{R}\times T_1 \\ (x_2,t_2) &\mapsto (x_1,t_1) = E(x_2,t_2)\,. \end{aligned} \tag{18}$$

The explicit form of the transformation can be easily computed by composing the two CATs, resulting in:

$$x_1 = \frac{x_2}{b(t_2)} \qquad W_1(t_1)dt_1 = \frac{W_2(t_2)}{b(t_2)^2}dt_2\,, \tag{19}$$

where $b(t_2) = \frac{u_2^{(2)}(t_2)}{u_2^{(1)}(t_1)}$ satisfies the non-linear SODE:

$$\ddot{b} + \dot{f}_2\dot{b} + \omega_2^2 b = \frac{W_2^2}{W_1^2}\frac{1}{b^3}\left[\omega_1^2 + \dot{f}_1\frac{\dot{u}_2^{(1)}}{u_2^{(1)}}(1-b^2\frac{W_1}{W_2})\right]\,, \tag{20}$$

and where $u_i^{(j)}$ refers to the i-th particular solution for system j; W_j, $\dot{f}_j$ and ω_j stand for the Wronskian and the LSODE coefficients for system j; and the dot means derivation with respect to the corresponding time variable.

Equation (20) constitutes a *generalization* of the well-known *Ermakov-Pinney equation.* That equation, together with the LSODE of system 2, is a *generalized Ermakov pair.* Also, any (quadratic) conserved quantity corresponding to the Schrödinger group of symmetries, which is shared by the two LSODE-systems, constitutes a *generalized Ermakov invariant.* Below, we consider specific examples in which the usual Ermakov invariant is constructed.

An important point needs to be made: equation (20) actually *defines* a generalized Arnold transformation, to be named (classical) *Arnold-Ermakov-Pinney transformation*, which transforms solutions of the LSODE 1 into solutions of the LSODE 2. However, it *does not describe* any of the two LSODE-systems themselves. Therefore, the Lie point symmetries of (20) need not to be those of the LSODEs in the general case.

The quantum version of the Arnold-Ermakov-Pinney transformation, $\hat{E}$, can be obtained computing the composition of a QAT and an inverse QAT with the diagram (17) in mind, to give:

$$\begin{aligned} \hat{E}:\ \mathcal{H}^{(2)}_{t_2} &\longrightarrow \mathcal{H}^{(1)}_{t_1} \\ \phi(x_2,t_2) &\longmapsto \varphi(x_1,t_1) = \hat{E}\left(\phi(x_2,t_2)\right) \\ &= E^*\left(\sqrt{b(t_2)}\,e^{-\frac{i}{2}\frac{m}{\hbar}\frac{1}{W_2(t_2)}\frac{\dot{b}(t_2)}{b(t_2)}x_2^2}\phi(x_2,t_2)\right)\,. \end{aligned} \tag{21}$$

The Quantum Arnold-Ermakov-Pinney transformation (QAEPT) is a unitary map importing solutions of a GCK Schrödinger equation from solutions of a different, auxiliary GCK Schrödinger equation which, in particular, might be the one corresponding to a harmonic oscillator, as in the next Section.

5. Examples

5.1. Ermakov System and interpretation of the Ermakov Invariant

Consider the particular case where LSODE-system 1 is a harmonic oscillator ($\omega_1(t_1) \equiv \omega_0$ and $\dot{f}_1 = 0$), which can be described by the Hamiltonian

$$H_{HO} = \frac{p_1^2}{2m} + \frac{1}{2} m\omega_0^2 x_1^2 , \tag{22}$$

and LSODE-system 2 is a time-dependent harmonic oscillator with frequency $\omega_2(t_2) \equiv \omega(t)$ and $\dot{f}_2 = 0$, with Hamiltonian

$$H = \frac{p^2}{2m} + \frac{1}{2} m\omega(t)^2 x^2 . \tag{23}$$

Then, expression (20) simplifies to:

$$\ddot{b} + \omega(t)^2 b = \frac{\omega_0^2}{b^3} . \tag{24}$$

That is just the Ermakov-Pinney (also known as Milne-Pinney) equation [7, 24, 25]. Obviously, for $\omega_0 = 0$ the Arnold-Ermakov-Pinney transformation reduces to the ordinary CAT, i.e. $E = A$.

Now, note that LSODE 1 Hamiltonian, H_{HO}, is conserved, and that it is so on both sides of the transformation E, given by (see (19)):

$$x_1 = \frac{x}{b} , \quad dt_1 = \frac{1}{b^2} dt . \tag{25}$$

Computing the momentum $p_1 = m\dot{x}_1 = m\frac{dx_1}{dt_1} = m\frac{dt}{dt_1}\frac{d}{dt}(\frac{x}{b}) = m(\dot{x}b - \dot{b}x)$, we can write H_{HO} in variables corresponding to system 2:

$$H_{HO} = \frac{1}{2m}(pb - m\dot{b}x)^2 + \frac{1}{2} m\omega_0^2 (\frac{x}{b})^2 \equiv I . \tag{26}$$

That is easily recognized as the usual Ermakov invariant I. Thus, we have found a way to characterize it through the CAT (or, more precisely, the CAEPT): I corresponds to the conserved quantity H_{HO} imported from the simple harmonic oscillator, which is used as an auxiliary system. Because the auxiliary system is arbitrary, I is conserved for any ω_0, provided (24) is satisfied.

The reader might wonder about the comparison of b (adimensional), satisfying (24) and providing the invariant (26), and the function α satisfying the usual Ermakov-Pinney equation $\ddot{\alpha} + \omega(t)^2 \alpha = \frac{1}{\alpha^3}$, which has the dimensions of the square root of time (see e.g. [26]). The relation of b and α is simply $b = \sqrt{\omega_0}\alpha$. In Quantum Mechanics, α is sometimes interpreted physically as the width of a wave packet, up to a factor $\sqrt{2m/\hbar}$ [26]. b may also be interpreted as the width of a wave packet for the quantum time-dependent harmonic oscillator over the natural length $\sqrt{2m\omega_0/\hbar}$ of the harmonic oscillator chosen as LSODE-system 1.

Using the explicit form of the inverse of (21) $\hat{E}^{-1}$ in this case, it is straightforward to arrive at solutions $\phi(x, t)$ of the Schrödinger equation of the time-dependent harmonic oscillator in terms of solutions of the Schrödinger equation for the simple harmonic oscillator $\varphi(x_1, t_1)$:

$$\phi(x, t) = \frac{1}{\sqrt{b}} e^{\frac{i}{2}\frac{m}{\hbar}\frac{\dot{b}}{b}x^2} \varphi\Big(\frac{x}{b}, \int \frac{1}{b^2} dt\Big) , \tag{27}$$

where b is any solution of (24). Although the explicit computations are left to the reader, note that, if $\varphi(x_1, t_1)$ is chosen to be, for instance, an eigenfunction of the quantum operator corresponding to (22), $\hat{H}_{HO}$, then the transformed wavefunction $\phi(x, t)$ is an eigenfunction of the quantum operator $\hat{I}$ corresponding to the invariant (26) (the explicit form of such operators is easily obtained from their classical counterpart by the canonical quantization prescription). That shows that $\hat{I}$ has discrete spectrum.

Moreover, $\hat{I}$ belongs to the $sl(2, \mathbb{R})$ subalgebra of the Schrödinger algebra of point variational symmetries of the system described by (23) (specifically, the one generating the compact subgroup). Even beyond that, the process can be repeated for any other operator representing an invariant in the simple harmonic oscillator (LSODE 1), showing the usefulness of the QAEPT to perform quick computations (see also Subsec. 5.2 for the case of creation and annihilation operators).

In the case here considered, the transformation $\hat{E}^{-1}$ together with (25) turns out to be very similar to the one used in BEC, known as scaling transformation, to transform the time-dependent potential (oscillator traps with time-dependent frequencies) into a time-independent harmonic oscillator potential [12, 13]. Also, in that case (i.e. for $\dot{f}_2 = 0$, $W_2 = 1$, $\dot{f}_1 = 0$, $W_1 = 1$) equation (21) reduces to the transformation given by Hartley and Ray [27] (this was already given by Lewis and Riesenfeld in [9]). However, the Quantum Arnold-Ermakov-Pinney transformation allows to choose in a suitable way the auxiliary system from which the solutions may be imported.

The Ermakov-Pinney equation entails a kind of nonlinear superposition principle, in the sense that its solutions can be written in terms of the solutions $y_1(t), y_2(t)$ of the corresponding linear equation (with $\omega_0 = 0$):

$$b(t)^2 = c_1 y_1(t)^2 + c_2 y_2(t)^2 + 2c_3 y_1(t) y_2(t)\,, \quad c_1 c_2 - c_3^2 = \omega_0^2\,. \tag{28}$$

The other way round, the general solution $y(t)$ of the linear equation can be written in terms of a particular solution $\rho(t)$ of the Ermakov-Pinney equation (24) as:

$$y(t) = c_1 \rho(t) \cos(\omega_0\, \theta(t) + c_2)\,, \tag{29}$$

where c_1, c_2 are arbitrary constants and $\theta(t) = \int^t \rho^{-2} dt'$. Note that this equation is just (25) for $t_1 = \theta(t)$, $\rho = b$, $x = y(t)$ and $x_1 = y(t)/b(t) = c_1 \cos(\omega_0 t_1 + c_2)$. As a result, the general solution of (24) can be determined from a particular solution $\rho(t)$ using (29) and (28). A similar situation holds for more general versions of the Ermakov-Pinney equation.

The process followed to find the Ermakov system is fairly easy to reproduce for more involved systems playing the role of LSODE-system 2, and the interpretation is exactly the same. For instance, consider a damped harmonic oscillator with time-varying frequency and damping (now we omit the time dependence) characterized by a GCK Hamiltonian:

$$H = \frac{p^2}{2m} e^{-f} + \frac{1}{2} m \omega^2 x^2 e^f\,. \tag{30}$$

As before, the auxiliary system is a simple harmonic oscillator. We arrive at a generalization of the Ermakov-Pinney equation:

$$\ddot{b} + \dot{f}\dot{b} + \omega^2 b = \frac{e^{-2f}}{b^3} \omega_0^2\,. \tag{31}$$

The AEP transformation is given by: $x_1 = \frac{x}{b}$, $dt_1 = \frac{e^{-f}}{b^2} dt$, so that we can compute $p_1 = m\dot{x}_1 = m\frac{dx_1}{dt_1} = m\frac{dt}{dt_1}\frac{d}{dt}(\frac{x}{b}) = me^f(\dot{x}b - \dot{b}x) = pb - m\dot{b}xe^f$ (in the last step we have taken into account that the canonical momentum for the GCK oscillator is $p = m\dot{x}e^f$). We then write H_{HO} in GCK variables:

$$H_{HO} = \frac{1}{2m}(pb - m\dot{b}xe^f)^2 + \frac{1}{2}m\omega_0^2(\frac{x}{b})^2 \equiv I\,, \tag{32}$$

which is an invariant for the GCK oscillator. This way, we have "imported" an invariant from an auxiliary system through the Arnold-Ermakov-Pinney transformation. Note that in [28] the authors also arrived at the result $I = H_{HO}$ using canonical transformations to link both systems (the reader may also note the resemblance of the canonical transformation there employed and the Arnold-Ermakov-Pinney transformation). The explicit construction of solutions of the Schrödinger equation proceeds as before and similar observations can be made.

5.2. Creation-Annihilation operators

The "importing" process discussed before can be repeated for any conserved quantity of the harmonic oscillator when it is used as an auxiliary system. Such an strategy may be quite relevant when going to the quantum theory. For instance, the classical version of the (conserved) annihilation (or annihilation) operators of the simple harmonic oscillator:

$$a = \left(\sqrt{\frac{m\omega_0}{2\hbar}}x_1 + \frac{i}{\sqrt{2\hbar m\omega_0}}p_1\right)e^{i\omega_0 t_1}\,, \tag{33}$$

can be used to find the corresponding conserved function for the GKC oscillator seen in the previous example. Performing the same transformation, we easily arrive at:

$$a = \left(\sqrt{\frac{m\omega_0}{2\hbar}}\frac{x}{b} + \frac{i}{\sqrt{2\hbar m\omega_0}}(pb - m\dot{b}xe^f)\right)e^{i\omega_0\int_0^t \frac{e^{-f}}{b^2}dt}\,. \tag{34}$$

We have used that $p = m\dot{x}e^f$ once more. A similar expression for a has been obtained recently in [26].

The QAEPT works in a similar way and the quantum version of the creation and annihilation operators are found straightforwardly:

$$\begin{aligned}
\hat{a}\phi(x,t) &= \hat{E}^{-1}\hat{a}_1\hat{E}\phi(x,t) = \hat{E}^{-1}\hat{a}_1E^*\left(\sqrt{b}\,e^{-\frac{i}{2}\frac{m}{\hbar}e^f\frac{\dot{b}}{b}x^2}\phi(x,t)\right) = \\
&= \frac{1}{\sqrt{b}}e^{+\frac{i}{2}\frac{m}{\hbar}e^f\frac{\dot{b}}{b}x^2}E^{*-1}\left(\left(\sqrt{\frac{m\omega_0}{2\hbar}}x_1 + \sqrt{\frac{\hbar}{2m\omega_0}}\frac{\partial}{\partial x_1}\right)e^{i\omega_0 t_1}E^*\left(\sqrt{b}\,e^{-\frac{i}{2}\frac{m}{\hbar}e^f\frac{\dot{b}}{b}x^2}\phi(x,t)\right)\right) = \\
&= \frac{1}{\sqrt{b}}e^{+\frac{i}{2}\frac{m}{\hbar}e^f\frac{\dot{b}}{b}x^2}\left(\left(\sqrt{\frac{m\omega_0}{2\hbar}}\frac{x}{b} + \sqrt{\frac{\hbar}{2m\omega_0}}b\frac{\partial}{\partial x}\right)e^{i\omega_0\int_0^t\frac{e^{-f}}{b^2}dt}\left(\sqrt{b}\,e^{-\frac{i}{2}\frac{m}{\hbar}e^f\frac{\dot{b}}{b}x^2}\phi(x,t)\right)\right) = \\
&= \left(\sqrt{\frac{m\omega_0}{2\hbar}}\frac{x}{b} + \frac{i}{\sqrt{2\hbar m\omega_0}}(-i\hbar b\frac{\partial}{\partial x} - m\dot{b}xe^f)\right)e^{i\omega_0\int_0^t\frac{e^{-f}}{b^2}dt}\phi(x,t)\,.
\end{aligned} \tag{35}$$

We would like to remark that there are many possible "importable" constructions for which analytical expressions can be found, such as the displacement operator $\hat{D}(a) = e^{a\hat{a}^\dagger - a^*\hat{a}}$, the radial squeezing operator $\hat{S}(\xi) = e^{\frac{1}{2}(\xi^*\hat{a}^2 - \xi(\hat{a}^\dagger)^2)}$, etc. The fundamental reason for the mere existence of these constructions is again the fact that classically linear systems share the same set of symmetries as the harmonic oscillator (and the free particle), which can be shown by the QAT. Some of these constructions can still be possible in classically non-linear systems, such as those presenting a potential of the form $\frac{1}{x^2}$ [28] (in our setup, that would amount to replace

LSODE 2 by the non-linear equation with a $\frac{1}{x^3}$ term). However, some of the symmetries may be lost in such systems with respect to the linear ones, so that linear $\hat{a}$ and $\hat{a}^\dagger$ may not exist as first-order operators, for instance, although they are still present as pseudo-differential operators [29].

Acknowledgments
Work partially supported by the Fundación Séneca, Spanish MICINN and Junta de Andalucía under projects 08814/PI/08, FIS2011-29813-C02-01 and FQM219-FQM1951, respectively. We thank the referee for valuable comments.

References

[1] Aldaya V, Cossío F, Guerrero J and López-Ruiz F F 2011 *J. Phys. A* **44** 065302
[2] Guerrero J, Aldaya V, López-Ruiz F F and Cossío F 2012 *Int. J. Geom. Meth. Mod. Phys.* **9** 1260011
[3] Guerrero J and López-Ruiz F F 2013, *Phys. Scr.* **87**, 038105
[4] Guerrero J, López-Ruiz F F, Aldaya V and Cossío F 2011 *J. Phys. A* **44** 445307
[5] López-Ruiz F F, Guerrero J and Aldaya V 2011 *J. Russ. Laser Res.* **32** 372
[6] Guerrero J, López-Ruiz F F, Aldaya V and Cossío F 2012, *J. Phys. A* **45**, 475303
[7] Ermakov V P 1880 *Univ. Izv. Kiev* **20** 1
[8] Ermakov V P 2008 *Appl. Anal. Discrete Math.* **2**, 123
[9] Lewis H R Jr. and Riesenfeld W B 1969 *J. Math. Phys.* **10** 1458
[10] Hawkins R M and Lidsey J E 2002 *Phys. Rev. D* **66** 023523
[11] Schuch D and Kaushal R S 2011 *J. Phys.: Conf. Series* **306** 012032
[12] Kagan Y, Surkov E L and Shlyapnikov G V 1996 *Phys. Rev. A* **54** R1753
[13] Castin Y and Dum R 1996 *Phys. Rev. Lett.* **77** 5315
[14] Leach P G L and Karasu A (Kalkanli) 2004 *J. Nonl. Math. Phys.* **11** 269
[15] Arnold V I, *Supplementary Chapters to the Theory of Ordinary Differential Equations* (Nauka, Moscow, 1978); English transl., *Geometrical Methods in the Theory of Ordinary Differential Equations*, (Springer-Verlag, New York-Berlin, 1983)
[16] Lie S 1883 *Arch. Math.* **VIII,IX** 187
[17] Mahomed F M, Qadir A 2009 *J. Nonlinear Math. Phys.* **16** 283
[18] Aminova A V and Aminov N A M 2010 *Sb. Math.* **201** 631
[19] Caldirola P 1941 *Nuovo Cimento* **18** 393
[20] Kanai E 1948 *Prog. Theor. Phys.* **3** 440
[21] Bateman H 1931 *Phys. Rev.* **38** 815
[22] Horvathy P A 1979 *Int. J. Theor. Phys* **18** 245
[23] Niederer U 1972 Helv. Phys. Acta **45** 802; Niederer U 1973 Helv. Phys. Acta **46** 191; Niederer U 1974 Helv. Phys. Acta **47** 167
[24] Milne W E 1930 *Phys. Rev.* **35** 86367
[25] Pinney E 1950 *Proc. A.M.S.* **1** 681
[26] Castaños O, Schuch D and Rosas-Ortiz O 2013, J. Phys. A **46**, 075304
[27] Hartley J G and Ray J R 1982, Phys. Rev. D **25**, 382
[28] Leach P G L and Andriopoulos K 2008, Phys. Scr. **77**, 015002
[29] Dodonov V V, Malkin I A and Man'ko V I 1974, Physica **74**, 597

Journal of Physics: Conference Series **538** (2014) 012016 doi:10.1088/1742-6596/538/1/012016

Quantum correlations expressed as information and entropic inequalities for composite and noncomposite systems

Margarita A Man'ko and Vladimir I Man'ko

P N Lebedev Physical Institute, Leninskii Prospect 53, Moscow 119991, Russia

E-mail: mmanko@sci.lebedev.ru

E-mail: manko@sci.lebedev.ru

Abstract. For noncomposite systems in classical and quantum domains, we obtain new inequalities such as the subadditivity and strong subadditivity conditions for Shannon entropies and information determined by the probability distributions and for von Neumann entropies of quantum states determined by the density operators. We extend the relations of Shannon and Tsallis entropies to the entropies of conditional probability distributions known for composite systems to the case of noncomposite systems. We give a review of the approach to construct the tomographic-probability distributions for qudit systems and present the entropic and information inequalities for spin tomograms, as well as the subadditivity and strong subadditivity conditions for tomograms of the both noncomposite and composite system states.

1. Introduction

Quantum correlations are the properties of quantum systems which can be associated with different phenomena. The correlations are determined by the structure of the density operator [1, 2]. The density operator of both composite and noncomposite quantum systems has only nonnegative eigenvalues, and this property provides the constraints determining the domain of parameters associated with matrix elements of the state density matrices in arbitrary representations. For composite systems, the density operator has different structure with respect to the decomposition in the form of convex sum of direct products of the density operators of the subsystem states.

The entanglement [3] of composite quantum systems containing two or more subsystems is one of the important manifestations of quantum correlations. The entanglement for two qubits can be detected by violation of the Bell inequality [4, 5].

Recently [6], it was demonstrated that the density operators can be mapped onto the probability vectors (quantum tomographic probabilities called state tomograms), which determine the operators. Recent reviews of this approach can be found in [7, 8]. Information and entropic properties of quantum systems were studied in the tomographic-probability representation in quantum mechanics in [9–12].

The other property reflecting the presence of quantum correlations in composite systems is quantum discord [13–16]; it corresponds to a difference in information properties of bipartite classical and quantum systems. The discord shows a difference between the properties of the

Shannon entropy [17] (determined by the probability distribution describing the classical state of a composite system) and the von Neumann entropy of the quantum bipartite system. Quantum correlations in noncomposite quantum systems were found to be described by the contextuality phenomenon [18–22].

Any probability distribution, in addition to the Shannon entropy, determines the so-called q-entropies [23, 24] containing an extra parameter dependence. The q-entropies are also defined for quantum states, and they are determined by the state density operator.

The Shannon and q-entropies obey some inequalities for composite systems in both classical and quantum domains. The strong subadditivity condition for the von Neumann entropy of three-partite system was proved in [25]. There are other aspects of the entropic inequalities studied in [26–32].

An entropic inequality called the subadditivity condition for the Shannon entropy of the classical bipartite system and the von Neumann entropy of the quantum bipartite system corresponds to the degree of correlations of the subsystem degrees of freedom. Recently [33–35], it was shown that analogous subadditivity conditions and information also exist for noncomposite systems, both classical and quantum. The subadditivity condition and information correspond to the degree of intrinsic correlations between different groups of the results obtained in the experiments where the classical and quantum observables are measured. An analogous result about the existence of the strong subadditivity condition for noncomposite classical and quantum systems was obtained in [35].

The aim of this paper is to review the new entropic and information inequalities found for noncomposite systems following [33–35] and introduce for these systems new concepts – analogs of the conditional entropies associated with composite systems and obtain new equalities characterizing the conditional entropies for Shannon entropy and q-entropies.

The main idea of our approach is to use the equivalence of N-dimensional Hilbert spaces of different quantum systems, e.g., quantum qubits. In view of this idea, one can apply the results known for a particular system, e.g., for the multipartite system of qudits to a different quantum system, e.g., to a single qudit. For this, we use the invertible map of integers $N = 0, 1, 2, 3, \ldots$ onto pairs of integers $0 \leftrightarrow (0,0)$, $1 \leftrightarrow (1,0)$, $2 \leftrightarrow (0,1)$, $3 \leftrightarrow (1,1) \ldots$, or onto triplets of integers $0 \leftrightarrow (0,0,0)$, $1 \leftrightarrow (1,0,0)$, $2 \leftrightarrow (0,1,0)$, $3 \leftrightarrow (0,0,1)$, $4 \leftrightarrow (1,1,0)$, $5 \leftrightarrow (1,0,1)$, $6 \leftrightarrow (0,1,1), \ldots$, etc. The map provides the possibility to introduce different basis vectors in the same Hilbert space, like $\mid m\rangle$, $\mid m_1, m_2\rangle$, $\mid m_1, m_2, m_3\rangle$, etc. An arbitrary state is presented by the density operator $\hat{\rho}$ described by the matrix elements $\rho_{mm'} = \langle m \mid \hat{\rho} \mid m'\rangle$, $\rho_{m_1m_2,m_1'm_2'} = \langle m_1m_2 \mid \hat{\rho} \mid m_1'm_2'\rangle, \ldots$ Thus, if one knows some properties of the matrix $\rho_{m_1m_2,m_1'm_2'}$, it is possible to visualize these properties for the matrix $\rho_{m,m'}$, in view of the map elaborated; it is a key point of our approach. We transfer the properties (inequalities) known for the density matrix $\rho_{m_1m_2,m_1'm_2'}$ of composite qudit systems to the density matrix of a single qudit $\rho_{m,m'}$. In principle, one can obtain from the inequalities known for multiqudit systems, like Bell inequalities or entropic inequalities, the same inequalities for single qudit systems. This approach was applied in [35–38].

This paper is organized as follows.

In section 2, we review the properties of sets of nonnegative numbers organized as tables with different labels by collective indices and obtain various inequalities for these nonnegative numbers. In section 3, we study new entropic inequalities for tomographic entropies of qudit states. In section 4, we present new entropic inequalities for density matrices of noncomposite systems, while in section 5 we discuss the quantum discord and its properties for noncomposite systems. We give our conclusions and prospectives in section 6.

2. Tables of nonnegative numbers

In this section, we describe a tool how to consider any set of N nonnegative numbers, the sum of which is equal to unity, as a set analogous to a vector or a set considered as a matrix. Also we show how such a set can be viewed as a table where each position in the table is labeled by three integers, or four integers, etc. The inverse procedure can also be suggested. This means that an arbitrary table which contains nonnegative numbers can be viewed as a vector with nonnegative number components. This vector can be identified with the probability vector or a point in the simplex. Also the table, where the nonnegative numbers are situated in the positions labeled by some collection of integers, can be treated as a joint probability distribution. When we identify the table with nonnegative numbers with the joint probability distribution, we take into account an additional information arising from the results of experiments where some random variables are measured.

As it was pointed out in [34] and employed in [33,35], there exist some entropic inequalities which can be associated not only with the probability distributions but with the sets of nonnegative numbers independently on the fact that these numbers are used as probabilities. Also in [35], it was stated that some properties of the von Neumann entropy can be considered as the properties of the complex number set where the complex numbers are associated with matrix elements of the quantum-state density matrix. Thus, in this paper we concentrate on the question: What inequalities are completely mathematical characteristics of sets of either nonnegative numbers or complex numbers, and what inequalities incorporate information that the numbers are associated with physical characteristics of the classical or quantum systems?

Thus, we can consider N nonnegative numbers and label these numbers by the integer number $k = 1, 2, \ldots, N$. We obtain a set of nonnegative numbers $\vec{p} = (p_1, p_2, \ldots, p_N)$, where $p_k \geq 0$ and $\sum_{k=1}^{N} p_k = 1$, and the vector notation $\vec{p}$ is introduced. If we add M zero numbers to this set, we get the set of $N' = N + M$ nonnegative numbers; the number M can be chosen arbitrary. This means that one can construct another vector $\vec{p}'$ with N' components. Some of these components are equal to zero.

Let us consider the same initial set of N nonnegative numbers. We can introduce a different labeling of these numbers if, e.g., the chosen integer has the form $N = N_1N_2$, where N_1 and N_2 are also integers. Then we organize these numbers as a table with labels (jk). Thus, we obtain $\mathcal{P}_{kj}$, where $k = 1, 2, \ldots, N_1$ and $j = 1, 2, \ldots, N_2$, satisfying $\sum_{k=1}^{N_1} \sum_{j=1}^{N_2} \mathcal{P}_{kj} = 1$. The numbers N_1 and N_2 can be chosen in a different way, if the number $N = n_1 n_2 \cdots n_s$ with factors n_α ($\alpha = 1, 2, \ldots, s$) equal to integers. In fact, we can combine the integers in this product, e.g., as $n_1 = N_1$, $n_2 n_3 \cdots n_s = N_2$. Thus, we can map the initial set of N nonnegative numbers on anyone of the tables $\mathcal{P}_{jk}$ constructed according to the described representation of the integer N in the form of the product of the integers. If the number N is prime number, one can add zeros and consider the integer N' for which one has the representation in the form of product of integers. Thus, one can always find many tables $\mathcal{P}_{jk}$, where indices (jk) play the role of positions, and the index α labels a particular table associated, e.g., with different numbers N'.

The consideration above presented demonstrates that any set of nonnegative numbers can be mapped onto many tables (matrices $\mathcal{P}_{jk}$). These rectangular matrices have matrix elements equal either to zero or to a number p_k from the initial vector $\vec{p}$. One can continue this construction of maps to associate the initial N nonnegative numbers p_k with, e.g., the table Π_{kjl} ($k = 1, 2, \ldots N_1$, $j = 1, 2, \ldots N_2$, and $l = 1, 2, \ldots N_3$). The positions in this table are labeled by three integers. We used the decomposition $N = N_1 N_2 N_3$ of integer N into product of three integers. As it is clear from the previous discussion, such a decomposition can always be found by constructing the integer $N' = N + M$ and using the table containing N' numbers. N numbers in this table Π_{kjl} are the initial N numbers p_k, and the other numbers are zeros.

It is clear that there exist many possibilities to construct different tables $\Pi_{kjl}^{(\alpha)}$ depending on the choice of the integer M and integers N_1, N_2, and N_3 providing the decomposition

$N' = N_1N_2N_3$. The index α labels different tables. It is also clear that for any initial set of nonnegative numbers p_k one can construct the map of these numbers onto table $T_{j_1,j_2,\ldots,j_k}$ containing all these numbers and the corresponding number of zero components. If the nonnegative numbers are experimentally measured probabilities, the tables constructed can be associated with probability distributions. The chosen table $\mathcal{P}_{kj}$ of nonnegative numbers provides two sets $\mathcal{P}_{1k}$ and $\mathcal{P}_{2j}$ of nonnegative numbers

$$P_{1k} = \sum_{j=1}^{N_2} \mathcal{P}_{kj}, \quad P_{2j} = \sum_{k=1}^{N_1} \mathcal{P}_{kj}. \tag{1}$$

Also the chosen table of nonnegative numbers Π_{kjl} provides the three other tables of nonnegative parameters

$$\mathcal{P}_{kj}^{(12)} = \sum_{l=1}^{N_3} \Pi_{kjl}, \quad \mathcal{P}_{jl}^{(23)} = \sum_{k=1}^{N_1} \Pi_{kjl}, \quad P_j^{(2)} = \sum_{k=1}^{N_1}\sum_{l=1}^{N_3} \Pi_{kjl}. \tag{2}$$

If the considered tables of nonnegative numbers are identified with joint probability distributions of physical composite systems, the above formulas correspond to constructing marginal distributions for bipartite and tripartite systems, respectively.

The constructed map of a set of nonnegative numbers p_k onto different kinds of tables, like $\mathcal{P}_{kj}$ or Π_{kjl}, etc., has an inverse in the following sense.

For a given table with N_1N_2 positions (matrix $\mathcal{P}_{kj}$), one can construct N-vector using the map of pairs of integers onto integers as $(11) \to 1, (12) \to 2, \ldots, (1N_2) \to N_2, (2,1) \to N_2+1, (2,2) \to N_2+2, \ldots, (N_1N_2) \to N$. Analogously, the table of nonnegative numbers Π_{kjl} can be mapped onto the N-vector. This observation can be used to rewrite the relations available for tables $\mathcal{P}_{kj}$ or Π_{kjl} in terms of numbers p_k. For example, it is known that joint probability distributions provide inequalities called the subadditivity conditions and strong subadditivity conditions, and these inequalities are valid for matrices $\mathcal{P}_{kj}$ and joint probability distributions Π_{kjl}, respectively. Being rewritten in the form of probability vector components p_k, they are valid for an arbitrary set of nonnegative numbers, which are equal to the probabilities.

Example of $N = 7$
For $N = 7$, one has the inequalities [35]

$$\begin{aligned}\left(-\sum_{k=1}^{7} p_k \ln p_k\right) &- (p_1+p_2+p_5+p_6)\ln(p_1+p_2+p_5+p_6)\\ -(p_3+p_4+p_7)\ln(p_3+p_4+p_7) &\leq -(p_1+p_2)\ln(p_1+p_2) - (p_3+p_4)\ln(p_3+p_4)\\ -(p_5+p_6)\ln(p_5+p_6) &- p_7\ln p_7 - (p_1+p_5)\ln(p_1+p_5) - (p_2+p_6)\ln(p_2+p_6)\\ &-(p_3+p_7)\ln(p_3+p_7) - p_4\ln p_4\end{aligned} \tag{3}$$

and

$$\begin{aligned}-\sum_{k=1}^{7} p_k \ln p_k &\leq -(p_1+p_2+p_5+p_6)\ln(p_1+p_2+p_5+p_6)\\ &-(p_3+p_4+p_7)\ln(p_3+p_4+p_7) - (p_1+p_3)\ln(p_1+p_3)\\ &-(p_2+p_4)\ln(p_2+p_4) - (p_5+p_7)\ln(p_5+p_7) - p_4\ln p_4.\end{aligned} \tag{4}$$

These inequalities are analogs of the strong subadditivity condition for three-partite system and the subadditivity condition for two-partite system, respectively. In spite of the fact that the

7 nonnegative numbers are not associated with any probabilities, the inequalities are satisfied. The inequalities were obtained by introducing an N'-dimensional vector with $N' = N + 1$ with the component $p_8 = 0$. The eight-dimensional vector obtained was mapped onto the rectangular matrix $\mathcal{P}_{kj}$ corresponding to the decomposition $N' = 2 \cdot 4$, and this provided the subadditivity condition (4). The strong subadditivity condition (3) corresponds to the decomposition $N' = 2 \cdot 2 \cdot 2$.

For four nonnegative numbers $\vec{p} = (a, b, c, d)$, an analog of the subadditivity condition discussed in [34] reads

$$-(a+b)\ln(a+b) - (c+d)\ln(c+d) - (a+c)\ln(a+c) - (b+d)\ln(b+d) \geq -a\ln a - b\ln b - c\ln c - d\ln d. \quad (5)$$

Inequalities for tables
The inequalities above discussed are examples of general inequalities for matrix tables $\mathcal{P}_{kj}$ and Π_{kjl} containing the nonnegative numbers; they are

$$-\sum_{j=1}^{m}\left(\sum_{k=1}^{n}\mathcal{P}_{jk}\right)\left(\ln\sum_{k'=1}^{n}\mathcal{P}_{jk'}\right) - \sum_{k=1}^{n}\left(\sum_{j=1}^{m}\mathcal{P}_{jk}\right)\left(\ln\sum_{j'=1}^{n}\mathcal{P}_{jj'}\right) \geq -\sum_{j=1}^{m}\sum_{k=1}^{n}\mathcal{P}_{jk}\ln\mathcal{P}_{jk}, \quad (6)$$

$$-\sum_{j=1}^{n_1}\sum_{k=1}^{n_2}\sum_{m=1}^{n_3}\Pi_{jkm}\ln\Pi_{jkm} - \sum_{k=1}^{n_2}\left(\sum_{j=1}^{n_1}\sum_{m=1}^{n_3}\Pi_{jkm}\right)\ln\left(\sum_{j'=1}^{n_1}\sum_{m'=1}^{n_2}\Pi_{j'km}\right)$$
$$\leq -\sum_{j=1}^{n_1}\sum_{k=1}^{n_2}\left(\sum_{m'=1}^{n_3}\Pi_{jkm'}\right)\ln\left(\sum_{m=1}^{n_3}\Pi_{jkm}\right) - \sum_{k=1}^{n_2}\sum_{m=1}^{n_3}\left(\sum_{j'=1}^{n_1}\Pi_{j'km}\right)\ln\left(\sum_{j=1}^{n_1}\Pi_{jkm}\right). \quad (7)$$

In the case where $\mathcal{P}_{kj}$ is the joint probability distribution for two random variables of the bipartite system, equation (6) is the usual subadditivity condition for Shannon entropies associated with the probability distribution. Inequality (7) is the strong subadditivity condition, if Π_{jkm} is the joint probability distribution of three random variables for the tripartite system. One should stress that the inequalities discussed are valid for an arbitrary probability vector $\vec{p}$ even of the system which is not composite. Such a system can have only a single random variable; nevertheless, there exist the inequalities for the probability distribution of this random variable identical to the subadditivity and strong subadditivity conditions. For the systems, which do not have subsystems, these inequalities can be interpreted as entropic properties corresponding to the existence of correlations between different results of measurements of the random variable.

3. Inequalities for tomographic entropies

For qudit states with the density matrix ρ, the tomographic-probability distribution (spin tomogram) $w(m, \vec{n})$ can be used as an alternative to the density matrix ρ [39, 40]. It is interpreted as the probability to obtain the spin projection m, where $m = -j, -j+1, \ldots, j$, on the quantization axes determined by the unit vector $\vec{n} = (\sin\theta\cos\varphi, \sin\theta\sin\varphi, \cos\theta)$. The tomogram can be obtained as a result of the experiment. Thus, for qudit states, nonnegative numbers p_k can be identified with the tomographic probabilities. For example, for the qudit state corresponding to spin $j = 3/2$ we have the four-vector $\vec{p} = (p_1, p_2, p_3, p_4)$, where $p_1 = w(-3/2, \vec{n})$, $p_2 = w(-1/2, \vec{n})$, $p_3 = w(1/2, \vec{n})$, and $p_4 = w(3/2, \vec{n})$.

The tomographic Shannon entropy is defined as (see, e.g., [41])

$$H(\vec{n}) = -\sum_{m=-3/2}^{3/2} w(m, \vec{n})\ln w(m, \vec{n}). \quad (8)$$

There are two tomographic-probability distributions

$$W_1(-2,\vec{n}) = w(-3/2,\vec{n}) + w(-1/2,\vec{n}), \qquad W_1(2,\vec{n}) = w(1/2,\vec{n}) + w(3/2,\vec{n}), \tag{9}$$
$$W_2(-1,\vec{n}) = w(-3/2,\vec{n}) + w(1/2,\vec{n}), \qquad W_2(1,\vec{n}) = w(-1/2,\vec{n}) + w(3/2,\vec{n}), \tag{10}$$

which are analogs of marginal probability distributions for the bipartite system. We can interpret these probability distributions as follows: $W_1(\vec{n})$ yields the probabilities to find the qudit state with the sum of spin projections equal to ± 2, and $W_2(\vec{n})$ provides the probabilities to find the qudit state with the sum of spin projections equal to ± 1. In spite the fact that the system is not bipartite, we have the subadditivity condition

$$-\sum_{m=-3/2}^{3/2} w(m,\vec{n}) \ln w(m,\vec{n}) \leq -W_1(1,\vec{n}) \ln W_1(1,\vec{n}) - W_1(-1,\vec{n}) \ln W_1(-1,\vec{n})$$
$$-W_2(2,\vec{n}) \ln W_2(2,\vec{n}) - W_2(-2,\vec{n}) \ln W_2(-2,\vec{n}). \tag{11}$$

Conditional tomographic probabilities
We introduce the conditional tomographic probability distribution for the qudit state following the procedure of constructing the distribution for the bipartite system. For the qudit-state tomogram with $j = 3/2$, we introduce two conditional probabilities

$$V_1(\vec{n}) = \frac{w(-3/2,\vec{n})}{w(-3/2,\vec{n}) + w(-1/2,\vec{n})}, \qquad V_2(\vec{n}) = \frac{w(-1/2,\vec{n})}{w(-3/2,\vec{n}) + w(-1/2,\vec{n})}, \tag{12}$$
$$\widetilde{V}_1(\vec{n}) = \frac{w(1/2,\vec{n})}{w(3/2,\vec{n}) + w(1/2,\vec{n})}, \qquad \widetilde{V}_2(\vec{n}) = \frac{w(3/2,\vec{n})}{w(3/2,\vec{n}) + w(1/2,\vec{n})}. \tag{13}$$

The tomographic conditional probability distributions (12) and (13) determine the tomographic conditional entropies. Since there exists the identity for four nonnegative numbers p_1, p_2, p_3, and p_4 (considered as the probabilities) of the form

$$-p_1 \ln \frac{p_1}{p_1+p_2} - p_2 \ln \frac{p_2}{p_1+p_2} - p_3 \ln \frac{p_3}{p_3+p_4} - p_4 \ln \frac{p_4}{p_3+p_4}$$
$$= (p_1+p_2)\ln(p_1+p_2) + (p_3+p_4)\ln(p_3+p_4) - p_1 \ln p_1 - p_2 \ln p_2 - p_3 \ln p_3 - p_4 \ln p_4, \tag{14}$$

one can rewrite this equality as the property of the conditional tomographic entropy

$$H(V,\widetilde{V}) = H(V \mid \widetilde{V}) + H(\widetilde{V}), \tag{15}$$

where the tomographic entropy of the qudit state reads

$$H(V,\widetilde{V}) = -\sum_{m=-3/2}^{3/2} w(m,\vec{n}) \ln w(m,\vec{n}). \tag{16}$$

Formula (15) can be generalized to yield an analog of the chain form of the conditional entropic equality for multipartite system.

The conditional tomographic entropy of the qudit state is defined as

$$H(V \mid \widetilde{V}) = H(V,\widetilde{V}) + W_1(2,\vec{n}) \ln W_1(2,\vec{n}) + W_1(-2,\vec{n}) \ln W_1(-2,\vec{n}). \tag{17}$$

Journal of Physics: Conference Series **538** (2014) 012016 doi:10.1088/1742-6596/538/1/012016

Conditional q-entropies
We define the q-entropy for four nonnegative numbers p_1, p_2, p_3, and p_4 as follows:

$$T_q(\vec{p}) = \frac{1}{1-q}\Big(\sum_{k=1}^{4} p_k^q - 1\Big). \tag{18}$$

Identifying the numbers p_k with tomographic probabilities of qudit state with $j = 3/2$, we obtain the Tsallis tomographic entropy [41]

$$T_q(\vec{n}) = \frac{1}{1-q}\Big(\sum_{m=-3/2}^{3/2} [w(m,\vec{n})]^q - 1\Big). \tag{19}$$

One can introduce the conditional tomographic q-entropy $T_q(V \mid \widetilde{V})$ for the qudit state with $j = 3/2$, using the formula presented in [34] for bipartite qudit states,

$$T_q(V \mid \widetilde{V}) = T_q(\vec{n}) - \frac{1}{1-q}\Big(W_1(-2,\vec{n})^q + W_1(2,\vec{n})^q - 1\Big). \tag{20}$$

For $q \to 1$, this entropy tends to the conditional Shannon entropy $H(V \mid \widetilde{V})$.

Formula (20) can be generalized to yield the chain form of q-entropic equality for the conditional entropies associated with multipartite systems.

In view of the inequalities derived for tomographic q-entropies of bipartite qudit systems in [34], we obtain analogous inequalities for q-entropy of the qudit state under consideration

$$T_q(V) \le T_q(\vec{n}), \qquad T_q(V \mid \widetilde{V}) \le T_q(V), \tag{21}$$

where

$$T_q(V) = \frac{1}{1-q}\Big(W_1(-2,\vec{n})^q + W_1(2,\vec{n})^q - 1\Big), \quad T_q(\widetilde{V}) = \frac{1}{1-q}\Big(W_2(1,\vec{n})^q + W_2(-1,\vec{n})^q - 1\Big).$$

4. Inequalities for density matrices

The quantum states of qudits with spin j $(2j+1 = N)$ are determined by density $N\times N$-matrices ρ_{km}. We can apply our tool elaborated for the set of complex numbers ρ_{km}. In this case, the tool is to map a pair of integers (k, m) onto the other collective pairs of integers, e.g., $k \to (l_k, j_k)$ and $m \to (l_m, j_m)$, i.e., $(k, m) \to (l_k j_k, l_m j_m)$. Continuing this procedure, we construct the map $(k, m) \to (l_k j_k n_k, l_m j_m n_m)$, etc. This means that the density matrix $\rho_{mm'}$ with $j = 3/2$ can be written in the form

$$\rho = \begin{pmatrix} \rho_{11,11} & \rho_{11,12} & \rho_{11,21} & \rho_{11,22} \\ \rho_{12,11} & \rho_{12,12} & \rho_{12,21} & \rho_{12,22} \\ \rho_{21,11} & \rho_{21,12} & \rho_{21,21} & \rho_{21,22} \\ \rho_{22,11} & \rho_{22,12} & \rho_{22,21} & \rho_{22,22} \end{pmatrix}. \tag{22}$$

Such a form provides a possibility to apply to this density matrix the positive maps which yield the nonnegative matrices

$$\rho(1)_{jk} = \sum_{s=1}^{2} \rho_{js,ks}, \qquad \rho(2)_{mn} = \sum_{s=1}^{2} \rho_{sm,sn}. \tag{23}$$

Complex numbers $\rho_{mm'}$ satisfy the criterion of nonnegativity of the matrix ρ, as well as the matrices with complex numbers $\rho(1)_{jk}$ and $\rho(2)_{mn}$. Then we have the subadditivity condition

$$-\mathrm{Tr}\,\rho\ln\rho \le -\mathrm{Tr}\,\rho(1)\ln\rho(1) - \mathrm{Tr}\,\rho(2)\ln\rho(2). \tag{24}$$

This inequality is valid for one qudit state, and the qudit is a system which does not contain subsystems.

Examples of spin-2 and spin-3 states
There are other examples [34] of such inequalities which are analogs of the strong subadditivity condition for qudit states with $j = 2$ and $j = 3$. In the case of 7×7-matrix ρ_{jk}, it has the form

$$-\mathrm{Tr}\,(\rho \ln \rho) - \mathrm{Tr}\,(R_2 \ln R_2) \leq -\mathrm{Tr}\,(R_{12} \ln R_{12}) - \mathrm{Tr}\,(R_{23} \ln R_{23})\,, \tag{25}$$

where the density matrix R_{12} has matrix elements expressed in terms of the density matrix ρ_{jk} as follows:

$$R_{12} = \begin{pmatrix} \rho_{11}+\rho_{22} & \rho_{13}+\rho_{24} & \rho_{15}+\rho_{26} & \rho_{17} \\ \rho_{31}+\rho_{42} & \rho_{33}+\rho_{44} & \rho_{35}+\rho_{46} & \rho_{37} \\ \rho_{51}+\rho_{62} & \rho_{53}+\rho_{64} & \rho_{55}+\rho_{66} & \rho_{57} \\ \rho_{71} & \rho_{73} & \rho_{75} & \rho_{77} \end{pmatrix}. \tag{26}$$

The density matrix R_{23} reads

$$R_{23} = \begin{pmatrix} \rho_{11}+\rho_{55} & \rho_{12}+\rho_{56} & \rho_{13}+\rho_{57} & \rho_{14} \\ \rho_{21}+\rho_{65} & \rho_{22}+\rho_{66} & \rho_{23}+\rho_{67} & \rho_{24} \\ \rho_{31}+\rho_{75} & \rho_{32}+\rho_{76} & \rho_{33}+\rho_{77} & \rho_{34} \\ \rho_{41} & \rho_{42} & \rho_{43} & \rho_{44} \end{pmatrix}, \tag{27}$$

while the matrix R_2 is

$$R_2 = \begin{pmatrix} \rho_{11}+\rho_{22}+\rho_{55}+\rho_{66} & \rho_{13}+\rho_{24}+\rho_{57} \\ \rho_{31}+\rho_{42}+\rho_{75} & \rho_{33}+\rho_{44}+\rho_{77} \end{pmatrix}. \tag{28}$$

In the case of 5×5-matrix ρ_{lk} corresponding to the qudit state with $j = 2$, we have inequality (25), where

$$R_{12} = \begin{pmatrix} \rho_{11}+\rho_{22} & \rho_{13}+\rho_{24} & \rho_{15} \\ \rho_{31}+\rho_{42} & \rho_{33}+\rho_{44} & \rho_{35} \\ \rho_{51} & \rho_{53} & \rho_{55} \end{pmatrix}, \qquad R_{23} = \begin{pmatrix} \rho_{11}+\rho_{55} & \rho_{12} & \rho_{13} & \rho_{14} \\ \rho_{21} & \rho_{22} & \rho_{23} & \rho_{24} \\ \rho_{31} & \rho_{32} & \rho_{33} & \rho_{34} \\ \rho_{41} & \rho_{42} & \rho_{43} & \rho_{44} \end{pmatrix}, \tag{29}$$

$$R_2 = \begin{pmatrix} \rho_{11}+\rho_{22}+\rho_{55} & \rho_{13}+\rho_{24} \\ \rho_{31}+\rho_{42} & \rho_{33}+\rho_{44} \end{pmatrix}. \tag{30}$$

In order to obtain these inequalities, we mapped the matrices ρ_{lk} on matrices R_{lk} acting on vectors in the eight-dimensional linear space. The matrix R_{lk} has matrix elements of the matrix ρ, and the other matrix elements are zeros.

The quantum subadditivity condition for qutrit state was obtained in [33]. The procedure developed is analogous to mapping of N-vectors on N'-vectors with adding zero components to the initial vector. After this map, we apply the known subadditivity and strong subadditivity conditions to the obtained nonnegative matrix R_{lk}. As a result, we get the above entropic inequalities for the density matrices of qudit states. It is obvious that analogous generalized inequalities can be obtained for arbitrary qudits with $j = 3/2, 2, 5/2, 3, 7/2, \ldots$ Also the strong subadditivity condition, which is usually associated with three-partite systems, can be written for the two-qudit system, in view of the suggested map of integers on the collection of integers.

5. Some inequalities and quantum discord

For a bipartite system with the joint probability distribution $\mathcal{P}_{jk}$, the Shannon information is defined as a difference of the sum of subsystem entropies and entropy of the composite system. The two-qudit state can be described by the joint tomographic probability distribution $w(m_1, m_2, u)$, which is given by diagonal matrix elements of the density matrix $\rho(1,2)$ in unitary rotated basis

$$w(m_1, m_2, u) = \langle m_1 m_2 \mid u^\dagger \rho(1,2) u \mid m_1 m_2 \rangle, \tag{31}$$

where u is the unitary matrix.

In the case $u = u_1 \otimes u_2$, where u_1 and u_2 are matrices of the irreducible representation of the rotation group, the tomogram $w(m_1, m_2, \vec{n}_1, \vec{n}_2)$ is the joint probability distribution of two random spin projections m_1 and m_2. The directions $\vec{n}_1$ and $\vec{n}_2$ are determined by Euler angles which fix the matrix elements of the representation matrices u_1 and u_2. The Shannon entropy

$$H_{12}(u) = -\sum_{m_1=-j_1}^{j_1} \sum_{m_2=-j_2}^{j_2} w(m_1, m_2, u) \ln w(m_1, m_2, u) \tag{32}$$

has the minimum for a specific value of the unitary matrix u [9], which is equal to the von Neumann entropy associated with the density matrix of the two qudit states, i.e.,

$$\min H(u) = -\mathrm{Tr}\,\rho(1,2) \ln \rho(1,2). \tag{33}$$

The density matrices $\rho(1) = -\mathrm{Tr}_2\,\rho(1,2)$ and $\rho(2) = -\mathrm{Tr}_1\,\rho(1,2)$ of the qudits give the von Neumann entropies $S(1) = -\mathrm{Tr}\,\rho(1) \ln \rho(1)$ and $S(2) = -\mathrm{Tr}\,\rho(2) \ln \rho(2)$. These entropies can be found as minima of the Shannon entropies

$$H_k(u) = -\sum_{m_k=-j_k}^{j_k} w_k(m_k, u) \ln w_k(m_k, u), \qquad k = 1, 2 \tag{34}$$

with respect to local unitary transforms $u = u_{10} \otimes u_{20}$. The subadditivity condition reads

$$S_{12} \leq S_1 + S_2. \tag{35}$$

In [34], we showed that inequality (35) can be written as

$$S_1 + S_2 \geq H_{12}(u_{10} \otimes u_{20}) \geq S_{12}. \tag{36}$$

The tomographic discord introduced as the difference

$$\mathcal{D} = (S_1 + S_2 - S_{12}) - I(u_{10} \otimes u_{20}) \geq 0 \tag{37}$$

characterizes quantum correlations in the bipartite system of qudits.

Quantum discord for spin-3/2 state

In view of the inequalities discussed above, we can introduce the quantum discord for one qudit state. For example, the qudit state corresponding to $j = 3/2$ with density matrix (22) determines the state tomogram $w(m, u) = \langle m \mid u\rho u^\dagger \mid m\rangle$, $m = -3/2, -1/2, 1/2, 3/2$. The von Neumann entropy of this state is the minimum of the Shannon entropy determined by the tomographic probability distribution; this means that $\min_u \left(-\sum_{m=-3/2}^{3/2} w(m,u) \right) = -\mathrm{Tr}\,\rho \ln \rho$. We rewrite the matrix in two forms: one form corresponds to an arbitrary matrix, and the other one specifies the spin-3/2 state,

$$\rho = \begin{pmatrix} \rho_{11} & \rho_{12} & \rho_{13} & \rho_{14} \\ \rho_{21} & \rho_{22} & \rho_{23} & \rho_{24} \\ \rho_{31} & \rho_{32} & \rho_{33} & \rho_{34} \\ \rho_{41} & \rho_{42} & \rho_{43} & \rho_{44} \end{pmatrix} \equiv \begin{pmatrix} \rho_{-3/2\ -3/2} & \rho_{-3/2\ -1/2} & \rho_{-3/2\ 1/2} & \rho_{-3/2\ 3/2} \\ \rho_{-1/2\ -3/2} & \rho_{-1/2\ -1/2} & \rho_{-1/2\ 1/2} & \rho_{-1/2\ 3/2} \\ \rho_{1/2\ -3/2} & \rho_{1/2\ -1/2} & \rho_{1/2\ 1/2} & \rho_{-1/2\ 3/2} \\ \rho_{3/2\ -3/2} & \rho_{3/2\ -1/2} & \rho_{3/2\ 1/2} & \rho_{3/2\ 3/2} \end{pmatrix}. \tag{38}$$

We consider two density matrices $\rho(1)$ and $\rho(2)$ by taking analogs of tracing with respect to subsystem degrees of freedom; the 2×2-matrices read

$$\rho(1) = \begin{pmatrix} \rho_{11} + \rho_{22} & \rho_{13} + \rho_{24} \\ \rho_{31} + \rho_{42} & \rho_{33} + \rho_{44} \end{pmatrix}, \qquad \rho_2 = \begin{pmatrix} \rho_{11} + \rho_{33} & \rho_{12} + \rho_{34} \\ \rho_{21} + \rho_{43} & \rho_{22} + \rho_{44} \end{pmatrix}. \tag{39}$$

Journal of Physics: Conference Series **538** (2014) 012016 doi:10.1088/1742-6596/538/1/012016

We constructed the 2×2-matrices from arbitrary density 4×4-matrices. In notation relevant to the spin-3/2 state, these density matrices read

$$\begin{aligned}\rho^{(3/2)}(1) &= \begin{pmatrix} \rho_{-3/2\ -3/2}+\rho_{-1/2\ -1/2} & \rho_{-3/2\ 1/2}+\rho_{-1/2\ 3/2} \\ \rho_{-1/2\ -3/2}+\rho_{-3/2\ -1/2} & \rho_{1/2\ 1/2}+\rho_{3/2\ 3/2} \end{pmatrix}, \\ \rho^{(3/2)}(2) &= \begin{pmatrix} \rho_{-3/2\ -3/2}+\rho_{1/2\ -1/2} & \rho_{-3/2\ -1/2}+\rho_{-1/2\ 3/2} \\ \rho_{-1/2\ -3/2}+\rho_{3/2\ -1/2} & \rho_{-1/2\ -1/2}+\rho_{3/2\ 3/2} \end{pmatrix};\end{aligned} \tag{40}$$

they determine the tomograms. While deriving (40), we used the other notation for the matrices $\rho^{(3/2)}(1) = \begin{pmatrix} r_{11} & r_{12} \\ r_{21} & r_{22} \end{pmatrix}$ and $\rho^{(3/2)}(2) = \begin{pmatrix} R_{11} & R_{12} \\ R_{21} & R_{22} \end{pmatrix}$. The two tomograms $w(\alpha, u_1)$ and $w(\beta, u_2)$ are diagonal matrix elements of the matrices $u_1\rho^{(3/2)}(1)u_1^\dagger$ and $u_2\rho^{(3/2)}(2)u_2^\dagger$, they are

$$w(\alpha, u_1) = \langle \alpha \mid u_1\rho^{(3/2)}(1)u_!^\dagger \mid \alpha\rangle, \qquad w(\beta, u_2) = \langle \beta \mid u_2\rho^{(3/2)}(2)u_2^\dagger \mid \beta\rangle. \tag{41}$$

The random variables α and β take two values ± 1. The von Neumann entropies of the "qubit" states with density matrices $\rho^{(3/2)}(1)$ and $\rho^{(3/2)}(2)$ read

$$S^{(3/2)}(1) = -\mathrm{Tr}\,\rho^{(3/2)}(1)\ln\rho^{(3/2)}(1), \qquad S^{(3/2)}(2) = -\mathrm{Tr}\,\rho^{(3/2)}(2)\ln\rho^{(3/2)}(2). \tag{42}$$

We introduce quantum discord as follows:

$$D = \mathrm{Tr}\,(\rho\ln\rho) + S^{(3/2)}(1) + S^{(3/2)}(2) - I(u_{10}\times u_{20}), \tag{43}$$

where the tomographic information $I(u)$ is

$$\begin{aligned} I(u) = &\sum_{m=-3/2}^{3/2} \langle m \mid u\rho u^\dagger \mid m \mid \ln\langle m \mid u\rho u^\dagger \mid m\rangle \\ &-\sum_\alpha \langle \alpha \mid u_1\rho^{3/2}(1)u_1^\dagger \mid \alpha\rangle \ln\Big(\sum_\alpha \langle \alpha \mid u_1\rho^{3/2}(1)u_1^\dagger \mid \alpha\rangle\Big) \\ &-\sum_\beta \langle \alpha \mid u_2\rho^{3/2}(2)u_2^\dagger \mid \beta\rangle \ln\Big(\sum_\beta \langle \beta \mid u_2\rho^{3/2}(2)u_1^\dagger \mid \beta\rangle\Big). \end{aligned} \tag{44}$$

One has the inequality

$$\begin{aligned} &S^{(3/2)}(1) + S^{(3/2)}(2) \geq \\ -&\sum_{m=-3/2}^{3/2} \langle m \mid u_{10}\otimes u_{20}\rho u_{10}^\dagger \otimes u_{20}^\dagger \mid m\rangle \ln\langle m \mid u_{10}\otimes u_{20}\rho u_{10}^\dagger \otimes u_{20}^\dagger \mid m\rangle \\ &\geq -\mathrm{Tr}\,\rho\ln\rho. \end{aligned} \tag{45}$$

This inequality is a new relation for the density matrix of the spin-3/2 state. In view of inequality (45), the quantum discord is nonnegative $D \geq 0$. The relations obtained are valid for an arbitrary density 4×4-matrix.

Example of qutrit state

Now we show new inequalities for qutrit state with the density matrix

$$\rho = \begin{pmatrix} \rho_{1\,1} & \rho_{1\,0} & \rho_{1\,-1} \\ \rho_{0\,1} & \rho_{0\,0} & \rho_{0\,-1} \\ \rho_{-1\,1} & \rho_{-1\,0} & \rho_{-1\,-1} \end{pmatrix}, \tag{46}$$

considering it as a particular 4×4-matrix $\tilde{\rho}$ [33] with matrix elements in the fourth row and column equal to zero. In this case, the 2×2-matrices associated with the 4×4-matrices are

$$\rho^{(1)}(1) = \begin{pmatrix} \rho_{1\,1} + \rho_{0\,0} & \rho_{1\,0} \\ \rho_{0\,1} & \rho_{-1\,-1} \end{pmatrix}, \qquad \rho^{(1)}(2) = \begin{pmatrix} \rho_{1\,1} + \rho_{-1\,-1} & \rho_{1\,0} \\ \rho_{0\,1} & \rho_{0\,0} \end{pmatrix}, \tag{47}$$

where we used index (1) to point out that the matrix is obtained for $j = 1$.

We have von Neumann entropies $S = -\mathrm{Tr}\,\rho \ln \rho = -\mathrm{Tr}\,\tilde{\rho} \ln \tilde{\rho}$, $S_1^{(1)} = -\mathrm{Tr}\,\rho_1^{(1)} \ln \rho_1^{(1)}$, and $S_2^{(1)} = -\mathrm{Tr}\,\rho_2^{(1)} \ln \rho_2^{(1)}$. The inequality $S \leq S_1^{(1)} + S_2^{(1)}$ was obtained for the density matrix of the qutrit state in [33], which is a new analog of the subadditivity condition. If we introduce tomographic probabilities corresponding to the matrix $\tilde{\rho}$ as diagonal elements of the matrix $u\tilde{\rho}u^\dagger$, i.e., $w(\alpha, u) = \langle \alpha \mid u\tilde{\rho}u^\dagger \mid \alpha \rangle$ and find the unitary 2×2-matrices u_{10} and u_{20} diagonalizing the matrices $\rho^{(1)}(1)$ and $\rho^{(1)}(2)$, we arrive at a stronger inequality for the qutrit density matrix, namely,

$$S_1^{(1)} + S_2^{(1)} \geq H(u_{10} \otimes u_{20}) \geq S, \tag{48}$$

where the tomographic entropy $H(u)$ for $u = u_{10} \otimes u_{20}$ is

$$H(u_{10} \otimes u_{20}) = -\sum_\alpha \langle \alpha \mid u_{10} \otimes u_{20}\tilde{\rho}u_{10}^\dagger \otimes u_{20}^\dagger \mid \alpha \rangle \ln \langle \alpha \mid u_{10} \otimes u_{20}\tilde{\rho}u_{10}^\dagger \otimes u_{20}^\dagger \mid \alpha \rangle. \tag{49}$$

The quantum discord for the qutrit state reads

$$D^{(1)} = S_1^{(1)} + S_2^{(1)} - S - I(u_{10} \otimes u_{20}). \tag{50}$$

The tomographic information $I(u_{10} \otimes u_{20})$ is determined by the equality for an arbitrary 4×4-matrix analogously to the case of $j = 3/2$. The discord for qutrit state $D^{(1)}$ is a nonnegative number. This property is a new characteristics of quantum correlations for qutrit states. It is clear that the procedure presented can be used to introduce quantum discord and new inequalities for arbitrary qudit states, as well as for multiqudit states.

6. Conclusions

To conclude, we point out the main results of our study.

We elaborated the method to extend all entropic and information equalities and inequalities known for composite (both classical and quantum) systems to the case of noncomposite and composite systems. The method is based on using an invertible map of N integers onto the pairs, triples, etc. of the integers. A particular case of entropic inequalities like the subadditivity condition and the strong subadditivity condition introduced for a single qudit state was studied. We showed examples of these inequalities for qudit states with $j = 2$ and 3; the subadditivity condition for qutrit state [33] was also presented.

We introduced the conditional probabilities and entropies for noncomposite qudit systems and constructed chain equalities for Shannon entropy and q-entropies for the qudit. Also the notion of quantum discord was given for a single qudit. We pointed out that the physical meaning of the new information and entropic equalities for noncomposite systems, we derived, needs clarification though it seems that the relations, we found, correspond to intrinsic correlations in the system, even if the system does not have the structure of separated subsystems. These aspects of information properties of noncomposite systems will be considered in a future publication.

Acknowledgments

This study was initiated by the memory of our discussions with Prof. Rui Vilela Mendes and Prof. Mary Beth Ruskai during the Madeira Math Encounters XXVI (October 3–11, 2003)

Quantum Information, Control and Computing. This work was supported by the Russian Foundation for Basic Research under Project No. 11-02-00456_a. We are grateful to the Organizers of the XVI Symposium "Symmetries in Science" (Bregenz, Austria, July 21–26, 2013) and especially to Prof. Dieter Schuch and Prof. Michael Ramek for invitation and kind hospitality.

References

[1] Landau L D 1927 *Z. Phys.* **45** 430–41
1965 *The damping problem in wave mechanics* ed Ter Haar D *Collected Papers of L. D. Landau* (New York: Gordon & Breach) pp 8–18
[2] von Neumann J 1927 *Nach. Ges. Wiss. Göttingen* **11** 245
von Neumann J 1932 *Mathematische Grundlagen der Quantenmechanik* (Berlin: Springer)
[3] Schrödinger E 1935 *Naturwissenschaften* **23** 807
[4] Bell J 1964 *Physics* **1** 195
[5] Clauser J F, Horne M A, Shimony A and Holt R 1969 *Phys. Rev. Lett.* **23** 880
[6] Mancini S, Man'ko V I and Tombesi P 1996 *Phys. Lett. A* **213** 1
[7] Man'ko M A, Man'ko V I, Marmo G, Simoni A and Ventriglia F 2013 *Nuovo Cimento C* **36** Ser. 3, p 163
[8] Man'ko O V and Chernega V N 2013 *JETP Lett.* **97** 557
[9] Man'ko M A, Man'ko V I and Vilela Mendes R 2006 *J. Russ. Laser Res.* **27** 507–32
[10] Man'ko M A and Man'ko V I 2012 *Beauty in Physics: Theory and Experiment* ed Bijker R (New York: AIP Conference Proceedings) **1488** pp 110–21
[11] Man'ko M A 2013 *Phys. Scr.* **T153** 014045
[12] Man'ko M A and Man'ko V I 2013 *J. Phys. Conf. Ser.* **442** 012008
[13] Oliver H and Zurek W H 2001 *Phys. Rev. Lett.* **88** 017901
[14] Henderson L and Vedral V 2001 *J. Phys. A: Math. Gen.* **34** 6899
[15] Man'ko V I and Yurkevich A 2013 *J. Russ. Laser Res.* **34** 463
[16] Isar A 2014 *J. Russ. Laser Res.* **35** 62
[17] Shannon C E 1948 *Bell Syst. Tech. J.* **27** 379
[18] Kochen S and Specker E P 1997 *J. Math. Mech.* **17** 59
[19] Klyachko A A, Ali Can H, Binicioğlu S and Shumovsky A S 2008 *Phys. Rev. Lett.* **101** 020403
[20] Vourdas A 2013 *J. Math. Phys.* **54** 082105
[21] Cabello A and Cuñha M T 2013 *Phys. Rev. A* **87** 022126
[22] Man'ko V I and Strakhov A 2013 *J. Russ. Laser Res.* **34** 267
[23] A. Rényi 1970 *Probability Theory* (Amsterdam: North-Holland)
[24] Tsallis C 2001 ed Abe S and Okamoto Y *Nonextensive Statistical Mechanics and Its Applications Lecture Notes in Physics* (Berlin: Springer) **560** pp 3–98
[25] Lieb E H and Ruskai M B 1973 *J. Math. Phys.* **14** 1938–41
[26] Ruskai M B 2004 [arXiv: quant-ph/0404126 v4]
[27] Carlen E A and Lieb E H 2008 *Lett. Math. Phys.* **83** 107–26
[28] Kim I H 2012 [arXiv:1210.5190]
[29] Ohya M and Petz D 2004 *Quantum Entropy and Its Use* (Heidelberg: Springer) 2nd ed.
[30] Ruskai M B 2007 Rep. Math. Phys. **60** 1
[31] Frank L R and Lieb E H 2012 [arXiv:1204.0825v1 quant-ph]
[32] Rastegin A E 2012 [arXiv:1210.6742 quant-ph]
[33] Chernega V N and Man'ko O V 2013 *J. Russ. Laser Res.* **34** 383
[34] Man'ko M A and Man'ko V I 2013 *J. Russ. Laser Res.* **34** 203–18
[35] Man'ko M A and Man'ko V I 2014 *Phys. Scr.* **T160** 014030
[36] Man'ko M A and Man'ko V I 2014 *J. Russ. Laser Res.* **35** iss. 3 [arXiv:1404.3650 quant-ph]
[37] Chernega V N, Man'ko O V and Man'ko V I 2014 *J. Russ. Laser Res.* **35** 383
[38] Man'ko V I and Markovich L A 2014 [arXiv:1404.1454 quant-ph]
[39] Dodonov V V and Man'ko V I 1997 *Phys. Lett. A* **229** 335
[40] Man'ko V I and Man'ko O V 1997 *J. Exp. Theor. Phys.* **85** 430
[41] Man'ko M A and Man'ko V I 2011 *Found. Phys.* **41** 330

Journal of Physics: Conference Series **538** (2014) 012017 doi:10.1088/1742-6596/538/1/012017

Classical integrals as quantum mechanical differential operators: a comparison with the symmetries of the Schrödinger Equation

M. C. Nucci and P. G. L. Leach [1]

Departimento di Matematica e Informatica, Universit'a di Perugia and Department of Mathematics and Statistics, University of Cyprus

E-mail: nucci@unipg.it

E-mail: leach@ucy.ac.cy

Abstract. Superintegrable systems are characterised by the possession of many symmetries and integrals. We use the simple harmonic oscillator as an example and examine the relationship between the Noetherian integrals of a given Lagrangian as quantum operators and the Lie symmetries of the corresponding Schrödinger Equation.

1. Introduction

We use the simple harmonic oscillator as the vehicle for our discussion and demonstration. It has the standard Lagrangian[2]

$$\tfrac{1}{2}\left(\dot{x}^2 - x^2\right). \tag{1}$$

The Noether point symmetries and their associated integrals are calculated according to the formulae [2]

$$\begin{aligned} \dot{f}(t,\,x) = \tau(t,\,x)\frac{\partial L(t,\,x,\,\dot{x})}{\partial t} &+ \eta(t,\,x)\frac{\partial L(t,\,x,\,\dot{x})}{\partial x} \\ +\left(\dot{\eta} - \dot{x}\dot{\tau}\right)\frac{\partial L(t,\,x,\,\dot{x})}{\partial \dot{x}} &+ \dot{\tau}L(t,\,x,\,\dot{x}) \end{aligned} \tag{2}$$

and

$$\begin{aligned} I = f(t,\,x) - \Big[&\tau(t,\,x)L(t,\,x,\,\dot{x}) \\ &+ \left(\eta(t,\,x) - \dot{x}\tau(t,\,x)\right)\frac{\partial L(t,\,x,\,\dot{x})}{\partial \dot{x}}\Big], \end{aligned} \tag{3}$$

where the point symmetry has the form $\Gamma = \tau(t,\,x)\partial_t + \eta(t,\,x)\partial_x$ and the function, $f(t,\,x)$, is the contribution consequent upon the possibility of a change in the endpoints when the variation of the Action Integral under the infinitesimal transformation generated by Γ is taken.

[1] permanent address: School of Mathematics, Statistics and Computer Science, University of KwaZulu-Natal, Private Bag X54001 Durban, Republic of South Africa 4000

[2] It is well known that there are many Lagrangians for the simple harmonic oscillator. For a sampling see [1]. The question to be addressed here is the relationship between the Noetherian integrals of a given Lagrangian as quantum operators and the corresponding Lie symmetries of the Schrödinger Equation.

The Noether point symmetries and associated integrals for the Lagrangian (1) are

$$\begin{aligned}
\Gamma_1 &= \partial_t & I_1 &= \tfrac{1}{2}\left(\dot{x}^2 + x^2\right) \\
\Gamma_2 &= \mathrm{e}^{2it}\left(\partial_t + ix\partial_x\right) & I_2 &= \tfrac{1}{2}\mathrm{e}^{2it}\left(\dot{x}^2 - 2i\dot{x}x - x^2\right) \\
\Gamma_3 &= \mathrm{e}^{-2it}\left(\partial_t - ix\partial_x\right) & I_3 &= \tfrac{1}{2}\mathrm{e}^{-2it}\left(\dot{x}^2 + 2i\dot{x}x - x^2\right) \\
\Gamma_4 &= \mathrm{e}^{it}\partial_x & I_4 &= \mathrm{e}^{it}\left(ix - \dot{x}\right) \\
\Gamma_5 &= \mathrm{e}^{-it}\partial_x & I_5 &= -\mathrm{e}^{-it}\left(ix + \dot{x}\right) \\
\Gamma_6 &= 0 & I_6 &= 1.
\end{aligned}$$

The Hamiltonian corresponding to the Lagrangian (1) is

$$H = \tfrac{1}{2}\left(p^2 + x^2\right), \tag{4}$$

where the conjugate momentum is $p = \partial L/\partial \dot{x} = \dot{x}$. The Schrödinger Equation for (4) is

$$2i\frac{\partial u(t,x)}{\partial t} = -\frac{\partial^2 u(t,x)}{\partial x^2} + x^2 u(t,x) \tag{5}$$

and it has the Lie point symmetries

$$\begin{aligned}
\Sigma_1 &= \partial_t, \\
\Sigma_2 &= \mathrm{e}^{2it}\left[\partial_t + ix\partial_x - i\left(\tfrac{1}{2} + x^2\right)u\partial_u\right], \\
\Sigma_3 &= \mathrm{e}^{-2it}\left[\partial_t - ix\partial_x + i\left(\tfrac{1}{2} - x^2\right)u\partial_u\right], \\
\Sigma_4 &= \mathrm{e}^{it}\left(\partial_x - xu\partial_u\right), \\
\Sigma_5 &= \mathrm{e}^{-it}\left(\partial_x + xu\partial_u\right), \\
\Sigma_6 &= u\partial_u \quad \text{and} \\
\Sigma_\infty &= \phi(t,x)\partial_u,
\end{aligned}$$

where $\phi(t,x)$ is any solution of (5).

The Schrödinger Equation for the physical problem comes with the condition that $u(t,x)$ vanish at spatial infinity. There is no such condition on $\phi(t,x)$ unless the problem is stated with the boundary condition being part of the problem.

The Lie point symmetries, Σ_i, $i = 1, 6$, are listed in parallel to the Noether point symmetries and integrals to make more readily a comparison of the properties reported below. In particular Σ_6 corresponds to the trivial integral I_6. We note that the coefficient functions for ∂_t and ∂_x are the same for the finite sets of symmetries.

The solution of the Schrödinger Equation is well known and is given by

$$u_n(t,x) = N_n \exp\left[-\left(n + \tfrac{1}{2}\right)it - \tfrac{1}{2}x^2\right] H_n(x), \tag{6}$$

where N_n is the normalisation constant for the nth-eigenstate and $H_n(x)$ is the nth Hermite polynomial which is the acceptable solution of the Hermite equation [3]

$$v''(x) - 2xv'(x) + 2nv(x) = 0. \tag{7}$$

2. Solution of the Schrödinger Equation using the Lie point symmetries of the equation

It is well known that the solution of (5) consistent with the boundary conditions can be obtained by means of separation of variables into time and space components and the use of the ladder operators introduced by Dirac [4], namely

$$a = D + x \quad \text{and} \quad a^\dagger = D - x, \tag{8}$$

where $a^\dagger$ is known as the creation operator and a is known as the annihilation operator since the former 'creates' states and the latter 'annihilates' states. What is obviously less well-known is that the Dirac operators are autonomous versions of two of the Lie point symmetries of (5) [5].

The Lie (point) symmetries of a differential equation constitute a Lie algebra under the operation of taking the Lie Bracket. In particular the Lie Brackets of the symmetries $\Sigma_i,\ i = 1, 6$ with Σ_∞ of the Schrödinger Equation, (5), produce another solution symmetry, *ie*,

$$[\Sigma_i,\, \Sigma_\infty]_{LB} = \tilde{\Sigma}_\infty, \tag{9}$$

where $\tilde{\Sigma}_\infty$ may be a constant multiple of Σ_∞ or a different solution symmetry[3]. Note that a solution symmetry (in this context) is a symmetry of the form $\phi(t,\, x)\partial_u$, where $\phi(t,\, x)$ is a solution of the Schrödinger Equation under consideration.

The Lie point symmetries of (5) all have the form

$$\Sigma = T(t)\partial_t + \Xi(t, x)\partial_x + \varepsilon(t, x)u\partial_u$$

apart from Σ_∞. The Lie Bracket of these two operators is [6]

$$[\Sigma,\, \Sigma_\infty]_{LB} = \left(T\phi_t + \Xi\phi_x - \varepsilon\phi\right)\partial_u. \tag{10}$$

The right side of (10) is a solution symmetry. One possibility is that it is trivial, *ie* zero. Then the expression on the right side of (10) is a linear first-order partial differential equation for $\phi(t,\, x)$. Consider Σ_5. The invariants of the first-order partial differential equation (10) can be calculated from the associated Lagrange's system,

$$\frac{\dot{t}}{0} = \frac{\dot{x}}{1} = \frac{\dot{u}}{-xu},$$

from which the common exponential term has been cancelled. The invariants are t and $u \exp\left[\frac{1}{2}x^2\right]$.

We set

$$u(t,\, x) = \exp\left[-\tfrac{1}{2}x^2\right] f(t)$$

and substitute it into (5) to obtain the first-order equation

$$2i\dot{f}(t) - f(t) = 0$$

for $f(t)$ with solution

$$f(t) = k \exp\left[-\tfrac{1}{2}it\right]$$

so that

$$u(t,\, x) = \exp\left[-\tfrac{1}{2}it - \tfrac{1}{2}x^2\right] \tag{11}$$

[3] A constant multiple, apart from zero, is of no consequence.

up to a constant of normalisation. The solution, (11), has no node and represents the ground-state solution. We denote it by u_0.

It is not necessary for the right side of (10) to be zero. Suppose it is some other solution, Φ. Then

$$T\phi_t + \Xi\phi_x - \varepsilon\phi = \Phi.$$

The action of Σ_5 on (11) is found by using the symmetry, Σ_∞, and the property that solutions are mapped into solutions. The Lie Bracket of the two symmetries, Σ_5 and Σ_∞, is

$$\left[e^{-it}\left(\partial_x + xu\partial_u\right),\, \phi(t,\,x)\partial_u\right]_{LB} = e^{-it}\left(\partial_x - x\right)\phi(t,\,x)\partial_u \tag{12}$$

so that, if we take the groundstate solution, u_0, in (11), the next eigenfunction is obtained from (12) when (11) is substituted, *ie*, the solution is mapped to the trivial solution. Consequently Σ_5 is the time-dependent creation operator and Σ_4 is the time-dependent annihilation operator. These Lie point symmetries of (5) are the origins of Dirac's famous operators.

Since the solution in (11) has no nodes and is annihilated by the action of Σ_4, it is the groundstate solution and we denote it by $u_0(t,\,x)$. Higher states may be obtained by the successive action of Σ_5. Thus

$$\begin{aligned} u_1(t,\,x) &= \left[\Sigma_5,\, u_0(t,\,x)\partial_u\right]_{LB} \\ &= \left[e^{-it}\left(\partial_x + xu\partial_u\right),\, \exp\left[-\tfrac{1}{2}it - \tfrac{1}{2}x^2\right]\partial_u\right]_{LB} \\ &= -2x\exp\left[-\tfrac{3}{2}it - \tfrac{1}{2}x^2\right]\partial_u, \\ u_2(t,\,x) &= \left[\Sigma_5,\, u_1(t,\,x)\partial_u\right]_{LB} \\ &= \left[e^{-it}\left(\partial_x + xu\partial_u\right),\, \left(-2x\exp\left[-\tfrac{3}{2}it - \tfrac{1}{2}x^2\right]\right)\partial_u\right]_{LB} \\ &= \left(4x^2 - 2\right)\exp\left[-\tfrac{5}{2}it - \tfrac{1}{2}x^2\right]\partial_u \end{aligned}$$

etc.

In a similar manner we find that Σ_1 is an eigenvalue operator and, if multiplied by i, is the energy operator; Σ_2 is a double annihilation operator and Σ_3 is a double creation operator.

3. Noetherian Integrals as differential operators

The Noetherian Integrals may be written as differential operators in the quantum scenario. They become

$$\begin{aligned} I_1 &= \tfrac{1}{2}\left(-\partial_x^2 + x^2\right), \\ I_2 &= \tfrac{1}{2}e^{2it}\left(-\partial_x^2 - (2x\partial_x + 1) - x^2\right), \\ I_3 &= \tfrac{1}{2}e^{-2it}\left(-\partial_x^2 + (2x\partial_x + 1) - x^2\right) \\ I_4 &= ie^{it}\left(x + \partial_x\right) \quad \text{and} \\ I_5 &= ie^{-it}\left(-x + \partial_x\right). \end{aligned}$$

(I_6 is an identity operator and we do not consider it further.)

The actions of each of these operators on the nth-eigenstate, given in (6), are given by

$$\begin{aligned} I_1u_n(t,\,x) &= \left(n + \tfrac{1}{2}\right)N_n\exp\left[-i\left(n + \tfrac{1}{2}\right)t - \tfrac{1}{2}x^2\right]H_n(x), \\ I_2u_n(t,\,x) &= -2n(n-1)N_n\exp\left[-i\left(n - \tfrac{3}{2}\right)t - \tfrac{1}{2}x^2\right]H_{n-2}(x), \\ I_3u_n(t,\,x) &= -N_n\exp\left[-i\left(n + \tfrac{5}{2}\right)t - \tfrac{1}{2}x^2\right]H_{n+2}(x), \\ I_4u_n(t,\,x) &= 2inN_n\exp\left[-i\left(n - \tfrac{1}{2}\right)t - \tfrac{1}{2}x^2\right]H_{n-1}(x), \\ I_5u_n(t,\,x) &= -iN_n\exp\left[-i\left(n + \tfrac{3}{2}\right)t - \tfrac{1}{2}x^2\right]H_{n+1}(x). \end{aligned}$$

In a less informative but conceptually clearer form these actions can be written as

$$\begin{aligned} I_1u_n(t,\,x) &= \left(n+\tfrac{1}{2}\right)u_n(t,\,x),\\ I_2u_n(t,\,x) &= -2n(n-1)N_nu_{n-2}(t,\,x),\\ I_3u_n(t,\,x) &= -\tfrac{1}{2}N_nu_{n+2}(t,\,x),\\ I_4u_n(t,\,x) &= 2inN_nu_{n-1}(t,\,x),\\ I_5u_n(t,\,x) &= -iN_nu_{n+1}(t,\,x). \end{aligned}$$

4. Variations on the Schrödinger Equation

We started from the standard Lagrangian, (1), and did all of the usual tricks to obtain the Schrödinger Equation, (5), its solutions and associated symmetries. We also have looked in the previous section at the action of the classical integrals when treated as operators on the solution. Now we examine what happens if we use these operators in a type of Schrödinger Equation. We should emphasise that one is looking at this from the point of view of symmetry and mathematics, not physics. Obviously there is no need to consider I_1 as that gives the standard quantal treatment.

The Schrödinger Equation corresponding to I_2 is

$$i\frac{\partial u}{\partial t} = \tfrac{1}{2}e^{2it}\left(-\partial_x^2 - 2\left(x\partial_x + 1\right) - x^2\right)u \tag{13}$$

which has the Lie point symmetries

$$\begin{aligned} \Gamma_1 &= \partial_t + ix\partial_x + \tfrac{1}{2}i\left(e^{2it} - 2x^2\right)u\partial_u\\ \Gamma_2 &= e^{2it}\partial_t + 2ixe^{2it}\partial_x\\ &+ i\left(-ie^{2it} + \tfrac{1}{2}e^{4it} - 2x^2 - 2e^{2it}x^2\right)u\partial_u\\ \Gamma_3 &= e^{-2it}\partial_t\\ \Gamma_4 &= \partial_x - xu\partial_u\\ \Gamma_5 &= e^{2it}\partial_x - \left(2 + e^{2it}\right)xu\partial_u \quad \text{and}\\ \Gamma_6 &= u\partial_u \end{aligned}$$

in addition to the usual infinite-dimensional abelian subalgebra of solution symmetries. The algebra of the finite subset is $sl(2,\,R) \oplus W_3$, *ie* the same algebra as in the case of the usual Schrödinger Equation, (5).

In the case of the Schrödinger Equation corresponding to I_3, namely

$$i\frac{\partial u}{\partial t} = \tfrac{1}{2}e^{-2it}\left(-\partial_x^2 + 2\left(x\partial_x + 1\right) - x^2\right)u, \tag{14}$$

a similar set of symmetries is obtained. Precisely the finite subalgebra comprises

$$\begin{aligned} \Delta_1 &= \partial_t - ix\partial_x - \tfrac{1}{2}i\left(2x^2 - e^{-2it}x\right)u\partial_u\\ \Delta_2 &= e^{2it}\partial_t\\ \Delta_3 &= \partial_t e^{-2it} - 2ie^{-2it}x\partial_x\\ &+ i\left(e^{-2it} - \tfrac{1}{2}ie^{-4it} - 2ix^2 - 2ie^{-2it}x^2\right)u\partial_u\\ \Delta_4 &= \partial_x + xu\partial_u\\ \Delta_5 &= e^{-2it}\partial_x + \left(2 + e^{-2it}\right)xu\partial_u \quad \text{and}\\ \Delta_6 &= u\partial_u. \end{aligned}$$

There is not much point in discussing the equations corresponding to I_4 and I_5 as they are linear first-order partial differential equations and have an infinite number of symmetries. The equations can be solved by means of the method of characteristics and one obtains

$$u = \exp\left[\tfrac{1}{2}x^2\right] f(x - it) \quad \text{and}$$
$$u = \exp\left[2it + \tfrac{1}{2}x^2\right] g(\exp[2it] - 2x),$$

and

$$u = \exp\left[-\tfrac{1}{2}x^2\right] f(x - it) \quad \text{and}$$
$$u = \exp\left[-2it - x\right] g\left(2x + \exp[-2it]\right),$$

respectively, for the equations

$$i\tfrac{\partial u}{\partial t} = \tfrac{\partial u}{\partial x} - xu \quad \text{and}$$
$$i\tfrac{\partial u}{\partial t} = e^{2it}\tfrac{\partial u}{\partial x} - \left(2 + e^{2it}\right) xu,$$

and

$$i\tfrac{\partial u}{\partial t} = \tfrac{\partial u}{\partial x} + xu \quad \text{and}$$
$$i\tfrac{\partial u}{\partial t} = e^{-2it}\tfrac{\partial u}{\partial x} + \left(2 + e^{-2it}\right) xu,$$

which are the equations corresponding to I_4 and I_5.

It is not at all obvious how to interpret these solutions in terms of the desired behaviour of the solution, $u(t, x)$, at Infinity. Personally, even for the solutions which have a suitable exponential terms, we cannot see how one could possibly select a solution for the so far arbitrary functions which would give anything like proper behaviour.

5. Discussion

Conventional Quantum Mechanics uses (5) derived from the Lagrangian, (1), via the quantisation of the corresponding Hamiltonian. Behind all of this there is a considerable body of theory applying to Classical Mechanics. On one hand, a matter not treated here, the Lagrangian is far from unique. All one needs to do is to glance at [1] to find a modest sampling. The critical point is that from the Lagrangian, whatever it is, one must be able to obtain the Newtonian Equation of Motion. The Theories of Lagrange and Hamilton are really mathematical constructs whereas the Equation of Newton is based upon a Law of Physics. Consequently one must be cautious in how one deals with the process of quantisation.

There is a question about the relevance of the solutions of the Schrödinger Equations obtained from the use of I_2 and I_3 since they do not seem to have any connection with the physically accepted solution obtained using I_1. Nevertheless their Schrödinger Equations have the same algebraic structure as that of the equation corresponding to I_1. Consequently one should expect a point transformation to relate the three equations pairwise. This is a task still to be performed!

The question of the application of these transformations to higher-dimensional systems is raised in [7]. (See also [8] in which the expression "quantum Arnold transformation" is introduced to describe this class of transformation.) It was in such an application in 1976 that it was realised for the classical case that it was simply a point transformation of one Hamiltonian to another. Admittedly the problem under consideration was the three-dimensional time-dependent linear oscillator, which could be considered to be rather special. One recalls the Jauch-Hill-Fradkin Tensor of the corresponding two- and three-dimensional simple harmonic oscillator and its interpretation in terms of the derivation of the orbital properties of the system [9,10]. Similar properties were found for the corresponding time-dependent tensor [11] . While it would seem that the solutions of the Schrödinger Equations obtained from the use of I_2 and I_3 appear to serve no purpose, it may be that there is some way of relating the quantal operators to the properties of the quantal oscillator in higher dimensions.

In considering multidimensional systems it is equally important that the algebraic properties of the system being investigated and the target system be reconcilable. If one is considering point transformations, then the number of point symmetries of both systems needs to be the same and not just the numbers for there needs to be algebraic consistency. This is the only type of symmetry considered in this paper, but there are applications in which the transformations may be more generalised and so one would not expect the conservation of point symmetry. A case in point is the reduction of the Kepler Problem to a simple harmonic oscillator plus a conservation law [12].

A final word of caution appears to be necessary. When one is using a transformation from one space to another, it is advised to make the transformation one-to-one. This then preserves the quantisation properties such as a discrete spectrum.

Acknowledgements

PGLL thanks the University of Cyprus for its kind hospitality while this work was being undertaken and the University of KwaZulu-Natal and the National Research Foundation of the Republic of South Africa is for their continued support. The opinions expressed in this paper should not be construed as being those of either of the latter two institutions.

We thank the referee for a very careful reading of the text and the making of some useful suggestions to clarify the meaning.

References

[1] Nucci M C and Leach P G L 2007 Lagrangians galore *Journal of Mathematical Physics* **48** 123510 (1–16)
[2] Noether E 1918 Invariante Variationsprobleme *Königliche Gesellschaft der Wissenschaften Göttingen Nachrichten Mathematik-physik Klasse* **2** 235–67
[3] Abramowitz M and Stegun I A 1965 *Handbook of Mathematical Functions* (New York: Dover) p 781
[4] Dirac P A M 1932 *The Principles of Quantum Mechanics*(Cambridge: Clarendon Press)
[5] Lemmer R L and Leach P G L 1999/1420 A classical viewpoint on quantum chaos *Arab Journal of Mathematical Sciences* **5** 1–17
[6] Andriopoulos K and Leach P G L 2005 Some group theoretical aspects of nonlinear quantal oscillators *Journal of Nonlinear Mathematical Physics* **12, S-1** 32–43
[7] López-Ruiz F F and Guerrero J 2014 Generalizations of the Ermakov system through the Quantum Arnold Transformation (submitted to these proceedings)
[8] Guerrero J and López-Ruiz F F 2013 The quantum Arnold transformation and the ErmakovPinney equation *Physica Scripta* **87** 038105
[9] Jauch J M and Hill E L 1940 On the problem of degeneracy in quantum mechanics *Physical Review* **57** 641–45
[10] Fradkin D M 1965 Three-dimensional isotropic oscillator and SU_3 *American Journal of Physics* **33** 207–11
[11] Günther N J and Leach P G L 1977 Generalized invariants for the time-dependent harmonic oscillator *Journal of Mathematical Physics* **18** 572–6
[12] Nucci M C and Leach P G L 2001 The harmony in the Kepler and related problems *Journal of Mathematical Physics* **42** 746–64

From probabilities to wave functions: A derivation of the geometric formulation of quantum theory from information geometry

Marcel Reginatto

Physikalisch-Technische Bundesanstalt, Bundesallee 100, 38116 Braunschweig, Germany

E-mail: marcel.reginatto@ptb.de

Abstract.
It is shown that the geometry of quantum theory can be derived from geometrical structure that may be considered more fundamental. The basic elements of this reconstruction of quantum theory are the natural metric on the space of probabilities (information geometry), the description of dynamics using a Hamiltonian formalism (symplectic geometry), and requirements of consistency (Kähler geometry). The theory that results is standard quantum mechanics, but in a geometrical formulation that includes also a particular case of a family of nonlinear gauge transformations introduced by Doebner and Goldin. The analysis is carried out for the case of discrete quantum mechanics. The work presented here relies heavily on, and extends, previous work done in collaboration with M. J. W. Hall.

1. Introduction

It is now known that quantum mechanics has a rich geometrical structure which allows for a geometric formulation of the theory. The geometric approach was introduced by Kibble [1] and later further developed by a number of other authors. One successful strategy has been to start from a known formulation of quantum mechanics and to identify geometrical features that can be used for the reformulation of the theory. This paper inverts this procedure: the aim is to *derive* the geometry of quantum theory from geometrical structure that may be considered more fundamental, and to examine the assumptions that are needed to do this. The work presented here relies heavily on, and extends, previous work of Reginatto and Hall [2, 3]. The basic elements of this geometrical reconstruction of quantum theory are the natural metric on the space of probabilities (information geometry), the description of dynamics using a Hamiltonian formalism (symplectic geometry), and requirements of consistency (Kähler geometry). One may associate a Hilbert space with the Kähler space, which leads to the standard version of quantum theory. The analysis is carried out for the case of discrete quantum mechanics; a similar approach has been carried out previously for continuous systems [2].

2. Information geometry

Consider a system with a discrete configuration space. If the configuration of the system is subject to uncertainty, the state of the system will be described by a probability $P = (P^1, ..., P^n)$ where n is the number of states. The probability that the system is in state i is P^i, where P^i

satisfies $P^i \geq 0$ and $\sum_i P^i = 1$. The space of probabilities can be visualized as points in the simplex $S_{n-1} = \{P \in \mathbf{R}_n^+ : \sum P^i = 1\}$.

There is a natural line element in this space, given by

$$ds^2 = G_{ij}\, dP^i\, dP^j = \frac{\alpha}{2P^i}\, \delta_{ij}\, dP^i\, dP^j, \tag{1}$$

where α is a constant. The value of α can not be determined *a priori*; it is usually set to $\frac{1}{2}$, but for the purposes of this paper it will be convenient not to follow this convention. Instead, α will be treated as a parameter. The line element of Eq. (1) defines a distance on the space of probabilities. This distance seems to have been introduced into statistics by Bhattacharyya [4, 5] as a way of providing a measure of divergence for multinomial probabilities [6]. Wootters has shown that this distance can be motivated using arguments based on the concept of the distinguishability of states and refers to this distance as the "statistical distance" [7].

The metric G_{ij} that appears in Eq. (1) is known as the *information metric*,

$$G_{ij} = \frac{\alpha}{2P^i}\, \delta_{ij}. \tag{2}$$

It has been shown by Čencov that the information metric is the only metric that is invariant under a family of probabilistically natural mappings known as *congruent embeddings by a Markov mapping* [8]. A simpler proof, which also makes use of these mappings, was given later by Campbell [9]. It will be useful to give a brief description of these mappings because their generalization (see below) play an important role in the derivation of the main result of this paper.

I follow the presentation of Ref. [9]. A Markov mapping is a particular type of linear transformation between a simplex S_{m-1} and a simplex S_{n-1} (with $m \leq n$) which preserves the probability; i.e., $\sum_{a=1}^{m} P^a = \sum_{b=1}^{n} \tilde{P}^b = 1$. For $m = n$, the mapping is just a permutation of the components P^i, but for $m < n$, the mapping relates spaces of different dimensions. A Markov mapping may be constructed in the following way. Let $A = \{A_1, ..., A_m\}$ be a partition of the set $\{1, 2, ..., n\}$ into disjoint sets. Associate a probability vector $Q_{(a)} = (q_{a1}, ..., q_{an})$ to each of the A_a, where the q_{ab} satisfy

$$q_{ab} = 0 \texttt{ if } b \notin A_a, \qquad q_{ab} > 0 \texttt{ if } b \in A_a, \qquad \sum_{b=1}^{n} q_{ab} = 1. \tag{3}$$

The probability vector $Q_{(a)}$ is therefore concentrated on A_a. Note that the $m \times n$ matrix $\mathbf{Q}$ with elements q_{ab} has the following properties: Each column has precisely one non-zero element and each row sums to one.

Define mappings between $f : S_{m-1} \to S_{n-1}$ and $g : S_{n-1} \to S_{m-1}$ by

$$\begin{aligned} \tilde{P}^b &= \sum_{a=1}^{m} P^a q_{ab}, \\ P^a &= \sum_{b \in A_a} \tilde{P}^b, \qquad a \in \{1, 2, ..., m\}. \end{aligned} \tag{4}$$

Following Čencov, the mapping f is known as a congruent embedding of S_{m-1} in S_{n-1} by a Markov mapping. The mapping g, which is also defined in terms of the partition A, has the property that the composition $g \circ f$ is the identity map on S_{m-1}.

I now consider a simple example of a Markov mapping for the case where $n = m + 1$. Set

$$\mathbf{Q} = \begin{pmatrix} 1 & 0 & \cdots & 0 & 0 & 0 \\ 0 & 1 & \cdots & 0 & 0 & 0 \\ \vdots & \vdots & \ddots & \vdots & \vdots & \vdots \\ 0 & 0 & \cdots & 1 & 0 & 0 \\ 0 & 0 & \cdots & 0 & k & (1-k) \end{pmatrix}, \tag{5}$$

with $0 < k < 1$. Then

$$P = (P^1, ..., P^m) \to \tilde{P} = (\tilde{P}^1, ..., \tilde{P}^m, \tilde{P}^{m+1}) := (P^1, ..., kP^m, (1-k)P^m). \tag{6}$$

A vector in the tangent space of the simplex transforms in a similar way,

$$V = (V^1, ..., V^m) \to \tilde{V} = (\tilde{V}^1, ..., \tilde{V}^m, \tilde{V}^{m+1}) := (V^1, ..., kV^m, (1-k)V^m). \tag{7}$$

To prove uniqueness of the information metric, Čencov [8] and Campbell [9] show that the only metric that preserves the inner product $< A, B >$ of two tangent vectors A, B under a Markov mapping is precisely the information metric. It is straightforward to show that the metric has this property. To see this for the simple example discussed above, simply compute

$$\begin{aligned} < \tilde{A}, \tilde{B} > &= \sum_{i=1}^{m+1} \left\{ \frac{\tilde{A}^i \tilde{B}^i}{\tilde{P}^i} \right\} = \sum_{i=1}^{m-1} \left\{ \frac{A^i B^i}{P^i} \right\} + \frac{kA^m \, kB^m}{kP^m} + \frac{(1-k)A^m \, (k-1)B^m}{(1-k)P^m} \\ &= \sum_{i=1}^{m} \left\{ \frac{A^i B^i}{P^i} \right\} =< A, B > . \end{aligned} \tag{8}$$

A proof of uniqueness is, as one would expect, not straightforward. The monograph of Caticha [10], which includes a useful review of information geometry for both the discrete and continuous cases, provides a nice, accessible proof of uniqueness which follows the presentation of Campbell.

3. Dynamics and symplectic geometry

I now set the probabilities in motion. Assume that the time evolution of the P^i is generated by an action principle and write the equations of motion using a Hamiltonian formalism. To do this, introduce additional coordinates S^i which are canonically conjugate to the P^i and a corresponding Poisson bracket for any two functions $F(P, S)$ and $G(P, S)$,

$$\{F, G\} = \sum_i \left(\frac{\partial F}{\partial P^i} \frac{\partial G}{\partial S^i} - \frac{\partial F}{\partial S^i} \frac{\partial G}{\partial P^i} \right). \tag{9}$$

As is well known, the Poisson bracket can be rewritten geometrically as

$$\{F, G\} = (\partial F / \partial P \, , \, \partial F / \partial S) \; \Omega \begin{pmatrix} \partial G / \partial P \\ \partial G / \partial S \end{pmatrix}, \tag{10}$$

where Ω is the corresponding symplectic form, given in this case by

$$\Omega = \begin{pmatrix} 0 & \mathbf{1} \\ -\mathbf{1} & 0 \end{pmatrix}, \tag{11}$$

where $\mathbf{1}$ is the $n \times n$ unit matrix. We thus have a symplectic structure and a corresponding *symplectic geometry*. The equations of motion for P^i and S^i are given by $\dot{P}^i = \{P^i, H\}$, $\dot{S}^i = \{S^i, H\}$ where H is the Hamiltonian that generates time translations.

Notice that the $2n$-dimensional phase space with coordinates P^i and S^i has a richer structure than the n-dimensional space of probabilities P^i; in particular, one may introduce the notion of *observables*, which are functions $O(P, S)$ of the coordinates, together with an algebra of observables defined in terms of the Poisson brackets of these functions [11]. However, not every function $O(P, S)$ qualifies as an observable because observables are also generators of infinitesimal transformations and these transformations must satisfy certain requirements. For example, the infinitesimal canonical transformation generated by any observable O must preserve the normalization and positivity of P. This implies

$$O(P, S + \chi) = O(P, S), \qquad \partial O/\partial S^i = 0 \text{ if } P^i = 0. \tag{12}$$

Note that the first condition implies gauge invariance of the theory under $S^i \to S^i + \chi$, where χ is a constant [11].

4. Kähler geometry

The $2n$-dimensional phase space with coordinates P^i and S^i is an extension of the n-dimensional space of probabilities P^i. It is natural to ask the following question: Is it possible to extend the metric G_{ij} in Eq. (2), which is only defined on the n-dimensional space of probabilities P^i, to the full $2n$-dimensional phase space of the P^i and S^i? This can be done, but certain conditions which ensure the compatibility of the metric and symplectic structures have to be satisfied. These conditions are equivalent to requiring that the space have a Kähler structure (see the Appendix of Ref. [2] for a proof).

A Kähler structure brings together metric, symplectic and complex structures in a harmonious way. To define such a space, introduce a complex structure $J^a_{\ b}$ and impose the following conditions [12],

$$\Omega_{ab} = g_{ac} J^c_{\ b}\,, \tag{13}$$
$$J^a_{\ c} g_{ab} J^b_{\ d} = g_{cd}\,, \tag{14}$$
$$J^a_{\ b} J^b_{\ c} = -\delta^a_{\ c}\,. \tag{15}$$

Eq. (13) is a compatibility equation between the symplectic structure Ω_{ab} and the metric g_{ab}, Eq. (14) is the condition that the metric should be Hermitian, and Eq. (15) is the condition that $J^a_{\ b}$ should be a complex structure.

The metric over the n-dimensional space of probabilities is the information metric, Eq. (2). Then, metric over the full space will be of the form

$$g_{ab} = \begin{pmatrix} \mathbf{G} & \mathbf{E} \\ \mathbf{E}^T & \mathbf{F} \end{pmatrix}, \tag{16}$$

where $\mathbf{G} = \mathtt{diag}(\frac{\alpha}{2P^i})$, and $\mathbf{E}$ and $\mathbf{F}$ are $n \times n$ matrices that need to be determined.

A simple matrix calculation using the Kähler conditions and the expression for Ω_{ab}, Eq. (11), leads to general forms for the metric g_{ab} and the complex structure $J^a_{\ b}$,

$$g_{ab} = \begin{pmatrix} \mathbf{G} & \mathbf{A}^T \\ \mathbf{A} & (\mathbf{1} + \mathbf{A}^2)\mathbf{G}^{-1} \end{pmatrix}, \qquad J^a_{\ b} = \begin{pmatrix} \mathbf{A} & (\mathbf{1} + \mathbf{A}^2)\mathbf{G}^{-1} \\ -\mathbf{G} & -\mathbf{G}\mathbf{A}\mathbf{G}^{-1} \end{pmatrix}. \tag{17}$$

where the $n \times n$ matrix $\mathbf{A}$ satisfies $\mathbf{G}\mathbf{A}\mathbf{G}^{-1} = \mathbf{A}^T$ but is otherwise arbitrary.

5. On the geometry of the space of probabilities in motion

At this point, it is useful to make some general remarks regarding the approach and the results obtained so far. The starting point is a system with a finite number of states (i.e., a discrete configuration space) and a probability $P = (P^1, ..., P^n)$ where n is the number of states. The first step of the procedure is to introduce the information metric, the natural metric on the space of probabilities. This leads to the most basic geometrical structure of the construction, *information geometry*. The second step is to consider dynamics and to introduce an action principle to derive equations of motion. This is done using a Hamiltonian formalism: introduce coordinates S^i canonically conjugate to the P^i, a Poisson bracket structure, and a Hamiltonian. This leads to additional geometrical structure, a *symplectic structure*. The third step is to extend the metric structure of information geometry, to define a geometry over the full space of the P^i and S^i. This can not be done in an arbitrary way. Consistency between the metric tensor and the symplectic form leads to a *Kähler geometry*.

Notice that the construction is very general: It applies to *any* system with a finite number of states that is described probabilistically and which admits equations of motion that can be derived from an action principle. In particular, the construction does not require any assumptions regarding classical or quantum physics.

The few assumptions that enter into the analysis lead to the beautiful result that *the natural geometry of the space of probabilities in motion is a Kähler geometry*.

6. Uniqueness of the Kähler metric via generalized Markov mappings

The Kähler conditions impose strong restrictions on the form of the metric g_{ab}, Eq. (17), but that they do not determine the metric uniquely because it depends on a matrix $\mathbf{A}$ which is to some extent arbitrary. Additional requirements are therefore needed to determine the form of g_{ab}.

As discussed before, Markov mappings play a crucial role in the proof of uniqueness of the information metric. In this section, it will be shown that the form of $\mathbf{A}$ can be determined by requiring invariance of the metric g_{ab} under a particular type of canonical transformation which extends the notion of a Markov mapping to the phase space with coordinates P^i and S^i. The simplest way to introduce a generalization of a Markov mapping is to treat it as a point transformation (i.e., a transformation of the P^i) and to use the well known fact that a point transformation can always be extended to a canonical transformation.

Consider first the simple case of Markov mappings between spaces of the same dimension; i.e., where $n = m$. In this case, the Markov mappings are just permutations of the components P^i and their extension to canonical transformations is trivial: carry out simultaneous permutations of the P^i and S^i. These transformations have two important properties: they are linear and they do not mix the P^i and the S^i.

The case that is non-trivial is the case where $n > m$. Here there are some subtle issues that need to be addressed, foremost that such canonical transformations will map spaces of different dimensions. This, however, is not a fundamental difficulty because there is a well developed theory of "nonclassical canonical transformations" which extends the concept of canonical transformations to allow for mappings of phase spaces of different dimensions [13, 14]. Although the canonical transformation that I derive in this section may be formulated within this formalism, a simpler approach is possible and therefore there will be no need to make use of the full theory of "nonclassical canonical transformations." The simpler approach followed here consists of extending the dimensionality of the space in a trivial way and imposing constraints.

For the purpose of this paper, it will be sufficient to work out the generalization for the case of the simple example of a Markov mapping that I discussed before, where the dimensionality of the space of probabilities is increased by one; i.e., $n = m + 1$. More generals cases can be derived by considering a series of successive transformations that are either permutations or which increase the dimensionality of the space by one at each step.

Consider then a system with states described by the coordinates P^j, S^j, $j = 1, ...m$, and a Hamiltonian $H(P^j, S^j)$ which describes the dynamics.

As a first step, increase the dimensionality of the space in a trivial way by adding coordinates P^{m+1}, S^{m+1} to the phase space. This increase in the dimensionality does not change the Hamiltonian. Therefore, the time evolution of the system remains the same and the additional coordinates P^{m+1}, S^{m+1} are *constants of the motion*. Consider now a point transformation relating old and new coordinates P^k, $\tilde{P}^k$ which satisfies the following relations,

$$\begin{aligned} P^i &= \tilde{P}^i, \\ P^m &= \tilde{P}^m + \tilde{P}^{m+1}, \\ P^{m+1} &= (1-k)\tilde{P}^m - k\tilde{P}^{m+1} \approx 0. \end{aligned} \tag{18}$$

where $i = 1, ..., m-1$, and the symbol "$\approx$" is used to indicate a weak equality (i.e., a constraint in the sense of an equality of numerical values, not of functions of the phase space coordinates). Using the last two relations of Eqs. (18), it is straightforward to show that $\tilde{P}^m$ and $\tilde{P}^{m+1}$ satisfy the constraints

$$\begin{aligned} \tilde{P}^m &\approx kP^m, \\ \tilde{P}^{m+1} &\approx (1-k)P^m, \end{aligned} \tag{19}$$

which agree with Eq. (6). Therefore, the relations defined in Eqs. (18) are equivalent to a Markov mapping. Notice that these constraints are preserved because P^{m+1} is a constant of the motion.

The second step is to extend this point transformation to a canonical transformation. In analogy to the case discussed above in which $n = m$, I will look for a linear canonical transformation which does not mix the P^i and the S^i. Notice that these two conditions lead to a *unique* canonical transformation (up to additive constants which are unimportant). To define the canonical transformation, introduce the generating function

$$K = \sum_{i=1}^{m-1} \left\{\tilde{P}^i S^i\right\} + \left(\tilde{P}^m + \tilde{P}^{m+1}\right) S^m + \left[(1-k)\tilde{P}^m - k\tilde{P}^{m+1}\right] S^{m+1}. \tag{20}$$

Derive the canonical transformation from the generating function in the standard way; i.e. $P^k = \partial K / \partial S^k$ and $\tilde{S}^k = \partial K / \partial \tilde{P}^k$. This leads to the following equations,

$$\begin{aligned} P^i &= \tilde{P}^i, & P^m &= \tilde{P}^m + \tilde{P}^{m+1}, & P^{m+1} &= (1-k)\tilde{P}^m - k\tilde{P}^{m+1}, \\ S^i &= \tilde{S}^i, & S^m &= k\tilde{S}^m + (1-k)\tilde{S}^{m+1}, & S^{m+1} &= \tilde{S}^m - \tilde{S}^{m+1}, \end{aligned} \tag{21}$$

and

$$\begin{aligned} \tilde{P}^i &= P^i, & \tilde{P}^m &= kP^m + P^{m+1}, & \tilde{P}^{m+1} &= (1-k)P^m - P^{m+1}, \\ \tilde{S}^i &= S^i, & \tilde{S}^m &= S^m + (1-k)S^{m+1}, & \tilde{S}^{m+1} &= S^m - kS^{m+1}. \end{aligned} \tag{22}$$

where $i = 1, ..., m-1$.

I have increased the dimensionality of the phase space by two dimensions and I have two constants of the motion, P^{m+1} and S^{m+1}. As shown in Appendix A, consistency requires both $P^{m+1} \approx 0$ and $S^{m+1} \approx 0$, which leads to

$$\begin{aligned} \tilde{P}^m &\approx kP^m, & \tilde{P}^{m+1} &\approx (1-k)P^m, \\ \tilde{S}^m &\approx S^m, & \tilde{S}^{m+1} &\approx S^m. \end{aligned} \tag{23}$$

It is also shown in Appendix A that the time evolution is preserved; i.e., the dynamics in the $2(n+1)$-dimensional phase space (with coordinates with tildes) reproduces precisely the dynamics in the $2n$-dimensional phase space (with coordinates without tildes).

Given the canonical transformation, Eq. (21), and its inverse, Eq. (22), one can examine the restrictions imposed on the matrix $\mathbf{A}$ by the requirement of invariance of the metric g_{ab} under this generalization of a Markov mapping. The calculation is summarized in Appendix B. The result is that the matrix $\mathbf{A}$ must be proportional to the $n \times n$ unit matrix, $\mathbf{A} = A\mathbf{1}$, where A is a constant. The line element depends on two parameteres only, α and A, and it takes the remarkably simple form

$$dl^2 = \sum_k \left[\frac{\alpha}{2P^k}(dP^k)^2 + 2A dP^k dS^k + \frac{2P^k}{\alpha}(1+A^2)(dS^k)^2 \right]. \tag{24}$$

7. Complex coordinates and wave functions

Up to now, I have made use of real coordinates P^i, S^i. Kähler geometry, however, is best expressed in terms of complex coordinates. I carry out a complex transformation that shows that the metric of Eq. (24) describes in fact a flat Kähler space.

Set $\mathbf{A} = A\mathbf{1}$ in Eqs. (17) and consider first the particular case $A = 0$. The tensors that define the Kähler structure take the form

$$\Omega_{ab} = \begin{pmatrix} 0 & 1 \\ -1 & 0 \end{pmatrix}, \; g_{ab} = \begin{pmatrix} \mathbf{G} & 0 \\ 0 & \mathbf{G}^{-1} \end{pmatrix}, \; J^a{}_b = \begin{pmatrix} 0 & \mathbf{G}^{-1} \\ -\mathbf{G} & 0 \end{pmatrix}. \tag{25}$$

Define now the *Madelung transformation*,

$$\psi^k = \sqrt{P^k}\exp(iS^k/\alpha), \qquad \bar{\psi}^k = \sqrt{P^k}\exp(-iS^k/\alpha). \tag{26}$$

A simple calculation shows that the tensors that define the Kähler geometry, expressed in terms of ψ^k, $\bar{\psi}^k$, take the standard form which is characteristic of flat-space [12],

$$\Omega_{ab} = \begin{pmatrix} 0 & i\alpha\mathbf{1} \\ -i\alpha\mathbf{1} & 0 \end{pmatrix}, \; g_{ab} = \begin{pmatrix} 0 & \alpha\mathbf{1} \\ \alpha\mathbf{1} & 0 \end{pmatrix}, \; J^a{}_b = \begin{pmatrix} -i\mathbf{1} & 0 \\ 0 & i\mathbf{1} \end{pmatrix}. \tag{27}$$

One may conclude that in this case ($A = 0$) there is a natural set of fundamental variables given by ψ^i and $\bar{\psi}^i$. In terms of these variables, the tensors that define the Kähler geometry take their simplest form. If the constant α is set equal to $\hbar$, these fundamental variables are precisely the *wave functions* of quantum mechanics. This is a remarkable result because it is based on geometrical arguments only. The derivation does not use any assumptions from quantum theory.

Consider now the more general case $A \neq 0$. The tensors that define the Kähler structure take the form

$$\Omega_{ab} = \begin{pmatrix} 0 & 1 \\ -1 & 0 \end{pmatrix}, \; g_{ab} = \begin{pmatrix} \mathbf{G} & A\mathbf{1} \\ A\mathbf{1} & (1+A^2)\mathbf{G}^{-1} \end{pmatrix}, \; J^a{}_b = \begin{pmatrix} A\mathbf{1} & (1+A^2)\mathbf{G}^{-1} \\ -\mathbf{G} & -A\mathbf{1} \end{pmatrix}. \tag{28}$$

In this case, define the *modified Madelung transformation*

$$\phi^k = \sqrt{P^k}\exp\left[i\left(\Lambda\, S^k/\alpha - \gamma \ln\sqrt{P^k}\right)\right], \quad \bar{\phi}^k = \sqrt{P^k}\exp\left[-i\left(\Lambda\, S^k/\alpha - \gamma \ln\sqrt{P^k}\right)\right], \tag{29}$$

where $\Lambda = 1/(1+A^2)$ and $\gamma = -A/(1+A^2)$. Once more, the tensors that define the Kähler geometry, expressed now in terms of ϕ^k, $\bar{\phi}^k$, take the standard form which is characteristic of flat-space,

$$\Omega_{ab} = \begin{pmatrix} 0 & i\alpha\Lambda^{-1}\mathbf{1} \\ -i\alpha\Lambda^{-1}\mathbf{1} & 0 \end{pmatrix}, \; g_{ab} = \begin{pmatrix} 0 & \alpha\Lambda^{-1}\mathbf{1} \\ \alpha\Lambda^{-1}\mathbf{1} & 0 \end{pmatrix}, \; J^a{}_b = \begin{pmatrix} -i\mathbf{1} & 0 \\ 0 & i\mathbf{1} \end{pmatrix}. \tag{30}$$

Journal of Physics: Conference Series **538** (2014) 012018 doi:10.1088/1742-6596/538/1/012018

This shows that the geometry of the Kähler space is the same whether $A = 0$ or $A \neq 0$. In fact, it is possible to map one case to the other using an A-dependent canonical transformation. It is clear then that both cases lead to the same theory (provided one sets $\alpha = \hbar$ when $A = 0$ or $\alpha\Lambda^{-1} = \hbar$ when $A \neq 0$), and in the following sections I will set $A = 0$ and use the complex coordinates (wave functions) ψ^i and $\bar{\psi}^i$.

The transformation that takes you from the coordinates of Eq. (26) to the coordinates of Eq. (29) is a particular case of a family of nonlinear gauge transformations introduced by Doebner and Goldin [15] (compare to their Eq. (2.2)). As pointed out by Doebner and Goldin, the theory that results from this particular family of nonlinear gauge transformations is physically equivalent to standard quantum mechanics. Here we arrive at the same conclusion, but now on the basis of the equivalence of the two cases $A = 0$ and $A \neq 0$ via a canonical transformation. One may therefore view the present derivation of the geometric formulation of quantum mechanics as providing a new route to this family of Doebner-Goldin nonlinear gauge transformations.

8. Group of unitary transformations and Hilbert space formulation

In the Hamiltonian formalism, any (smooth) function of the canonically conjugate variables generates an infinitesimal transformation. The group associated with these generators is the group of canonical transformations. The group of transformations of the theory developed here, however, can not be the full group of canonical transformations: only a subset of such generators is permissible because the infinitesimal transformations are subject to certain restrictions. In this section, I show that the group of transformations of the theory is the group of unitary transformations and that one may introduce a Hilbert space formulation of the theory. Both of these results are needed to establish the equivalence of the geometric formulation derived here to standard quantum mechanics.

To understand the role of the unitary group, one can argue as follows. Since the Kähler structure includes a symplectic structure, the group of symplectic transformations, Sp($2n$,R), will play an important role in the theory. But the group of transformations of the theory can not be the symplectic group due to the requirements that the transformations have to satisfy. The first requirement is that they preserve the normalization of the probability, $\sum_i P^i = \sum_i \psi^i \bar{\psi}^i = 1$. The second requirement is that the metric be form invariant under the transformations; i.e., that the line element $dl^2 = 2\alpha \sum_j d\bar{\psi}^j d\psi^j$ of the Kähler space be preserved by the transformations. Requiring normalization of the probability and metric invariance leads to the group of rotations on the $2n$-dimensional sphere, O($2n, R$). Unitary transformations are the only symplectic transformations which are also rotations; i.e., Sp($2n$,R) $\cap$ O($2n, R$)= $U(n)$ [16]. Therefore, we are led to the group of unitary transformations. One can show that time evolution is described by a one-parameter group of unitary transformations (see Appendix C for a proof).

One can now introduce a Hilbert space formulation. There is a standard construction that associates a complex Hilbert space with any Kähler space. Given two complex vectors ψ^i and φ^i, define the Dirac product by [1]

$$\begin{aligned} \langle\psi|\varphi\rangle &= \frac{1}{2\alpha}\sum_i \left\{ \left(\psi^i, \bar{\psi}^i\right) \cdot [g + i\Omega] \cdot \begin{pmatrix} \varphi^i \\ \bar{\varphi}^i \end{pmatrix} \right\} \\ &= \frac{1}{2}\sum_i \left\{ \left(\psi^i, \bar{\psi}^i\right) \left[\begin{pmatrix} 0 & \mathbf{1} \\ \mathbf{1} & 0 \end{pmatrix} + i \begin{pmatrix} 0 & i\mathbf{1} \\ -i\mathbf{1} & 0 \end{pmatrix} \right] \begin{pmatrix} \varphi^i \\ \bar{\varphi}^i \end{pmatrix} \right\} \\ &= \sum_i \bar{\psi}^i \varphi^i \end{aligned} \tag{31}$$

In this way one arrives at the Hilbert space formulation of quantum mechanics.

9. Concluding remarks

The geometry of quantum theory can be derived from information geometry, the natural geometry on the space of probabilities, using only a few assumptions. The derivation has a number of interesting features:

- Doubling of the dimensionality of the space (i.e., $\{P^i\} \rightarrow \{P^i, S^i\}$) from dynamical considerations,
- Complex structure from consistency between metric and symplectic structures,
- Wave functions as the natural complex coordinates of the Kähler space,
- Representation in terms of canonical transformations of a particular case of a family of Doebner-Goldin nonlinear gauge transformations,
- Time evolution described by a one-parameter group of unitary transformations,
- Hilbert space formulation expressed in terms of geometrical quantites associated with the Kähler space.

The derivation presented here relies heavily on, and extends, a geometrical reconstruction of quantum theory by Reginatto and Hall which takes information geometry as its starting point [2, 3]. Mehrafarin [17] and Goyal [18, 19] have also developed reconstructions of quantum theory using information-geometrical approaches. A detailed comparison to their approaches has not been carried out yet; however, one of the main differences is in the handling of *dynamics*, which plays a crucial role here. In particular, the use of an action principle to describe the dynamics of probabilities leads in a natural way to geometrical structure that goes beyond information geometry. A generalization of Markov mappings has proven to be very useful for deriving a unique Kähler geometry; these types of transformations may be of interest for other classes of problems where probabilities play an important role.

Finally, a few remarks concerning the limit $\alpha \rightarrow 0$ (i.e., the limit $\hbar \rightarrow 0$ in quantum mechanics). In this case, neither the Madelung transformation, Eq. (26), nor the modified Madelung transformation, Eq. (26), are well defined so I consider the description in terms of the real coordinates P^i, S^i. Of the three tensors that describe the Kähler structure, Eq. (25), only the symplectic structure Ω_{ab} remains well defined: both the metric g_{ab} and the complex structure $J^a{}_b$ become singular when $\alpha = 0$. Therefore, in the limit $\alpha \rightarrow 0$, the metric g_{ab} and complex structure $J^a{}_b$ are no longer available and the Kähler structure is lost. The only mathematical structure that remains is the symplectic structure, which can be used to define equations of motion for the canonically conjugate variables P^i, S^i given an appropriate Hamiltonian. One interesting consequence of taking the limit $\alpha \rightarrow 0$ is that the argument in Appendix C is no longer valid (because without a metric one can no longer require metric invariance). A more careful analysis of the limit $\hbar \rightarrow 0$ will be presented in a future publication.

10. Acknowledgments

I am very grateful to the organizers of the Symmetries in Science XVI symposium for the invitation to attend a wonderful and stimulating meeting. I also want to thank Michael J. W. Hall, Ariel Caticha, Gerald A. Goldin and Heinz-Dietrich Doebner for their comments and suggestions regarding the work presented in this paper.

Appendix A. Generalized Markov mappings: constants of the motion and dynamics

To derive the canonical transformation that generalizes the Markov mapping of Eq. (6), the dimensionality of the original phase space was increased by two in a trivial way. This led to two constants of the motion, P^{m+1} and S^{m+1}. P^{m+1} was set to $P^{m+1} \approx 0$, with corresponding constraints for the $\tilde{P}^k$ of the form

$$\begin{aligned} \tilde{P}^m &\approx kP^m, \\ \tilde{P}^{m+1} &\approx (1-k)P^m. \end{aligned} \tag{A.1}$$

These are precisely the conditions that are needed to get a generalization of the Markov mapping of Eq. (6). When $k = 1/2$, $\tilde{P}^m \approx \tilde{P}^{m+1}$, which is expected because in this case there should be invariance under the re-labeling $m \leftrightarrow m+1$. To fix the value of S^{m+1}, notice that $S^{m+1} \approx c$ leads to constraints for the $\tilde{S}^k$ of the form

$$\begin{aligned} \tilde{S}^m &\approx S^m + (1-k)c, \\ \tilde{S}^{m+1} &\approx S^m - kc. \end{aligned} \tag{A.2}$$

Argue once more that there should be invariance under the re-labeling $m \leftrightarrow m+1$ in the case when $k = 1/2$. But this can only be satisfied if $c = 0$. On can conclude therefore that the constants of the motion must satisfy

$$\begin{aligned} P^{m+1} &= (1-k)\tilde{P}^m - k\tilde{P}^{m+1} \approx 0, \\ S^{m+1} &= \tilde{S}^m - \tilde{S}^{m+1} \approx 0. \end{aligned} \tag{A.3}$$

The corresponding constraints for the unprimed coordinates are of the form

$$\begin{aligned} \tilde{P}^m &\approx kP^m, \qquad \tilde{P}^{m+1} \approx (1-k)P^m, \\ \tilde{S}^m &\approx S^m, \qquad \tilde{S}^{m+1} \approx S^m. \end{aligned} \tag{A.4}$$

I check now that the dynamics in the $2(n+1)$-dimensional phase space (with coordinates with tildes) reproduces precisely the dynamics in the original $2n$-dimensional phase space (with coordinates without tildes). Using Eqs. (21-22), one can show that

$$\begin{aligned} \dot{\tilde{P}}^i &= \frac{\partial H}{\partial S^i}, \qquad \dot{\tilde{P}}^m = \frac{\partial H}{\partial S^m}k, \qquad \dot{\tilde{P}}^{m+1} = \frac{\partial H}{\partial S^m}(1-k), \\ \dot{\tilde{S}}^i &= -\frac{\partial H}{\partial P^i}, \qquad \dot{\tilde{S}}^m = -\frac{\partial H}{\partial P^m}, \qquad \dot{\tilde{S}}^{m+1} = -\frac{\partial H}{\partial P^m}. \end{aligned} \tag{A.5}$$

These equations lead to

$$\begin{aligned} \dot{P}^i &= \dot{\tilde{P}}^i = \frac{\partial H}{\partial S^i}, \qquad \dot{P}^m = \dot{\tilde{P}}^m + \dot{\tilde{P}}^{m+1} = \frac{\partial H}{\partial S^m} \\ \dot{S}^i &= \dot{\tilde{S}}^i = -\frac{\partial H}{\partial P^i}, \qquad \dot{S}^m = k\,\dot{\tilde{S}}^m + (1-k)\,\dot{\tilde{S}}^{m+1} = -\frac{\partial H}{\partial P^m}, \end{aligned} \tag{A.6}$$

which are the correct equations of motion for the original space.

Appendix B. Invariance of the Kähler metric under generalized Markov mappings

The metric of the Kähler space is given by

$$g_{ab} = \begin{pmatrix} \mathbf{G} & \mathbf{A}^T \\ \mathbf{A} & (\mathbf{1}+\mathbf{A}^2)\mathbf{G}^{-1} \end{pmatrix}, \tag{B.1}$$

where $\mathbf{G} = \mathtt{diag}(\frac{\alpha}{2P^i})$ and the $n \times n$ matrix $\mathbf{A}$ satisfies $\mathbf{G}\mathbf{A}\mathbf{G}^{-1} = \mathbf{A}^T$. For the calculations in this Appendix it is convenient to introduce the matrix $\mathbf{B}$ with matrix elements given by

$$B_{jk} = \sqrt{P_j/P_k}\,A_{jk}. \tag{B.2}$$

It is straightforward to show that $\mathbf{B}$ is a symmetric matrix, $B_{jk} = B_{kj}$.

To restrict the form of $\mathbf{B}$, it will be sufficient to consider the invariance of the metric under the particular generalized Markov mapping which corresponds to the inverse canonical transformation of Eq. (22). After taking into consideration the constraints, Eq. (23), the generalized Markov mapping can be written in the form

$$\begin{aligned} P &= (P^i, P^m) \to \tilde{P} = (\tilde{P}^i, \tilde{P}^m, \tilde{P}^{m+1}) := (P^i, kP^m, (1-k)P^m), \\ S &= (S^i, S^m) \to \tilde{S} = (\tilde{S}^i, \tilde{S}^m, \tilde{S}^{m+1}) := (S^i, S^m, S^m), \end{aligned} \tag{B.3}$$

where $i = 1, ..., m-1$.

As a first step, look at the contribution to the line element dl^2 from the mixed terms $dP^k dS^k$. In terms of the coordinates without tildes,

$$dl^2 = \sum_{i=1}^{m-1} \left\{ B_{ii} dP^i dS^i + \sqrt{\frac{P^i}{P^m}} B_{im} dP^i dS^m + \sqrt{\frac{P^m}{P^i}} B_{mi} dP^m dS^i \right\} + B_{mm} dP^m dS^m. \tag{B.4}$$

There is a corresponding expression for the coordinates with tildes, and with the help of Eq. (B.3) it can be rewritten in terms of coordinates without tildes. This leads to

$$\begin{aligned} dl^2 &= \sum_{i=1}^{m-1} \left\{ \tilde{B}_{ii} dP^i dS^i + \left[\sqrt{\frac{P^i}{kP^m}} \tilde{B}_{im} + \sqrt{\frac{P^i}{(1-k)P^m}} \tilde{B}_{i(m+1)} \right] dP^i dS^m \right\} \\ &+ \sum_{i=1}^{m-1} \left\{ \left[\sqrt{\frac{k^3 P^m}{P^i}} \tilde{B}_{mi} + \sqrt{\frac{(1-k)^3 P^m}{P^i}} \tilde{B}_{(m+1)i} \right] dP^m dS^i \right\} \\ &+ \left[k\tilde{B}_{mm} + \sqrt{\frac{k^3}{1-k}} \tilde{B}_{m(m+1)} + \sqrt{\frac{(1-k)^3}{k}} \tilde{B}_{(m+1)m} + (1-k)\tilde{B}_{(m+1)(m+1)} \right] dP^m dS^m. \end{aligned} \tag{B.5}$$

Equate terms in Eqs. (B.4) and (B.5) proportional to the same $dP^a dS^b$, where $a, b = 1, ..., m$. This leads to the four relations

$$\begin{aligned} B_{ii} &= \tilde{B}_{ii}, \\ B_{im} &= \sqrt{\frac{1}{k}} \tilde{B}_{im} + \sqrt{\frac{1}{(1-k)}} \tilde{B}_{i(m+1)}, \\ B_{mi} &= \sqrt{k^3} \tilde{B}_{mi} + \sqrt{(1-k)^3} \tilde{B}_{(m+1)i}, \\ B_{mm} &= k\tilde{B}_{mm} + \sqrt{\frac{k^3}{1-k}} \tilde{B}_{m(m+1)} + \sqrt{\frac{(1-k)^3}{k}} \tilde{B}_{(m+1)m} + (1-k)\tilde{B}_{(m+1)(m+1)}. \end{aligned} \tag{B.6}$$

Since the matrix $\mathbf{B}$ is symmetric, $B_{im} = B_{mi}$, which leads to

$$\sqrt{\frac{1}{k}} \tilde{B}_{im} + \sqrt{\frac{1}{(1-k)}} \tilde{B}_{i(m+1)} = \sqrt{k^3} \tilde{B}_{mi} + \sqrt{(1-k)^3} \tilde{B}_{(m+1)i}. \tag{B.7}$$

By symmetry, $\tilde{B}_{im} = \tilde{B}_{i(m+1)}$ at $k = 1/2$, but this relation can only be satisfied if $\tilde{B}_{im} = \tilde{B}_{i(m+1)} = 0$, which in turn implies $B_{im} = B_{mi} = 0$. Since B_{im} and B_{mi} are independent of k, it follows that they must always be zero. This shows that the off-diagonal elements of the matrix $\mathbf{B}$ are zero.

Now look at the contribution to the line element dl^2 from terms proportional to $dP^a dP^a$ and $dS^a dS^a$. The terms proportional to $dP^a dP^a$ give the two relations

$$\begin{aligned} B_{ii} &= \tilde{B}_{ii}, \\ B_{mm} &= k\tilde{B}_{mm} + (1-k)\tilde{B}_{(m+1)(m+1)}, \end{aligned} \tag{B.8}$$

while the terms proportional to $dS^a dS^a$ give the two relations

$$\begin{aligned} 1 + B_{ii}^2 &= 1 + \tilde{B}_{ii}^{\,2}, \\ 1 + B_{mm}^2 &= k(1 + \tilde{B}_{mm}^{\,2}) + (1-k)\left(1 + \tilde{B}_{(m+1)(m+1)}^{\,2}\right). \end{aligned} \tag{B.9}$$

Combining Eqs. (B.8) and (B.9) leads to

$$\begin{aligned} B_{ii} &= \tilde{B}_{ii}, \\ B_{mm} &= \tilde{B}_{mm} = \tilde{B}_{(m+1)(m+1)}. \end{aligned} \tag{B.10}$$

Notice that this result is valid for arbitrary values of k. Since there is nothing special about the particular labels m and $(m+1)$, *all* the diagonal elements of the matrices $\mathbf{B}$ and $\tilde{\mathbf{B}}$ must be equal. Then

$$\begin{aligned} \mathbf{B} &= B\mathbf{1}_{m\times m}, \\ \tilde{\mathbf{B}} &= B\mathbf{1}_{(m+1)\times(m+1)}, \end{aligned} \tag{B.11}$$

where $\mathbf{1}_{n\times n}$ is the $n \times n$ unit matrix and B still has to be determined.

To carry out this last step, use the relations

$$\begin{aligned} B(P,S) &= B(P^i, P^m, S^i, S^m), \\ \tilde{B}(\tilde{P},\tilde{S}) &= B(P(\tilde{P}), S(\tilde{S})) = B(\tilde{P}^i, \tilde{P}^m + \tilde{P}^{m+1}, \tilde{S}^i, k\tilde{S}^m + (1-k)\tilde{S}^{m+1}). \end{aligned} \tag{B.12}$$

The functional form of $B(P,S)$ must be the same as the functional form of $\tilde{B}(\tilde{P},\tilde{S})$, and these expressions must be both invariant under permutations and independent of k. The only functional form that seems to satisfy all these conditions appears to be $B(P,S) = B(\sum_i P^i)$. But $\sum_i P^i = 1$, therefore one can conclude that B is a *constant* and the matrix

$$\mathbf{B} = B\mathbf{1} \tag{B.13}$$

is a constant matrix proportional to the unit matrix. This in turn implies that

$$\mathbf{A} = A\mathbf{1} \tag{B.14}$$

where A is a constant.

Appendix C. Group of unitary transformations

The generators of infinitesimal transformations (for example, Hamiltonians that generate the time evolution of the system) have to satisfy certain requirements; therefore, not every function of the coordinates qualifies as a Hamiltonian. To determine the allowed functional form of Hamiltonians, it is convenient to set $A = 0$ and work with the complex coordinates ψ^k, $\bar{\psi}^k$ introduced in Section 7. In this Appendix, I follow the derivation presented in Ref. [3] and show that Hamiltonians must be of the form $H = \sum_{j,k} M_{jk}\bar{\psi}^j\psi^k$ (up to an additive constant), where M_{jk} is a Hermitian matrix. Thus, time evolution is described by a one-parameter group of unitary transformations.

To show this, one needs to consider three requirements. The first requirement is that the transformations preserve the normalization of the probability, $\sum_i P^i = \sum_i \psi^i \bar{\psi}^i = 1$. The second requirement is that the metric be form invariant under those transformations; i.e., that the transformations preserve the line element $d\sigma^2 = 2\alpha \sum_j d\bar{\psi}^j d\psi^j$ of the Kähler space. Requiring normalization of the probability and metric invariance leads to the group of rotations on the $2n$-dimensional sphere. Such rotations are linear with respect to ψ^j and $\bar{\psi}^j$. For an infinitesimal transformation, it follows that

$$\dot{\psi}^j = -i\frac{\partial H}{\partial \bar{\psi}^j}, \qquad \dot{\bar{\psi}}^j = i\frac{\partial H}{\partial \psi^j}, \tag{C.1}$$

are linear in ψ and $\bar{\psi}$, where H is the Hamiltonian that generates the motion. Then H must be of the form

$$H = E(t) + \sum_{j,k}\left[M_{jk}\bar{\psi}^j\psi^k + N_{jk}\psi^j\psi^k + \bar{N}_{jk}\bar{\psi}^j\bar{\psi}^k\right] \tag{C.2}$$

where $E(t)$ is a arbitrary function of time, M is Hermitian, and N is symmetric.

The third and final requirement is to consider rotations on the $2n$-dimensional sphere that are compatible with the equations of motion. Conservation of probability requires that the Hamiltonian be invariant (up to an additive constant) under $S^j \to S^j + \chi$, since to first-order [11]

$$0 = \epsilon \sum_j \dot{P}^j = \epsilon \sum_j \frac{\partial H}{\partial S^j} = H(P, S+\epsilon) - H(P, S). \tag{C.3}$$

This condition, when written in terms of complex coordinates, is equivalent to invariance of the Hamiltonian under $\psi \to \psi e^{i\chi}$. Using the notation $Q := \sum N_{jk}\psi^j\psi^k$, Eq. (C.2) leads to the equality

$$\left[Qe^{2i\chi} + \bar{Q}e^{-2i\chi}\right] = 0 \tag{C.4}$$

which must be valid for all χ. Differentiating with respect to χ gives the additional equality

$$2i\left[Qe^{2i\chi} - \bar{Q}e^{-2i\chi}\right] = 0. \tag{C.5}$$

Combining these two expressions leads to $Q = 0$. Since this must hold for all ψ, it follows that $N_{jk} \equiv 0$, i.e., the Hamiltonian has the Hermitian form

$$H = E(t) + \sum_{j,k} M_{jk}\bar{\psi}^j\psi^k \tag{C.6}$$

as desired.

References

[1] Kibble T W B 1979 *Commun. math. Phys.* **65** 189
[2] Reginatto M and Hall M J W 2012 *AIP Conf. Proc.* **1443** 96
[3] Reginatto M and Hall M J W 2013 *AIP Conf. Proc.* **1553** 246
[4] Bhattacharyya A 1943 *Bull. Calcutta Math. Soc.* **35** 99
[5] Bhattacharyya A 1946 *Sanhkyā* **7** 401
[6] Good I J 1990 *J. Stat. Comput. Simulation* **36** 179
[7] Wootters W K 1981 *Phys. Rev. D.* **23** 357
[8] Čencov N N 1981 *Statistical decision rules and optimal inference* (Transl. Math. Monographs **53**) (Providence, R. I.: Amer. Math. Soc.)
[9] Campbell L L 1986 *Proc. Amer. Math. Soc.* **98** 135
[10] Caticha A 2012 *Entropic Inference and the Foundations of Physics* (Sao Paulo, Brazil: Brazilian Chapter of the International Society for Bayesian Analysis – ISBrA)

[11] Hall M J W 2004 *J. Phys. A* **37** 7799
[12] Goldberg S I 1982 *Curvature and Homology* (New York: Dover Publications)
[13] Scheifele G 1970 *Celest. Mech.* **2** 296
[14] Kurcheeva I V 1977 *Celest. Mech.* **15** 353
[15] Doebner H-D and Goldin G A 1996 *Phys. Rev. A* **54**, 3764
[16] Arnold V I 1978 *Mathematical methods of classical mechanics* (Berlin: Springer)
[17] Mehrafarin M 2005 *Int. J. Theor. Phys.* **44** 429
[18] Goyal P 2008 *Phys. Rev. A* **78** 052120
[19] Goyal P 2010 *New Journal of Physics* **12** 023012

Relations between nonlinear Riccati equations and other equations in fundamental physics

Dieter Schuch

Institut für Theoretische Physik, J.W. Goethe-Universität Frankfurt am Main, Max-von-Laue-Str. 1, D-60438 Frankfurt am Main, Germany

E-mail: Schuch@em.uni-frankfurt.de

Abstract. Many phenomena in the observable macroscopic world obey nonlinear evolution equations while the microscopic world is governed by quantum mechanics, a fundamental theory that is supposedly linear. In order to combine these two worlds in a common formalism, at least one of them must sacrifice one of its dogmas. Linearizing nonlinear dynamics would destroy the fundamental property of this theory, however, it can be shown that quantum mechanics can be reformulated in terms of nonlinear Riccati equations. In a first step, it will be shown that the information about the dynamics of quantum systems with analytical solutions can not only be obtainable from the time-dependent Schrödinger equation but equally-well from a complex Riccati equation. Comparison with supersymmetric quantum mechanics shows that even additional information can be obtained from the nonlinear formulation. Furthermore, the time-independent Schrödinger equation can also be rewritten as a complex Riccati equation for any potential. Extension of the Riccati formulation to include irreversible dissipative effects is straightforward. Via (real and complex) Riccati equations, other fields of physics can also be treated within the same formalism, e.g., statistical thermodynamics, nonlinear dynamical systems like those obeying a logistic equation as well as wave equations in classical optics, Bose–Einstein condensates and cosmological models. Finally, the link to abstract "quantizations" such as the Pythagorean triples and Riccati equations connected with trigonometric and hyperbolic functions will be shown.

1. Introduction

In the 19th century electrodynamics (as formulated by James C. Maxwell) and thermodynamics (condensed in three laws derived from phenomenological experience) were a huge success. They projected an image of nature as being continuous (strongly favoured by physicists like Ernst Mach) as opposed to being built from discrete components as assumed in an atomistic approach (favoured by scientists like Ludwig Boltzmann). The ideas of Democritus (who coined the term *atom* for a part of our material world that is no longer divisible) and, in a more abstract way, of Plato and Pythagoras who assumed smallest units as building blocks of our world, seemed to belong purely to the realm of philosophy.

Towards the end of the 19th century and the beginning of the 20th, the situation changed. First, in order to explain the spectrum of the black-body radiation, Max Planck had to "discretize" a physical quantity called *action* that, for example, can be expressed as a product of position times momentum or energy times time, i.e., a product of two physical quantities that cannot be measured exactly at the same time. (Something that Heisenberg later on proved to be impossible in principle.) Planck first considered the corresponding minimal quantum of action, denoted by "h" (for the German "Hilfsgrösse"), only as a kind of temporary auxiliary quantity that could be abandoned when a more natural explanation for the spectrum could be found. It turned out this was not the case. In fact, by introducing his quantum of action, conservative Planck actually initiated a revolution in the physicists' view of the world in a way he never expected, not to mention intended! And at the beginning of the 20th century even more peculiar effects were observed. For instance light, that (after Maxwell) was finally considered to be a continuous wave, behaved like discrete particles in certain experiments such as the photoelectric effect. Contrarily, electrons that (in the meantime) were assumed to be particles, displayed wave-like behaviour in some experiments and produced interference patterns. The dichotomy of light versus matter, or continuous versus discrete, was only resolved in the mid-twenties of the last century by Schrödinger and Heisenberg (and finally Dirac) with the development of quantum theory.

Though physically equivalent, Schrödinger's wave mechanics turned out to be more successful and receptive to the physics community than Heisenberg's matrix mechanics that used a less familiar mathematical description than Schrödinger's partial differential equation. Since both formulations are closely-related to classical Hamiltonian mechanics they also have similar properties, in particular, there is no direction of time in the evolution of the system and energy is a constant of motion (at least in the cases that are usually discussed in textbooks and can be solved analytically in closed form).

However, as everyone can observe daily in the surrounding world, nature actually behaves quite differently. There is a direction of time in almost every evolutionary process (and we cannot reverse it even if we would sometimes like to). Also mechanical energy is not a conserved quantity but dissipated into heat by effects like friction. There are ways of explaining and including these phenomena into the theories mentioned earlier but, for ordinary people, they are not really convincing. Nevertheless, quantum theory (with all its technological developments) is undoubtedly the most successful theory so far, and not only in physics but also from an economic viewpoint.

The situation in physics took a different twist near the end of the 20th century with the development of nonlinear (NL) dynamics. This theory is able to describe evolutionary processes like population growth with given resources or weather patterns and other such complex systems as they occur in real life. At the same time it can also take into account phenomena like irreversibility of evolution and dissipation of energy.

So why not combine the two theories to get the best of both worlds?

From a formal, mathematical point of view, there are several aspects that must be considered. Here, only two will be mentioned that are relevant to the subsequent discussion.

1. C. N. Yang, in his lecture on the occasion of Schrödinger's centennial celebration [1], stated that in his opinion, the major difference between classical and quantum physics is the occurrence of the imaginary unit $i = \sqrt{-1}$ in quantum mechanics since here it enters physics in a fundamental way (not just as a tool for computational convenience) and "complex numbers became a conceptual element of the very foundation of physics". The very meaning of the fundamental equations of wave mechanics and matrix mechanics "would be totally destroyed if one tries to get rid of i by writing them in terms of real and imaginary parts". This, however, should not be a problem in the unification of quantum theory and NL dynamics since, also in the latter, complex numbers play an eminent role (e.g. in the complex quadratic family leading to the Mandelbrot set).

2. The contrast between reversible time-evolution in quantum theory and irreversible evolution, as possible in NL dynamics, appears more problematic. But, even if one would restrict the unification only to systems with reversible evolution the major problem seems to be that quantum theory is a linear one, i.e., it is essentially based on linear differential equations. On the other hand, NL dynamics, by definition, uses NL differential equations (or discretized versions

of such). Why then should linearity be so important for quantum theory? As mentioned above, quantum theory can explain the wave properties of material systems such as diffraction patterns that can essentially be considered as superposition of different solutions of the same equation. But this superposition principle is usually only attributed to linear differential equations!

So is there any way out of this dilemma?

In principle, NL dynamics cannot be linearized otherwise it would lose not only its *linguistic* meaning but also its *physical* properties. Therefore, the only solution seems to be a NL version of quantum theory which takes into account all the conventional properties of this theory (including a kind of superposition principle) while displaying formal compatibility with NL dynamics and which might (hopefully) provide additional information that cannot be obtained easily (or even be expected) from the linear form of quantum theory.

For this purpose, the next section will consider the time-dependent Schrödinger equation (TDSE) for cases where it possesses exact analytic solutions in the form of Gaussian wave packets (WPs). (In other words, for potentials at most quadratic in position variable, here explicitly for the one-dimensional harmonic oscillator (HO) with $V = \frac{m}{2}\omega^2 x^2$ with constant frequency ω, its generalization for $\omega = \omega(t)$ and, in the limit $\omega \to 0$, for the free motion $V = 0$.) It will be shown that the information about the dynamics of these systems can equally-well be obtained from a complex, quadratically-nonlinear, inhomogeneous Riccati equation. An interesting feature of this equation is that it can be linearized, therefore being able to provide a kind of superposition principle. Conversely, due to the nonlinearity, this equation is very sensitive to the initial conditions, which is not at all obvious in the linearized form. Due to these attributes, it is predestined to bridge the gap between linear theories like quantum mechanics and NL ones. Several methods of treating the NL Riccati equation will be mentioned and formal similarities with supersymmetric (SUSY) quantum mechanics will be shown. During the course, a dynamical invariant will also be introduced with the dimension of action (the quantity that is quantized according to Planck). Further, it will be shown how factorization of the corresponding operator can lead to generalized creation/annihilation operators allowing for the construction of generalized coherent states (CSs) with TD width, i.e., TD position (and momentum) uncertainties as known from squeezed states. An application of the generalized creation/annihilation operators to obtain unusual solutions of the TDSE for the free motion will be mentioned.

The linearization of the complex Riccati equation to a complex Newton-type equation will allow for a geometric interpretation of the linearized variable as well as of the afore-mentioned invariant.

In section 3, it will be shown that a, formally, similar treatment is also possible for the time-independent (TI) SE but now without restriction to quadratic potential since $V(x)$ corresponds to $\omega(t)$ in the TD case and enters only as an inhomogeneity into the Riccati equation. Since the discussion in section 2 holds also for TD $\omega(t)$, the same applies in the TISE for any $V(x)$. With this treatment it is possible to obtain complex solutions with space-dependent phase of the wave function. Using a formalism well-known from SUSY quantum mechanics, complex (non-hermitian) potentials with real eigenvalues can be obtained in a new way. As a first example, a "complexified" Pöschl–Teller potential will be presented that is derived from the free motion SE (see [2] for further details).

Section 4 will briefly show that a generalization is possible that includes dissipation and irreversibility. Several effective models describing these systems and their interrelations will be mentioned. Using a specific NL modification of the TDSE, it will be shown how the equations of the WP dynamics change in this case, particularly the complex Riccati equation describing the TD of the WP width.

The transformation between the linear SE (TD and TI) and the NL Riccati equation allows for connections with many other fields of physics where NL Riccati equations play an important role. This will be shown in section 5 using, as examples, systems with Bose–Einstein (BE) or Fermi–Dirac statistics in statistical thermodynamics and systems obeying the logistic equation in NL dynamics. Changing to complex Riccati equations, systems in classical optics as well as BECs will be considered and, by analogy, even cosmological problems can be treated in this way.

Complex Riccati (or related Bernoulli) equations also provide the answer to the question of finding Pythagorean triples, i.e. right-angled triangles with integer length of their sides. Furthermore, the geometry of these triangles with their description in terms of trigonometric functions allows one to find Riccati equations from their derivatives. Admitting imaginary variables as well, this can be generalized to include hyperbolic functions.

Section 6 will summarize the results and point out some areas for future developments.

2. Complex Riccati equations equivalent to the time-dependent Schrödinger equation

In the following the one-dimensional TDSE for problems with exact analytic solutions in the form of Gaussian WPs will be considered. This applies for potentials that are at most quadratic in position variable. In particular the HO ($V = \frac{m}{2}\omega^2 x^2$) with constant frequency, $\omega = \omega_0$, or the parametric oscillator with TD frequency, $\omega = \omega(t)$, will be discussed where the corresponding expressions for the free motion ($V(x) = 0$) are always obtained in the limit $\omega \to 0$.

The WP solution of the TDSE

$$i\hbar \frac{\partial}{\partial t}\Psi(x,t) = \left\{ -\frac{\hbar^2}{2m}\frac{\partial^2}{\partial x^2} + \frac{m}{2}\omega^2 x^2 \right\} \Psi(x,t) \tag{1}$$

(where $\hbar = \frac{h}{2\pi}$ with $h =$ Planck's constant) can be written as

$$\Psi(x,t) = N(t) \exp\left\{ i \left[y(t)\, \tilde{x}^2 + \frac{<p>}{\hbar}\, \tilde{x} + K(t) \right] \right\} \tag{2}$$

with the shifted coordinate $\tilde{x} = x - <x> = x - \eta(t)$, where the mean value $<x> = \int_{-\infty}^{+\infty} \Psi^* x \Psi dx = \eta(t)$ corresponds to the classical trajectory, $<p> = m\dot{\eta}$ represents the classical momentum and the coefficient of the quadratic term in the exponent, $y(t) = y_R(t) + i y_I(t)$, is a complex function of time. The (possibly TD) normalization factor $N(t)$ and the purely TD function $K(t)$ in the exponent are not relevant to the following discussion.

The equations of motion for $\eta(t)$ and $y(t)$, or $\left(\frac{2\hbar}{m} y = \mathcal{C}\right)$, that are obtained by inserting WP (2) into Eq. (1) are relevant for our purpose and are given by

$$\ddot{\eta} + \omega^2 \eta = 0 \;, \tag{3}$$

and

$$\dot{\mathcal{C}} + \mathcal{C}^2 + \omega^2 = 0 \;, \tag{4}$$

where overdots denote derivatives with respect to time. The Newtonian equation (3) simply expresses that the maximum of the WP, located at $x = <x> = \eta(t)$, follows the classical trajectory. The equation for the quantity $\frac{2\hbar}{m} y(t) = \mathcal{C}$ has the form a of a *complex* NL Riccati equation and describes the time-dependence of the WP width that is related to the position uncertainty via $y_I = \frac{1}{4<\tilde{x}^2>}$ with $<\tilde{x}^2> = <x^2> - <x>^2$ being the mean square deviation of position.

2.1. Direct solution of the Riccati equation

There are different ways of treating this (inhomogeneous) Riccati equation, elucidating different aspects of the equation. First, it can be solved directly by transforming it into a (homogeneous) NL (complex) Bernoulli equation if a particular solution $\tilde{\mathcal{C}}$ of the Riccati equation is known. The general solution of Eq. (4) is then given by $\mathcal{C} = \tilde{\mathcal{C}} + \mathcal{V}(t)$ where $\mathcal{V}(t)$ fulfils the Bernoulli equation

$$\dot{\mathcal{V}} + 2\,\tilde{\mathcal{C}}\mathcal{V} + \mathcal{V}^2 = 0\,. \tag{5}$$

The coefficient of the linear term depends on the particular solution $\tilde{\mathcal{C}}$. Equation (5) can be linearized via $\mathcal{V} = \frac{1}{\kappa(t)}$ to yield

$$\dot{\kappa} - 2\,\tilde{\mathcal{C}}\,\kappa \;=\; 1\,, \tag{6}$$

which can be solved straightforwardly. For constant $\tilde{\mathcal{C}}$, $\kappa(t)$ can be expressed in terms of exponential or hyperbolic functions (for real $\tilde{\mathcal{C}}$) or trigonometric functions (for imaginary $\tilde{\mathcal{C}}$). In this case, $\mathcal{C}$ can be written as

$$\mathcal{C}(t) \;=\; \tilde{\mathcal{C}} \;+\; \frac{e^{-2\tilde{\mathcal{C}}t}}{\frac{1}{2\tilde{\mathcal{C}}}\left(1 - e^{-2\tilde{\mathcal{C}}t}\right) \;+\; \kappa_0} \tag{7}$$

where the enumerator is obviously the derivative of the denominator.

For $\tilde{\mathcal{C}}$ being TD, $\kappa(t)$ and hence $\mathcal{V}$ can be expressed in terms of $\mathcal{I}(t) = \int^t dt'\, \mathrm{e}^{-\int^{t'} dt'' 2\tilde{\mathcal{C}}(t'')}$. So the general solution of Eq. (4) can be written as

$$\mathcal{C}(t) \;=\; \tilde{\mathcal{C}} \;+\; \frac{d}{dt}\,\ln\left[\kappa_0 + \mathcal{I}(t)\right] \tag{8}$$

with the logarithmic derivative representing the solution of the Bernoulli equation. It defines a one-parameter family of solutions depending on the (complex) initial value $\kappa_0 = \mathcal{V}_0^{-1}$ as a parameter. To realize the strong qualitative influence of the initial value κ_0 on the solution of the Riccati equation (which is not surprising since this is a NL differential equation), a comparison with supersymmetric (SUSY) quantum mechanics [3, 4, 5] could be quite useful.

In SUSY quantum mechanics, the Hamiltonian can be represented by a 2 $\times$ 2 diagonal matrix where the potentials V_i of the Hamiltonian operators $H_i \;=\; -\frac{\hbar^2}{2m}\frac{d^2}{dx_2} + V_i$ (i= 1,2) on the diagonal are determined by a (real) Riccati equation for the so-called "superpotential" $W(x)$,

$$V_{1/2} \;=\; \frac{1}{2}\left[W^2 \;\mp\; \frac{\hbar}{\sqrt{m}}\,\frac{d}{dx}W\right] \tag{9}$$

and differ only by the sign of the term depending on the derivative of $W(x)$. Again, one can try to solve this Riccati equation by reducing it to a Bernoulli equation using a particular solution $\tilde{W}$. In the case of the HO, this particular solution is given by $\tilde{W} \;=\; \sqrt{m}\omega_0\, x$, leading to the two potentials

$$\tilde{V}_{1/2} \;=\; \frac{m}{2}\,\omega_0^2\, x^2 \;\mp\; \frac{\hbar}{2}\,\omega_0\,, \tag{10}$$

essentially the parabolic harmonic potential, only shifted by minus/plus the ground state energy $\frac{\hbar}{2}\omega_0$. The general solution can again be written in the form $W(x) \;=\; \tilde{W}(x) \;+\; \Phi(x)$ where $\Phi(x)$ must now fulfil the Bernoulli equation

$$\frac{\hbar}{\sqrt{m}}\,\frac{d}{dx}\Phi \;+\; 2\tilde{W}\,\Phi \;+\; \Phi^2 \;=\; 0 \tag{11}$$

(written here for the plus sign of the derivative). This can be solved in the same way by linearization to finally yield the general solution

$$W(x) \;=\; \tilde{W}(x) \;+\; \frac{\hbar}{\sqrt{m}}\,\frac{d}{dx}\,\ln\left[\varepsilon + \mathcal{I}(x)\right] \,. \tag{12}$$

The integral $\mathcal{I}(x)$ is formally identical to the one in the TD case, only t must be replaced by x and $\tilde{C}$ by $\tilde{W}$. Also this solution depends on a (this time real) parameter ε (corresponding to $\Phi^{-1}(0)$). This generalized $W(x)$ gives rise to a one-parameter family of isospectral potentials (e.g., for i = 1)

$$V_1(x;\varepsilon) \;=\; \tilde{V}_1(x) \;-\; \frac{\hbar^2}{m}\,\frac{d^2}{dx^2}\,\ln\left[\varepsilon + \mathcal{I}(x)\right] \,, \tag{13}$$

i.e., potentials with different shapes but the same energy spectrum, namely that of the HO, only with ground state energy equal to zero (apart from $\varepsilon = 0$, where this state is missing).

The shape of the potentials is now, unlike the parabolic harmonic potential, no longer symmetric under the exchange $x \to -x$; there is even a second minimum showing up for negative x whose depth increases with decreasing ε (for details, see [5]). Only for $\varepsilon \to \infty$, the ln-term vanishes and the parabolic potential $\tilde{V}_1(x)$ is re-gained. So, obviously the parameter ε can have drastic qualitative consequences for the solution of the Riccati equation. The same can also happen in the above-mentioned TD case when κ_0 is varied.

2.2. Treatment via Ermakov equation

There is a second way of treating the complex Riccati equation (4). Transforming it into a real NL differential equation which, in combination with the Newtonian equation (3), leads to a dynamical invariant with the dimension of an action. For this purpose a new (real) variable $\alpha(t)$ is introduced via $\mathcal{C}_I = \left(\frac{2\hbar}{m}y_I\right) = \frac{1}{\alpha^2}$, i.e., $\alpha \propto \sqrt{<\tilde{x}^2>}$ and is thus directly proportional to the WP width. Inserting this into the imaginary part of Eq. (4) allows one to determine the real part of the variable as $\mathcal{C}_R = \left(\frac{2\hbar}{m}y_R\right) = \frac{\dot{\alpha}}{\alpha}$, which, when inserted into the real part of (4) together with the above definition of $\mathcal{C}_I$, finally turns the complex Riccati equation into the real NL so-called Ermakov equation [1] for $\alpha(t)$,

$$\ddot{\alpha} + \omega^2\alpha = \frac{1}{\alpha^3} \,. \tag{14}$$

It was shown by Ermakov [9] in 1880, i.e., 45 years before quantum mechanics was formulated by Schrödinger and Heisenberg, that from the pair of equations (3) and (14), coupled via ω^2, by eliminating ω^2 from the equations, a dynamical invariant, the Ermakov-invariant,

$$I_L = \frac{1}{2}\left[(\dot{\eta}\alpha - \eta\dot{\alpha})^2 + \left(\frac{\eta}{\alpha}\right)^2\right] \;=\; \text{const.} \tag{15}$$

can be obtained (this invariant was rediscovered by several authors, also in a quantum mechanical context; see, e.g., [10, 11, 12]).

This invariant has (at least) two remarkable properties: i) it is also a constant of motion for $\omega = \omega(t)$, in the case where the corresponding Hamiltonian does not have this property; ii) apart from a missing constant factor m, i.e., mass of the system, it has the dimension of an *action*,

[1] This equation was studied already in 1874 by Adolph Steen [6]. However, Steen's work was ignored by mathematicians and physicists for more than a century because it was published in Danish in a journal usually not containing many articles on mathematics. An English translation of the original paper [7] is available and generalizations can be found in [8].

not of an energy. The missing factor m is due to the fact that Ermakov used the *mathematical* Eq. (3) whereas, in a *physical* context, Newton's equation of motion, i.e., Eq. (3) multiplied by m, is relevant.

Using the definitions of $\mathcal{C}_I$ and $\mathcal{C}_R$, the invariant can be rewritten as

$$I_L = \frac{1}{2}\alpha^2\left[(\dot{\eta} - \mathcal{C}_R\eta)^2 + (\mathcal{C}_I\eta)^2\right] = \frac{1}{2}\alpha^2\left[(\dot{\eta} - \mathcal{C}\eta)(\dot{\eta} - \mathcal{C}^*\eta)\right] . \tag{16}$$

In this form it is obvious that the invariant (and consequently an operator that can be obtained from it by canonical quantization) can be factorized in a similar way as the Hamiltonian operator H_{op} of the HO or an operator related to it via

$$\hat{H}_{op} = \frac{H_{op}}{\hbar\omega_0} = \left(a^+ a + \frac{1}{2}\right) , \tag{17}$$

where a^+a is the so-called number operator and the creation and annihilation operators are defined by

$$a^+ = -i\sqrt{\frac{m}{2\hbar\omega_0}}\left(\frac{p_{op}}{m} + i\,\omega_0\,x\right) = \frac{1}{\sqrt{2\hbar\omega_0}}\left(-\frac{\hbar}{\sqrt{m}}\frac{\partial}{\partial x} + \sqrt{m}\omega_0\,x\right) , \tag{18}$$

$$a = i\sqrt{\frac{m}{2\hbar\omega_0}}\left(\frac{p_{op}}{m} - i\,\omega_0\,x\right) = \frac{1}{\sqrt{2\hbar\omega_0}}\left(\frac{\hbar}{\sqrt{m}}\frac{\partial}{\partial x} + \sqrt{m}\omega_0\,x\right) , \tag{19}$$

where $p_{op} = \frac{\hbar}{i}\frac{\partial}{\partial x}$ and a is the adjoint operator of a^+.

The number that is the eigenvalue of a^+a is the number of quanta of the action $\hbar$ since $\frac{H_{op}}{\omega_0}$ has the dimension of an *action*! With the help of a, the ground state wave function can be obtained and from this, by successive application of a^+, the excited states can be created. Via superposition of all these states, Schrödinger obtained a stable Gaussian WP (with constant width) [13]. Generalizations of Schrödinger's approach were achieved for the description of coherent light beams emitted by lasers in terms of what is now called coherent state (CS).

One of at least three different definitions of CSs is that these are eigenstates of the annihilation operator a with (complex) eigenvalue z, $a\,|z> \;=\; z\,|z>$. Comparing the CS $|z>$ for the HO with the minimum uncertainty WP solution in the form of Eq. (2), it shows that $\omega_0 = \frac{2\hbar}{m}y_I = \frac{1}{\alpha_0^2}$. So, in definitions (18) and (19), $i\,\omega_0$ can be replaced by $i\,\frac{2\hbar}{m}y_I = i\mathcal{C}_I$. Therefore, for the more general case of WPs or CSs with TD width ($\mathcal{C}_R \neq 0$), $i\,\omega_0$ must be replaced by the full complex quantity $\mathcal{C}$ in a and by $\mathcal{C}^*$ in the adjoint operator a^+. If one then substitutes $\frac{1}{\sqrt{\omega_0}} = \alpha_0$ in front of the brackets with $\alpha(t)$, the generalized creation and annihilation operators take the form

$$a^+(t) = -i\sqrt{\frac{m}{2\hbar}}\,\alpha(t)\left(\frac{p_{op}}{m} - \mathcal{C}^* x\right) , \tag{20}$$

$$a(t) = i\sqrt{\frac{m}{2\hbar}}\,\alpha(t)\left(\frac{p_{op}}{m} - \mathcal{C}x\right) . \tag{21}$$

These operators can even be turned into constants of motion if an additional phase factor is taken into account. But in the case of CSs, as discussed here, this factor can be absorbed into the phase of the CS and will therefore be omitted in the following (for further details see [14]).

Employing the above definition of the CS, but now with our generalized annihilation operator, i.e., $a(t)\,|z> \;=\; z\,|z>$, the CS (in position representation, $< x|z >= \Psi_z(x,t)$) can be obtained as

$$\Psi_z(x,t) = \left(\frac{m}{\pi\hbar}\right)^{\frac{1}{4}} \lambda^{-\frac{1}{2}} \exp\left\{i\left[y\tilde{x}^2 + \frac{<p>\tilde{x}}{\hbar} + \frac{<p><x>}{2\hbar}\right]\right\} , \tag{22}$$

which is in complete agreement with our WP definition (2) (here, also $N(t)$ and $K(t)$ are now specified where the complex quantity $\lambda(t)$ is defined via $\mathcal{C} = \left(\frac{2\hbar}{m}y\right) = \frac{\dot{\lambda}}{\lambda}$ and will be discussed in the next subsection).

From the mean values of position and momentum, $< x >_z = \eta$ and $< p >_z = m\dot{\eta}$, calculated with these CSs, the real and imaginary parts of the complex eigenvalue $z = z_R + iz_I$ can be determined to be

$$z_R = \sqrt{\frac{m}{2\hbar}}\left(\frac{\eta}{\alpha}\right) \quad , \quad z_I = \sqrt{\frac{m}{2\hbar}}\,(\dot{\eta}\alpha - \eta\dot{\alpha}) \tag{23}$$

which looks familiar when compared with the Ermakov invariant (15). Indeed, the absolute square of z is, up to a constant factor, identical to I_L,

$$I_L = \frac{\hbar}{m}\left(z_I^2 + z_R^2\right) = \frac{\hbar}{m}zz^* = \frac{\hbar}{m}|z|^2 \,. \tag{24}$$

In the same way that the anihilation operator of the HO can be used to obtain the corresponding ground state wave function and the creation operator to obtain the excited states, the generalized annihilation and creation operators (20) and (21) can be used to obtain solutions of the TDSE where the potential of the problem enters via the complex TD variable $\mathcal{C}$ (or y) of the corresponding Riccati equation.

So, e.g., for the free motion (V = 0) application of $a(t)$ on the ground state wave function $|0>$ leads (in position representation) via $a(t)|0> = 0$ to

$$< x|0> = \Psi_0(x,t) = \left(\frac{m}{\pi\hbar}\right)^{\frac{1}{4}} \lambda^{-\frac{1}{2}} \exp\left\{iyx^2\right\} \tag{25}$$

where $y(t)$ is obtained from the Riccati equation (4) for $\omega = 0$. Note that this ground state (in contrast with the TI HO) has a position dependent phase via iy_Rx^2.

The corresponding excited states can be obtained by applying $a(t)$ on this ground state according to $\Psi_n(x,t) = (a^+)^n|0>$ in position-space representation as

$$\Psi_n(x,t) = \left[\left(\frac{m}{\pi\hbar}\right)^{\frac{1}{2}} \frac{1}{\alpha}\frac{1}{n!2^n}\right]^{1/2} \mathcal{H}\left(\frac{x}{x_0}\right) \exp\left\{-\frac{1}{2}\left(\frac{x}{x_0}\right)^2 + i\frac{\dot{\alpha}}{\alpha}x^2\right\} \tag{26}$$

with $x_0 = \sqrt{\frac{\hbar}{m}}\alpha(t)$ being TD due to $\alpha(t)$. Hence, the dimensionless variable $\xi = \frac{x}{x_0}$ that usually appears in this context in the Hermite polynomials $\mathcal{H}(\xi)$ is now TD! Note: the $\Psi_n(x,t)$ are solutions of the TDSE for $V = 0$ but they are no eigenfunctions of the operator $H_{op} = -\frac{\hbar^2}{2m}\frac{\partial^2}{\partial x^2}$ (but of $I_{L,op}$!). Similar solutions for the TD free motion SE have also been found by Guerrero et al [15] in a different way.

2.3. Linearization of the complex Riccati equation

Another property of the Riccati equation, particularly interesting in a quantum mechanical context, is the existence of a superposition principle for this NL differential equation [16, 17]. This is related to the fact that the Riccati equation can always be linearized. In our case, this can be achieved using the ansatz

$$\left(\frac{2\hbar}{m}y\right) = \mathcal{C} = \frac{\dot{\lambda}}{\lambda} \,, \tag{27}$$

with complex $\lambda(t)$, leading to

$$\ddot{\lambda} + \omega^2(t)\lambda = 0 \tag{28}$$

which has the form of the Newton-type equation (3) of the corresponding problem, but now for a complex variable.

First, a kind of geometric interpretation of the motion of λ in the complex plane shall be given. Expressed in Cartesian coordinates, λ can be written as $\lambda = u+iz$, or in polar coordinates as $\lambda = \alpha e^{i\varphi}$. Inserting the polar form into Eq. (27) leads to

$$\left(\frac{2\hbar}{m}y\right) = \frac{\dot{\alpha}}{\alpha} + i\,\dot{\varphi} \tag{29}$$

where the real part is already identical to $\frac{2\hbar}{m}y_R$, as defined above.

The quantity α defined in $\frac{2\hbar}{m}y_I$ as being proportional to the position uncertainty is identical to the absolute value of λ if it can be shown that

$$\dot{\varphi} = \frac{1}{\alpha^2}\,. \tag{30}$$

This, however, can be proven by simply inserting real and imaginary parts of (29) into the imaginary part of the Riccati equation (4). Comparing relation (30), that can also be written in the form

$$\dot{z}u - \dot{u}z = \alpha^2\dot{\varphi} = 1\,, \tag{31}$$

with the motion of a particle under the influence of a central force in two-dimensional physical space, it shows that this relation corresponds to the "conservation of angular momentum", but here for the motion in the *complex* plane!

Relation (31) also shows that real and imaginary parts, or phase and amplitude, respectively, of the complex quantity are not independent of each other but uniquely coupled. This coupling, which, as mentioned in the Introduction, is typical for quantum systems (but not only for these) is due to the quadratic nonlinearity in the Riccati equation. We will find an analogous situation also in the TI case, discussed in the next section.

The complex variable $\lambda = u + iz$ also allows for a kind of geometric interpretation of the Ermakov invariant. For this purpose it will be used that the imaginary part of λ is directly proportional to the classical trajectory, i.e., $z = \frac{m}{\alpha_0 p_0}\eta(t)$ (this can be proven using the corresponding TD Green function, for details see, e.g., [18]). Therefore, the second quadratic term of the invariant (15) is proportional to $(\frac{z}{\alpha})^2 = \sin^2\varphi$. Consequently, the first term must be $(\frac{u}{\alpha})^2 = \cos^2\varphi$ to yield a constant value for I_L. So the invariant can be written as

$$I_L = \frac{1}{2}\left(\frac{\alpha_0 p_0}{m}\right)^2 \left[\left(\frac{u}{\alpha}\right)^2 + \left(\frac{z}{\alpha}\right)^2\right] = \text{const.}\,. \tag{32}$$

Defining a new variable $\mathcal{Y}(\varphi) = (\frac{z}{\alpha}) = \sin\varphi$ that depends on the angle φ instead of time t, the first term in the the square bracket can be expressed as

$$\mathcal{Y}'(\varphi) = \frac{d}{d\varphi}\mathcal{Y} = \cos\varphi = \left(\frac{u}{\alpha}\right)\,. \tag{33}$$

In this form, the invariant (32) is formally equivalent to the Hamiltonian of a HO with angle-dependent variable $\mathcal{Y}(\varphi)$ (instead of TD variable $\eta(t)$) and unit frequency $\omega = 1$, leading to the corresponding equation of motion

$$\mathcal{Y}'' + \mathbf{1}^2\mathcal{Y} = 0\,. \tag{34}$$

However, the time-dependence is implicitly contained in the time-dependence of the angle φ, i.e., $\varphi = \varphi(t)$. Therefore, expressing Eq. (34) as a differential equation with respect to time instead of with respect to angle φ and using $\frac{d}{d\varphi} = \alpha^2\frac{d}{dt}$ with $\mathcal{Y}(\varphi(t)) = \hat{\mathcal{Y}}(t)$ yields

$$\frac{d^2}{dt^2}\hat{\mathcal{Y}}(t) + 2\,\frac{\dot{\alpha}}{\alpha}\,\frac{d}{dt}\hat{\mathcal{Y}}(t) + \dot{\varphi}^2\hat{\mathcal{Y}}(t) = 0\,. \tag{35}$$

For $\dot{\alpha} = 0$, i.e. $\alpha =$ constant, with $z(t) \propto \eta(t)$ and $\dot{\varphi} =$const. $= \omega_0$, Eq. (35) just turns into $\ddot{\eta} + \omega_0^2\eta = 0$, i.e., Eq. (3) for TI frequency $\omega = \omega_0$ and a WP solution with constant width.

For $\dot{\alpha} \neq 0$, i.e. $\alpha = \alpha(t)$, an additional first derivative term appears in (35) that looks like a linear velocity dependent friction force in the Langevin equation (see below) with friction coefficient $2\,\frac{\dot{\alpha}}{\alpha}$. At first sight this looks contradictory since we are not dealing so far with dissipative systems with irreversible time-evolution. A closer look shows that Eq. (35) is actually still invariant under time reversal since the coefficient of the second term also contains a time-derivative (in $\frac{\dot{\alpha}}{\alpha}$). So, together with $\frac{d}{dt}\hat{\mathcal{Y}}$, this term also does not change its sign under time-reversal (unlike in the Langevin equation, where the friction coefficient γ is usually assumed to be constant). Eq. (35) takes into account that not only the angle of the system described by $\lambda(t)$, but also its radius may change in time.

3. Complex Riccati equations related to the time-independent Schrödinger equation

We have seen in the TD case that the real and imaginary parts, or phase φ and amplitude α, of the complex variable $\lambda(t) = \alpha\, e^{i\varphi}$ which fulfils the linear equation (28), obtained via Eq. (27) from the Riccati equation (4), are not independent of each other but coupled via the conservation law (30). A similar situation exists when considering the TISE, but now in the space-dependent case.

This can be shown using Madelung's hydrodynamic formulation of quantum mechanics [19] where the wave function is written in polar form as

$$\Psi(\mathbf{r},t) = \varrho^{1/2}(\mathbf{r},t)\ \exp\left(\frac{i}{\hbar}S(\mathbf{r},t)\right) \tag{36}$$

with the square root of the probability density $\varrho = \Psi^*\Psi$ as amplitude and $\frac{1}{\hbar}S$ as phase ($\mathbf{r}$ is the position vector in three dimensions).

Inserting this form into the TDSE (1) (now in three dimensions, and replacing $\frac{\partial}{\partial x}$ by the nabla operator ∇), leads to a modified Hamilton–Jacobi equation for the phase,

$$\frac{\partial}{\partial t}S + \frac{1}{2m}(\nabla S)^2 + V - \frac{\hbar^2}{2m}\frac{\Delta\varrho^{1/2}}{\varrho^{1/2}} \ = 0\ , \tag{37}$$

and a continuity equation for the amplitude,

$$\frac{\partial}{\partial t}\varrho + \frac{1}{m}\nabla(\varrho\ \nabla S) = 0\ . \tag{38}$$

Already here, the coupling of phase and amplitude can be seen clearly since the Hamilton–Jacobi equation for the phase S contains a term (misleadingly called "quantum potential", V_{qu}) depending on ϱ, and the continuity equation for the density ϱ contains ∇S. It can be shown that also in the TI case this coupling is not arbitrary but related to a conservation law.

In 1994, G. Reinisch [20] did this in a NL formulation of TI quantum mechanics. Since in this case $\frac{\partial}{\partial t}\varrho = 0$ and $\frac{\partial}{\partial t}S = -E$ are valid, the continuity equation (38) (we now use the notation $\varrho^{1/2} = |\Psi| = a$) turns into

$$\nabla(a^2\nabla S) \ = \ 0 \tag{39}$$

and the modified Hamilton–Jacobi equation into

$$-\frac{\hbar^2}{2m}\Delta a + (V - E)a = -\frac{1}{2m}(\nabla S)^2 a\ . \tag{40}$$

Equation (39) is definitely fulfilled for $\nabla S = 0$, turning (40) into the usual TISE for the real wave function $a = |\Psi|$ with position-independent phase S. (N.B.: the kinetic energy term divided by a is just identical to V_{qu}!)

However, Eq. (39) can also be fulfilled for $\nabla S \neq 0$ if only the conservation law

$$\nabla S = \frac{C}{a^2} \tag{41}$$

is fulfilled with constant (or, at least, position-independent) C.

This relation now shows explicitly the coupling between phase and amplitude of the wave function and is equivalent to Eq. (30) in the TD case. Inserting (41) into the rhs of Eq. (40) changes this into the Ermakov equation

$$\Delta a + \frac{2m}{\hbar^2}(E - V)a = (\frac{1}{\hbar}\nabla S)^2 \, a = \left(\frac{C}{\hbar}\right)^2 \frac{1}{a^3} \, , \tag{42}$$

equivalent to Eq. (14) in the TD case.

Returning to the method described in [20], so far the energy E occurring in Eq. (42) is still a free parameter that can take any value. However, solving this equation numerically for arbitrary values of E generally leads to solutions a that diverge for increasing x. Only if the energy E is appropriately tuned to any eigenvalue E_n of the TISE (see Eq. (44), below) does this divergence disappear and normalizable solutions can be found. So, the quantization condition that is usually obtained from the requirement of truncation of an infinite series in order to avoid divergence of the wave function is, in this case, obtained from the requirement of non-diverging solutions of the NL Ermakov equation (42) by variation of the parameter E. This has been numerically verified in the case of the one-dimensional HO and the Coulomb problem and there is the conjecture that this property is "universal" in the sense that it does not depend on the potential V (see [20, 21]).

The corresponding complex Riccati equation is now given by

$$\nabla\left(\frac{\nabla\Psi}{\Psi}\right) + \left(\frac{\nabla\Psi}{\Psi}\right)^2 + \frac{2m}{\hbar^2}(E - V) = 0 \tag{43}$$

with the complex variable $C = \left(\frac{\nabla\Psi}{\Psi}\right) = \frac{\nabla a}{a} + i\frac{1}{\hbar}\nabla S$ which corresponds to $\left(\frac{2\hbar}{m}y\right) = \frac{\dot{\lambda}}{\lambda} = \frac{\dot{\alpha}}{\alpha} + i\,\dot{\varphi}$ in the TD problem.

It is possible to show straightforwardly that Eq. (43) can be linearized to yield the usual TISE

$$-\frac{\hbar^2}{2m}\Delta\Psi + V\Psi = E\Psi \, , \tag{44}$$

but in this form the information on the coupling of phase and amplitude, expressed by Eq. (41) and originating from the quadratic NL term in Eq. (43), gets lost.

However, particularly this relation (41) now allows for the construction of complex solutions of the form

$$\Psi(\mathbf{r}) \;=\; a(\mathbf{r})\exp\left[-\frac{i}{\hbar}\int^{\mathbf{r}} \frac{C}{a^2(\mathbf{r}')}d\mathbf{r}'\right] \tag{45}$$

once the solution of the Ermakov equation (42) is known. Using the techniques known from SUSY quantum mechanics it is possiple to start from a real potential with real eigenvalue E to construct a complex partner potential, but still with real eigenvalues, opening a new way to obtain non-hermitian Hamiltonians with real spectrum.

The procedure shall be outlined briefly for a one-dimensional example. In this case, the TISE

$$\left(-\frac{\hbar^2}{2m}\frac{d^2}{dx^2}+V_1\right)\Psi = H_1\ \Psi = E\Psi \tag{46}$$

can be expressed in terms of the two linear operators

$$A = -\frac{\hbar}{\sqrt{m}}\frac{d}{dx}+W(x) \quad \text{and} \quad B = \frac{\hbar}{\sqrt{m}}\frac{d}{dx}+W(x) \tag{47}$$

as

$$AB = -\frac{\hbar^2}{2m}\frac{d^2}{dx^2}+W^2 \ - \ \frac{\hbar}{\sqrt{m}}\frac{d}{dx}W = 2(H_1-E) \tag{48}$$

where the potential V_1 is related to $W(x)$ via

$$V_1 = \frac{1}{2}\left(W^2 - \frac{\hbar}{\sqrt{m}}\frac{d}{dx}\right) . \tag{49}$$

The partner Hamiltonian H_2 can be obtained by reversing the order of the operators A and B, i.e.,

$$BA = -\frac{\hbar^2}{2m}\frac{d^2}{dx^2}+W^2 \ + \ \frac{\hbar}{\sqrt{m}}\frac{d}{dx}W \ = 2(H_2-E) \tag{50}$$

where H_2 differs from H_1 only by an additional term in the potential, i.e.,

$$V_2 = \frac{1}{2}\left(W^2 + \frac{\hbar}{\sqrt{m}}\frac{d}{dx}W\right) = V_1 + \frac{\hbar}{\sqrt{m}}\frac{d}{dx}W . \tag{51}$$

In the same way as usual in SUSY quantum mechanics, the function $W(x)$ is related to the wave function $\Psi(x)$ of Eq. (46) via its logarithmic derivative (where Ψ here is not necessarily the ground state wave function and E not necessarily the ground state energy $E=0$),

$$W = -\frac{\hbar}{\sqrt{m}}\frac{\frac{d}{dx}\Psi}{\Psi} = -\frac{\hbar}{\sqrt{m}}\frac{d}{dx}\ln\Psi . \tag{52}$$

Particularly in our case, we chose $\Psi(x)$ as a complex function as given in (45). Therefore, the additional term in V_2 will also be complex, turning it into a complex potential though V_1 and E are real!

As a first example the case $V=0$, i.e., the free motion shall be mentioned. In this case it can be shown that the Ermakov equation (42) (for proper choice of the occurring parameters and $\hbar = m = 1$) has the solution

$$a(x) = \sqrt{\cosh(\kappa x)} \tag{53}$$

with $E = -\frac{\kappa}{2}$, $\kappa > 0$, which can be used to construct the complex wave function $\Psi(x)$ via (45) and the function $W(x)$ via (52) as

$$W(x) = -\frac{\kappa}{2}\tanh(\kappa x) + \frac{i}{2}\frac{1}{\cosh(\kappa x)} . \tag{54}$$

Since $V_1 = 0$, V_2 is just the derivative of $W(x)$ and given by

$$V_2 = -\frac{\kappa^2}{2}\frac{1}{\cosh^2(\kappa x)} - i\frac{\kappa}{2}\frac{\sinh(\kappa x)}{\cosh^2(\kappa x)} = \left[1+\frac{i}{\kappa}\sinh(\kappa x)\right]V_{PT}(x) , \tag{55}$$

where V_{PT} is the well known Pöschl–Teller potential.

Whereas the real part of V_2, i.e. $V_{PT}(x)$, is symmetric under the exchange $x \rightarrow -x$, the imaginary part is antisymmetric under this operation. Further details will be discissed in a forthcoming publication [2].

4. Riccati equations in dissipative Schrödinger equations

Since real physical systems are always in contact with some kind of environment and this coupling usually introduces the phenomena irreversibility and dissipation (but not necessarily always both simultaneously!), the questions arise of how this can be taken into account in the formalism of classical (Hamiltonian) mechanics and, especially, in a quantum mechanical context.

In the trajectory picture, dissipation can be included by adding a phenomenological friction force that is proportional to velocity (or momentum) to the Newtonian equation of motion, thus turning it into an irreversible evolution equation, i.e.,

$$m\ddot{x} = m\dot{v} = -m\gamma v - \frac{\partial}{\partial x}V \tag{56}$$

which is the Langevin equation without stochastic contribution (γ is the friction coefficient). This equation is used to describe Brownian motion from a trajectory point of view. An equivalent description of this phenomenon can also be given in terms of (classical) probability distributions ϱ_{cl} via the Fokker–Planck equations that contain irreversible diffusion terms. Particularly in position space, this can be written in the form of the Smoluchowski equation for $\varrho_{cl}(x,t)$,

$$\frac{\partial}{\partial t}\varrho_{\rm cl} + \frac{\partial}{\partial x}\left(\frac{F(x)}{m\gamma}\varrho_{\rm cl}\right) - \frac{\mathsf{kT}}{m\gamma}\frac{\partial^2}{\partial x^2}\varrho_{\rm cl} \ = \ 0\ , \tag{57}$$

with the conservative force $F(x)$, Boltzmann's constant k and temperature T. Comparison with Einstein's theory of Brownian motion [22] shows that the coefficient of the diffusion term fulfils the Einstein relation $D = \frac{\mathsf{kT}}{m\gamma}$.

However, Eqs. (56) and (57) do not fit consistently into the Lagrange/Hamilton formalism that is invariant under canonical transformations and provides a basis for quantization. So, these phenomenological equations do not yet provide the answer to the questions posed at the beginning of this section.

There are numerous approaches in the literature trying to answer these questions in different ways, most of them with some problematic aspects. In the following, only one approach will be discussed, that uses an effective modified TDSE that, for potentials at most quadratic in position variable, also leads to Gaussian WP solutions like the ones discussed in section 2 and thus allows for direct comparison with the corresponding conservative case. However, this approach can be uniquely linked with other similar approaches and approaches using modified Lagrangian/Hamiltonian functions (with subsequent canonical quantization) or even system-plus-reservoir approaches like the one by Caldeira and Leggett. Since a detailed discussion of these interrelations has already been given in the proceedings of the last Symmetries in Science Symposium [23], this shall not be repeated here.

For the following it will be sufficient to consider a modified TDSE that is able to take into account the effect of an environment without introducing any individual degrees of freedom of it, but changing the observable dynamics in a way, that particularly the averaged equation of motion for the position, according to Ehrenfest, contains a linear velocity dependent friction force as it occurs in the Langevin equation (56). Approaches adding a so-called "friction potential" (that usually introduces some kind of nonlinearity) to the Hamiltonian operator in a way that it provides the desired averaged equation of motion, contain a lot of ambiguity and therefore lead to sometimes rather inconsistant results.

We therefore tried (see [24]) an approach that is not based on the dissipative aspect of the trajectory picture but on the irreversible aspect of the probability picture. For this purpose, we add a time-reversal symmetry-breaking diffusion term to the continuity equation for ϱ, thus turning it, as previously discussed, into an irreversible Smoluchowski equation

$$\frac{\partial}{\partial t}\varrho + \frac{\partial}{\partial x}\left(\varrho\, v_-\right) - D\frac{\partial^2}{\partial x^2}\varrho \ = \ 0 \tag{58}$$

but now for the quantum mechanical $\varrho(x,t)$. For the continuity equation, Madelung [25] and Mrowka [26] had shown that with a bilinear ansatz for the probability current density, $j = \varrho\left(\frac{1}{m}\frac{\partial}{\partial x}S\right) = \varrho v_- = \left(\frac{\hbar}{2mi}\right)\left(\Psi^*\frac{\partial}{\partial x}\Psi - \Psi\frac{\partial}{\partial x}\Psi^*\right)$, this can be separated into the TDSE and its complex conjugate where the separation "constant" is proportional to the potential V. However, due to the diffusion term in (58), Ψ and Ψ^*-dependent terms can no longer be separated in general.

For separation the condition

$$-D\frac{\frac{\partial^2}{\partial x^2}\varrho}{\varrho} = F_1(\Psi) + F_2(\Psi^*) \tag{59}$$

must be fulfilled.

This can be achieved by introducing the additional separation condition

$$-D\frac{\frac{\partial^2}{\partial x^2}\varrho}{\varrho} = \gamma\left(\ln\,\varrho - <\ln\,\varrho>\right)\;, \tag{60}$$

where $\ln\,\varrho = \ln\,\Psi + \ln\,\Psi^*$. This leads to an additional *complex* logarithmic term [24] in the SE for $\Psi_{NL}(x,t)$,

$$i\hbar\frac{\partial}{\partial t}\Psi_{NL}(x,t) = \{H_L + \gamma\frac{\hbar}{i}\left(\ln\Psi_{NL} - <\ln\Psi_{NL}>\right)\}\Psi_{NL}(x,t) = \{H_L + W_{\text{SCH}}\}\Psi_{NL}(x,t)\;, \tag{61}$$

where H_L is the usual linear Hamiltonian. The additional NL term W_{SCH} can be written as real and imaginary contributions in the form

$$W_{\text{SCH}} = W_R + iW_I = \frac{\gamma}{2}\frac{\hbar}{i}\left(\ln\frac{\Psi_{NL}}{\Psi^*_{NL}} - <\ln\frac{\Psi_{NL}}{\Psi^*_{NL}}>\right) + \frac{\gamma}{2}\frac{\hbar}{i}\left(\ln\,\varrho_{NL} - <\ln\,\varrho_{NL}>\right) \tag{62}$$

where the real part only depends on the phase of the wave function and provides the friction force in the averaged equation of motion. The imaginary part does not contribute to dissipation but introduces irreversibility into the evolution of the wave function. It corresponds to the diffusion term in the Smoluchowski equation but still allows for normalizability due to the subtraction of the mean value of $\ln\,\varrho$. Furthermore, it has properties that are usually attributed to a stochastic term occurring in the Langevin equation (details are not relevant to the following discussion and will be discussed elsewhere).

The imaginary part breaks the time-reversal symmetry on the level of the probability density, introduces a non-unitary time evolution and turns the Hamiltonian into a non-hermitian one while still guaranteeing normalizable wave functions and real energy mean values since its mean value vanishes.

By itself, the real part would provide dissipation but retain a unitary time-evolution of the wave function, whereas the imaginary part on its own would provide irreversibility via a non-unitary time-evolution but no dissipation. Consequently only the combination of real and imaginary parts provides all the desired properties of the quantum system under consideration. The reason for this is the coupling of phase and amplitude of the wave function as expressed by Eq. (41) since W_R depends on the phase and W_I on the amplitude.

4.1. Modification of the Gaussian WP dynamics

Inserting WP (2) into the NLSE (61) yields, for $\eta(t) = <x>_{NL}$ and hence for the motion of the WP maximum, the Newtonian equation including the friction force

$$\ddot{\eta} + \gamma\dot{\eta} + \omega^2\eta = 0 \tag{63}$$

and for the complex variable $(\frac{2\hbar}{m}y)_{NL} = \mathcal{C}_{NL}$ the modified Riccati equation

$$\dot{\mathcal{C}}_{NL} + \gamma\, \mathcal{C}_{NL} + \mathcal{C}_{NL}^2 + \omega^2(t) \;=\; 0 \tag{64}$$

with an additional linear term depending on the friction coefficient γ.

The relation between the imaginary part of this variable and the position uncertainty or Ermakov variable, respectively, remains unchanged as in the conservative case

$$\mathcal{C}_{NL,I} \;=\; \frac{\hbar}{2m < \tilde{x}^2 >_{NL}} = \frac{1}{\alpha_{NL}^2} \tag{65}$$

but the real part differs by a contribution from the friction coefficient, i.e.,

$$\mathcal{C}_{NL,R} \;=\; \frac{\dot{\alpha}_{NL}}{\alpha_{NL}} - \frac{\gamma}{2}\,. \tag{66}$$

Inserting (65) and (66) into the Riccati equation (64) turns this into the Ermakov equation

$$\ddot{\alpha}_{NL} + \left(\omega^2 - \frac{\gamma^2}{4}\right)\alpha_{NL} = \frac{1}{\alpha_{NL}^3}\;, \tag{67}$$

i.e., in comparison with the conservative case, only ω^2 has been replaced by $\Omega^2 = \left(\omega^2 - \frac{\gamma^2}{4}\right)$. Again, from Eqs. (63) and (67) a dynamical invariant can be obtained that now has the form [27] [28] [29]

$$I_{NL} \;=\; \frac{1}{2}\, e^{\gamma t}\left[\left(\dot{\eta}\,\alpha_{NL} - \left(\dot{\alpha}_{NL} - \frac{\gamma}{2}\alpha_{NL}\right)\eta\right)^2 + \left(\frac{\eta}{\alpha_{NL}}\right)^2\right]. \tag{68}$$

Riccati equation (64) can also be linearized, now using the ansatz $\mathcal{C}_{NL} = \frac{\dot{\tilde{\lambda}}}{\tilde{\lambda}} = \frac{\dot{\lambda}}{\lambda} - \frac{\gamma}{2}$ to yield the Newtonian equation with linear friction term

$$\ddot{\tilde{\lambda}} + \gamma\dot{\tilde{\lambda}} + \omega^2\tilde{\lambda} \;=\; 0 \tag{69}$$

for the complex variable $\tilde{\lambda}(t) = \lambda e^{-\gamma t/2} = \alpha_{NL} e^{-\gamma t/2 + i\varphi}$. Inserting the polar form into the imaginary part of Eq. (64) leads to the unchanged conservation law

$$\dot{\varphi} = \frac{1}{\alpha_{NL}^2} \tag{70}$$

as in the conservative case.

Comparing Eqs. (63), (64) and (69) with the corresponding Eqs. (3), (4) and (28) in the conservative case, in the linear second-order differential equations for η and λ, a linear term with first derivative has been added while, in the first-order Riccati equation, an additional term linear in $\mathcal{C}_{NL} = \left(\frac{2\hbar}{m}y\right)$ appears. All the additional terms depend on the coefficient γ of the friction force and vanish in the limit $\gamma \to 0$.

Similar modifications can also be obtained for the TISE (for details see [23]).

5. Riccati equations in other fields of physics

5.1. Riccati equations in statistical thermodynamics

Let us return to solution (7) of the Riccati equation (4), but now for real $\mathcal{C}$, and rewrite it in the form

$$\mathcal{C}(t) = \tilde{\mathcal{C}} + \frac{2\tilde{\mathcal{C}}\; e^{-2\tilde{\mathcal{C}}t}}{\kappa_0\; 2\tilde{\mathcal{C}} + \left(1 - e^{-2\tilde{\mathcal{C}}t}\right)} = \tilde{\mathcal{C}} + \frac{2\tilde{\mathcal{C}}}{\kappa_0\; 2\tilde{\mathcal{C}}\; e^{2\tilde{\mathcal{C}}t} + \left(e^{2\tilde{\mathcal{C}}\;t} - 1\right)}\;. \tag{71}$$

For the choice $\kappa_0 = 0$, this turns into

$$\mathcal{C}(t) = \tilde{\mathcal{C}} + \frac{2\tilde{\mathcal{C}}}{\left(e^{2\tilde{\mathcal{C}}t} - 1\right)} = \tilde{\mathcal{C}} \, \coth \tilde{\mathcal{C}}t \ . \tag{72}$$

Replacing time t by $t \rightarrow \frac{1}{\mathrm{kT}} = \beta$ (with k = Boltzmann's constant and T = temperature) and the constant particular solution by $\tilde{\mathcal{C}} = \frac{\hbar}{2}\omega$, one obtains

$$\frac{\hbar}{2}\omega + \frac{\hbar\,\omega}{e^{\hbar\omega/\mathrm{kT}} - 1} = \frac{\hbar}{2}\omega \, \coth\left(\frac{\hbar\omega}{2\mathrm{kT}}\right) = \ < E >_{th} \tag{73}$$

which is the expression known from statistical thermodynamics for the average energy of a single oscillator in thermal equilibrium [30]. The first term on the lhs is just the ground state energy of the HO, the second is equal to Planck's distribution function for the black body radiation. This type of relation between Eq. (73) and the Riccati equation has also been found by Rosu et al [31]. Equation (73) can also be expressed in terms of the partition function $Z = \sum_n e^{-n\hbar\omega\beta} = \frac{1}{1 - e^{\hbar\omega\beta}}$ as

$$< E >_{th} \ = \frac{\hbar}{2}\omega + \frac{\frac{\partial}{\partial\beta} Z^{-1}}{Z^{-1}} = \frac{\hbar}{2}\omega - \frac{\partial}{\partial\beta}\ln Z \ . \tag{74}$$

The Riccati equation corresponding to solution (72) in the form (73) can be written as

$$\mathcal{C}' + \mathcal{C}^2 - \tilde{\mathcal{C}}^2 = 0 \tag{75}$$

with $\mathcal{C} = \mathcal{C}(\beta)$ depending on the variable $\beta = \frac{1}{\mathrm{kT}}$, prime denoting derivative with respect to this variable and $\tilde{\mathcal{C}}$, as mentioned above, being the ground state energy, $\tilde{\mathcal{C}} = \frac{\hbar}{2}\omega$. The inverse quantity of $\mathcal{C}$, multiplied by $-\tilde{\mathcal{C}}^2$, i.e., $\mathcal{K} = -\tilde{\mathcal{C}}^2\mathcal{C}^{-1}$ fulfils

$$-\mathcal{K}' + \mathcal{K}^2 - \tilde{\mathcal{C}}^2 = 0 \ , \tag{76}$$

i.e., also a Riccati equation but now with negative sign for the derivative term and the solution

$$\mathcal{K}\left(\frac{1}{\mathrm{kT}}\right) = \frac{\hbar}{2}\omega - \frac{\hbar\omega}{e^{\hbar\omega/\mathrm{kT}} + 1} = \frac{\hbar}{2}\omega \, \tanh\left(\frac{\hbar\omega}{2\mathrm{kT}}\right) \ . \tag{77}$$

From the ground state energy, in this case, a term is subtracted that represents a Fermi–Dirac distribution whereas in the solution for $\mathcal{C}$, a term was added to the ground state energy that represents a Bose–Einstein distribution. So both quantum statistics can be obtained from the solution of the Bernoulli equations derived from the Riccati equations for $\mathcal{C}\left(\frac{1}{\mathrm{kT}}\right)$ and (essentially) its inverse quantity.

5.2. The logistic equation as Riccati/Bernoulli equation

The last Riccati equation for $\mathcal{K}\left(\frac{1}{\mathrm{kT}}\right) = \frac{\hbar\omega}{2} \, \tanh\left(\frac{\hbar\omega}{2}\frac{1}{\mathrm{kT}}\right)$ can be rewritten in terms of more neutral, general variables using $\frac{1}{\mathrm{kT}} \rightarrow x$ and $\hbar\omega \rightarrow 1$, changing $\mathcal{K}\left(\frac{1}{\mathrm{kT}}\right) \rightarrow g(x)$ with $g(x) = \frac{1}{2} \, \tanh\left(\frac{x}{2}\right)$, fulfilling the Riccati equation

$$g' + g^2 - \frac{1}{4} = 0 \tag{78}$$

with $' = \frac{d}{dx}$.

Just adding the constant $\frac{1}{2}$ to $g(x)$, one obtains the function $f(x)$ as

$$f(x) = \frac{1}{2} + g(x) = \frac{1}{2}\left(1 + \tanh\left(\frac{x}{2}\right)\right) = \frac{1}{1 + e^{-x}} \tag{79}$$

which describes the standard sigmoid function that fulfills the (homogeneous, Bernoulli-type) logistic equation

$$-f' + f^2 - f = 0 \tag{80}$$

or

$$-f' = f(1 - f) \; . \tag{81}$$

This equation has wide applications in NL dynamics from modelling population growth in ecology to modelling tumor growth in medicine, from autocatalytic reaction models in chemistry to fields like linguistics, economics and many others.

5.3. Complex Riccati equation in classical optics

The wave equation of classical optics can be written (see, e.g., [32]) as

$$\Delta\phi - \frac{n^2}{c^2}\frac{\partial^2}{\partial t^2}\phi = 0 \tag{82}$$

with the scalar field $\phi(\mathbf{r}, t)$, $c =$ velocity of light in vacuum, $n = \frac{c}{v} =$ diffraction index (with $v =$ velocity of light in medium). In general $n = n(\mathbf{r})$, i.e., the diffraction index depends on the medium and can therefore be a function of space. In this case the solutions of Eq. (82) are no longer plane waves but have a similar, but more general form. Using the ansatz

$$\phi(\mathbf{r}, t) = \exp\left[A(\mathbf{r}) + ik_0(L(\mathbf{r}) - ct)\right] \tag{83}$$

with $L(\mathbf{r})$ being the optical wave length or phase, also called Eikonal, which is, as well as $A(\mathbf{r})$, the quantity that determines the amplitude of ϕ, a function of the spatial variable $\mathbf{r}$. Inserting this ansatz into the wave equation (82) leads to two coupled differential equations for $A(\mathbf{r})$ and $L(\mathbf{r})$,

$$\nabla(\nabla A) + (\nabla A)^2 - (k_0\nabla L)^2 + (k_0 n)^2 = 0 \tag{84}$$

$$\nabla(k_0\nabla L) + 2(\nabla A)(k_0\nabla L) = 0 \tag{85}$$

where the inhomogeneity in Eq. (84) originates from the time-derivative via

$$-\frac{n^2}{c^2}\frac{\partial^2}{\partial t^2}\phi = k_0^2\, n^2\phi \; . \tag{86}$$

Since $\phi(\mathbf{r}, t)$ is a complex function, Eqs. (84) and (85) are the real and imaginary parts of the, originally, complex differential equation (82). Similar to the situation in sections 2 and 3, the linear complex equation can be considered as a linearized form of a complex NL Riccati equation whose variable is the logarithmic derivative of the one fulfilling the linear equation.

In this sense, the complex variable in this case can be written as $C(\mathbf{r}) = (\nabla A) + i\; k_0(\nabla L)$ (corresponding to $\frac{\nabla a}{a} + i\;\frac{1}{\hbar}\nabla S$ in the case of the TISE) and the complex Riccati equation (corresponding to the wave equation (82)) reads

$$\nabla C + C^2 + (k_0\; n(\mathbf{r}))^2 = 0 \; . \tag{87}$$

The position-dependent inhomogeneity $(k_0\; n(\mathbf{r}))^2$ corresponds to the term $\frac{2m}{\hbar^2}(E - V(\mathbf{r}))$ in the TISE. It should be noted that the case, where a complex potential has been discussed in section 3, corresponds to the optical situation with complex diffraction index where the imaginary part describes absorption. A more detailed discussion will be given elsewhere [2].

5.4. Complex Riccati equation for Bose–Einstein condensates

In the mean field approximation, a Bose–Einstein condensate (BEC) can be described by a macroscopic WP for the condensate, $\Psi(\mathbf{r}, t)$, which obeys the cubic NL Gross–Pitaevskii equation

$$i\hbar \frac{\partial}{\partial t}\Psi = \left\{ -\frac{\hbar^2}{2m}\Delta + V(\mathbf{r}, t) + g|\Psi|^2 \right\} \Psi \tag{88}$$

where $V(\mathbf{r}, t)$ represents the trapping potential and can be given (e.g., for a Paul trap) by $V(\mathbf{r}, t) = \frac{m}{2}\omega^2(t)r^2$ with $r = |\mathbf{r}|$ = absolute value of the vector $\mathbf{r}$ and TD frequency $\omega = \omega(t)$; g parametrizes the strength of the atomic interaction.

Although Eq. (88) cannot be solved analytically, the dynamics of the BEC characterized by this equation can be described in terms of so-called moments M_n ($n = 1 - 4$) (for details, see e.g., [33]), where M_1 represents the norm, M_2 the width, M_3 the radial momentum and M_4 the energy of the WP. It can be shown that these moments satisfy a set of coupled first-order differential equations (where $\frac{d}{dt}M_1$ corresponds to the conservation of probability or particle number). This set can be reduced to a single equation for M_2 which, using a new variable $X = \sqrt{M_2}$, can be expressed in the form of an Ermakov equation,

$$\ddot{X} + \omega^2(t)X = \frac{k^2}{X^3} \ . \tag{89}$$

As we have seen already before (e.g. in section 2), he real Ermakov equation is equavalent to a complex NL Riccati equation for the variable $\mathcal{C} = \frac{\dot{X}}{X} + i\ \frac{k}{X^2}$. Again, as shown in section 2, this can be used to define generalized creation/annihilation operators that allow one to obtain coherent states following the dynamics that corresponds to the abovementioned moments.

To include dissipative effects, one could add another NL term like the logarithmic one from Eq. (61) to the Gross–Pitaevskii equation which would correspond to adding a linear term to the Riccati equation. So, one simply has to solve this modified Riccati equation (or the corresponding Ermakov equation) to obtain all moments M_n for the dissipative BEC [34].

5.5. Complex Riccati equation in cosmology

Another example shall be given where a physical system can be described by a (complex) Riccati equation, or its equivalent, a real Ermakov equation, now not for a microscopic but for a really macroscopic system. For this purpose, we switch to cosmology and apply the cosmological principle, i.e., the assumption of a spatially-homogeneous and isotropic universe, to Einstein's field equations. In this case, the Robertson–Walker metric applies [35] that contains the scale factor $a(t)$ (so to say, the radius of the universe) and the curvature k that can attain the values 0 for a flat or $\pm$ 1 for a closed or open universe. This finally leads to the Friedmann–Lemaître equations

$$\left(\frac{\dot{a}}{a}\right)^2 = H^2 = \frac{2}{3}\varrho - \frac{k}{a^2} \tag{90}$$

and

$$\dot{\varrho} = -3\ H(\varrho + p) \tag{91}$$

with $H = \frac{\dot{a}}{a}$ = Hubble parameter; overdots denote derivatives with respect to cosmic proper time, ϱ = energy density and p = pressure (where $4\pi G = c = 1$ and the cosmological constant $\Lambda = 0$ have been set).

Assuming the matter source as a self-interacting scalar field $\Phi = \Phi(t)$, the energy density ϱ and pressure p can be written as

$$\varrho = \frac{1}{2}\left(\frac{d}{dt}\Phi\right)^2 + U(t) \ , \tag{92}$$

$$p = \frac{1}{2}\left(\frac{d}{dt}\Phi\right)^2 - U(t) \ . \tag{93}$$

Taking the time-derivative of (90) and inserting (91) using (92) and (93) leads to

$$\frac{d}{dt}\left(\frac{\dot{a}}{a}\right) = -\left(\frac{d}{dt}\Phi\right)^2 + \frac{k}{a^2} \ . \tag{94}$$

Introducing a new time variable (conformal time τ) via $\frac{d}{dt} = a\,\frac{d}{d\tau}$, Eq. (94) can be rewritten as an Ermakov equation,

$$\frac{d^2}{d\tau^2}a + \left(\frac{d}{d\tau}\Phi\right)^2 a = \frac{k}{a^3} \ , \tag{95}$$

which is equivalent to a Riccati equation for the complex quantity $\mathcal{C} = \left(\frac{\frac{d}{d\tau}a}{a} + i\,\frac{1}{a^2}\right)$. The above derivation of the Ermakov equation in comparison with the afore-mentioned BEC had been given by Lidsey in [36]. Everything said about the complex Riccati or real Ermakov equation in section 2 can also be applied to this system, e.g., corresponding creation and annihilation operators and coherent states can be defined (see [14]).

5.6. Complex Riccati/Bernoulli equations, "quantized" triangles and Pythagorean triples

Looking for the origin of the idea of quantization one might first think of the Greek philosopher Democritus (ca. 460 - ca. 371 BC) and his idea of dividing our world into minute components that are not further divisible, the "atoms".

A similar idea was formulated in a more abstract way by another equally famous Greek philosopher living around the same time. In his work *Timaios*, Plato (428/27 - 348/47 BC) gives his view of how the world is built up in terms of right-angled triangles, essentially his "quanta". Werner Heisenberg, equally fascinated with, and puzzled by, this text, summarizes this idea [37] in the general sense as follows: "Matter is made up of right-angled triangles which, after being paired to form isosceles triangles or squares, are joined together to build the regular bodies of stereometry: cube, tetrahedron, octahedron and icosahedron. These four solids are then supposed to be the basic units of the four elements earth, fire, air and water".

So, geometric objects like right-angled triangles should be the quanta of nature. But one could go even further and ask if particular right-angled triangles might play a special role in quantization. Yet another Greek philosopher enters the scene. We remember Pythagoras' theorem from school, i.e., $a^2 + b^2 = c^2$ where a and b are the two shorter sides (catheti) and c is the longest side (hypotenuse) of a right-angled triangle. Pythagoras (around 570 - 500 BC) and his disciples were well-known for their dogma "everything is number", with number meaning *integer*. So let us look for right-angled triangles where the lengths of all three sides are integers (a kind of "second quantization") fulfilling the Pythagorean theorem. The most common example for such a Pythagorean triple is $(3, 4, 5)$ with $9 + 16 = 25$. But, asked for a few more examples of the kind, even mathematically-affiliated persons have difficulties finding any – though an infinite number of triples exists! Moreover, there is even a rather simple rule of finding these triples. This rule was probably already known to the Babylonians more than 3500 years ago [38] but, at least, Diophantus of Alexandria (around 250 AD) knew of it.

What does this have to do with our complex Riccati equations? For this purpose the case of the homogeneous Bernoulli equation (5) will be considered. As shown above, with a particular solution of the Riccati equation its inhomogeneity can always be removed. The resulting additional linear term in the Bernoulli equation (at least for constant coefficient $\tilde{\mathcal{C}}$) can also be removed. So we are dealing with a complex equation of the form

$$\frac{d}{dt}\mathcal{C} + \mathcal{C}^2 = 0 \ . \tag{96}$$

Then $-\frac{d}{dt}\mathcal{C}$ is also a complex quantity, $\mathcal{C}^2$, where its real and imaginary parts as well as its absolute value again define a right-angled triangle (in the complex plane) and each side contains contributions from $\mathcal{R}$ *and* $\mathcal{I}$, i.e., $\Re\{\mathcal{C}^2\} = \mathcal{R}^2 - \mathcal{I}^2$, $\Im\{\mathcal{C}^2\} = 2\,\mathcal{R}\mathcal{I}$ and $|\mathcal{C}|^2 = \mathcal{R}^2 + \mathcal{I}^2$.

If we now assume that $\mathcal{R}$ and $\mathcal{I}$ are integers (with $\mathcal{R} > \mathcal{I}$), all three sides of the right-angled triangle created by $\mathcal{C}^2$ in the complex plane are also integers. As examples, we choose: (a) $\mathcal{R} = 2$, $\mathcal{I} = 1$: $\mathcal{R}^2 - \mathcal{I}^2 = 3$, $2\,\mathcal{R}\mathcal{I} = 4$, $\mathcal{R}^2 + \mathcal{I}^2 = 5$ with $9 + 16 = 25$; (b) $\mathcal{R} = 3$, $\mathcal{I} = 2$: $\mathcal{R}^2 - \mathcal{I}^2 = 5$, $2\,\mathcal{R}\mathcal{I} = 12$, $\mathcal{R}^2 + \mathcal{I}^2 = 13$ with $25 + 144 = 169$.

All possible Pythagorean triples can be obtained in this way just by applying all integers $\mathcal{R}$ and $\mathcal{I}$ with $\mathcal{R} > \mathcal{I}$. In a physical context this means that whenever a physical quantity obeys a complex Riccati equation and this quantity can be "quantized" in the sense that its real and imaginary parts can be expressed as multiples of some units, its evolution (in time, space or depending on other variables like inverse temperature etc.) can also be expressed in terms of the same units.

5.7. Scale invariance, trigonometric and hyperbolic functions and Riccati equations

In the last sub-section it was shown how particular right-angled triangles can be related to (complex) Riccati/Bernoulli equations. Yet another relation between right-angled triangles and Riccati equations can be established. For this purpose, we divide all sides of these triangles by the length of the hypotenuse so that this now has a unit length and the length of the two catheti are given (in the case of the complex numbers) by the cosine (real part) and sine (imaginary part) of the angle between the hypothenuse and the real axis. The same also applies to any right-angled triangle in a unit circle in the real plane, accordingly.

As mentioned before, in NL dynamics it is not the absolute changes that are relevant but relative ones, thus providing scale invariance of the corresponding laws. On the other hand, the variables of the Riccati equations are simply logarithmic derivatives of the quantities that fulfil the corresponding linearized second order differential equation, i.e., the enumerator of the Riccati variable is the derivative of the denominator. Looking at the catheti of our right-angled triangles, their ratio fulfils this requirement (not only for the triangles in the unit circle, but for any right-angled triangle with constant length of its sides). So the ratio of the two catheti is either $\tan\varphi = \frac{\sin\varphi}{\cos\varphi}$ or $\cot\varphi = \frac{\cos\varphi}{\sin\varphi}$, where $\sin\varphi$ and $\cos\varphi$ are (up to the sign) the derivatives of each other. Therefore, there should also be a connection to Riccati equations; and it obviously exists. For example, the derivative of $\tan x$ with respect to x is $\frac{d}{dx}\tan x = \frac{1}{\cos^2 x}$. Using in addition the relation $\cos^2 x + \sin^2 x = 1$, this can be rewritten as the Riccati equation

$$-\frac{d}{dx}\tan x + \tan^2 x + 1 = 0\,, \tag{97}$$

with unit inhomogeneity. In the same way, one obtains from $\frac{d}{dx}\cot x = -\frac{1}{\sin^2 x}$

$$\frac{d}{dx}\cot x + \cot^2 x + 1 = 0\,. \tag{98}$$

Changing from the real argument x to an imaginary argument ix changes the trigonometric functions into hyperbolic ones. As we have seen in 5.1, these functions and corresponding Riccati equations actually occur in statistical thermodynamics where the transition from t to $\frac{\hbar}{kT}$ actually represents a transition to an imaginary time. This concept of imaginary (or complex) time is sometimes used for computational reasons to transform the TDSE with oscillatory solutions into a diffusion-type equation and also for other purposes in, e.g., theoretical chemistry [39].

Looking at the hyperbolic functions, the situation is very similar to the trigonometric ones. Also in $\tanh\varphi = \frac{\sinh\varphi}{\cosh\varphi}$ and $\coth\varphi = \frac{\cosh\varphi}{\sinh\varphi}$, the enumerators and denominators are derivatives of each other and, now with the relation $\cosh^2 x - \sinh^2 x = 1$, the following Riccati equations can be obtained:

$$\frac{d}{dx}\tanh x + \tanh^2 x - 1 = 0 \tag{99}$$

$$\frac{d}{dx}\coth x + \coth^2 x - 1 = 0 \; . \tag{100}$$

The inhomogeneity now has a negative sign which would correspond in the TD case in section 2 to a change from an attractive to a repulsive quadratic potential.

6. Conclusions and perspectives

It has been shown in section 2 that the information about the dynamics of a quantum mechanical WP, i.e. the equations of motion for the time-evolution of its maximum and width, are not only be obtainable from the solution of the TDSE, but equally-well from a complex NL Riccati equation. The classical particle aspect that is reflected by the Newtonian equation determining the motion of the wave packet maximum can be obtained, using that the complex Riccati equation can be linearized by the ansatz $\mathcal{C} = \frac{\dot{\lambda}}{\lambda}$ to a Newtonian equation for the complex quantity $\lambda(t)$, which can be written as $\lambda = u + z = \alpha\, e^{i\varphi}$. Knowing that $z \propto \eta(t)$, the imaginary part of λ immediately provides (up to a constant factor) the classical trajectory $\eta(t)$.

Reformulation of the complex Riccati equation in terms of a real variable, $\alpha(t)$, that is proportional to the WP width (and is identical to the absolute value of $\lambda(t)$), the solution of the resulting NL so-called Ermakov equation for α provides the information about the dynamics of this WP property. Both equations describing particle and wave aspects of the WP can be combined to yield a dynamical invariant with the dimension of "action". Factorization of an operator corresponding to this invariant allows for the definition of generalized creation/annihilation operators leading not only to CSs with TD width, but also to unconventional solutions of the TDSE.

Formal similarities exist between this TD problem and SUSY quantum mechanics where (real) position-dependent Riccati equations play a similar role and the construction of isospectral potentials demonstrates the importance of parameters like initial conditions.

In section 3 it has been shown that the Riccati formalism established for the TDSE can also be applied to the TISE if the TD complex quantity $\mathcal{C} = \frac{\dot{\lambda}}{\lambda}$ is replaced by the space-dependent complex quantity $\frac{\nabla\Psi}{\Psi}$ for a complex wave function Ψ. In a certain way, this looks like a complex version of SUSY quantum mechanics where not only the logarithmic derivative of the ground state is considered as a variable for a Riccati equation, but also any (even complex) excited state may fulfil a now complex Riccati equation. Applying methods known from SUSY actually allows one to start from a real potential with complex wave function to obtain a system with complex potential, i.e., non-hermitian Hamiltonian, but still with real eigenvalue.

The flexibility of the Riccati formalism was illustrated in section 4 where it was shown that effects like dissipation and irreversibility can easily be included into it. In the TD case, only an additional term linear in $\mathcal{C}$ appears in Eq. (4). Since, in the solution via the homogeneous Bernoulli equation (5), a linear term already occurs only the coefficient of this term changes.

Finally in section 5, the formal similarity between this formulation of quantum mechanics and several other fields of physics that also can be written in the form of (real or complex) Riccati equations was demonstrated. Replacing time t with (actual imaginary) "time" $\frac{\hbar}{kT}$ leads to well-known expressions from statistical thermodynamics as solutions of Riccati equations that are also able to distinguish between bosonic and fermionic properties.

By simply adding a constant to the solution of the Fermi–Dirac equation, the solution of an equation well-known in NL dynamics can be found, namely the logistic equation which is actually also a Riccati/Bernoulli equation. This equation, originally found in the context of population dynamics, has wide applications and not only in physics.

While the previous examples were still dealing with real Riccati equations, in the next example it was shown that also the wave equation of classical optics can be rewritten as a complex Riccati equation where the (position-dependent) diffraction index essentially corresponds to the (position-dependent) potential of the Riccati equation obtained from the TISE. The case of the complex potential would therefore correspond to a complex diffraction index with the imaginary part describing absorption.

Also the dynamics of a BEC can be described by a complex Riccati or the corresponding Ermakov equation. We then went from the microscopic scale to the really macroscopic one, to cosmology, and demonstrated that the Friedmann–Lemaître equations, which under certain assumptions describe the dynamics of our universe, can be written in terms of a real Ermakov or equivalent complex Riccati equation.

Finally, returning to the abstact mathematical roots we show how complex Riccati/Bernoulli equations are related to the Pythagorean triples and thus to the most fundamental "quantization" problem. Generalization of this problem from very particular right-angled triangles to arbitrary ones where the lengths of their sides can be expressed in terms of trigonometric functions also displayed the connection of these functions with Riccati equations. Including imaginary arguments for these functions, Riccati equations can also be formulated for hyperbolic functions. This then opens a new approach to the theory of solitons which will be discussed elsewhere.

There are many more examples like electrodynamics, quantum optics, etc., where the Riccati equation allows for a unifying formulation with the fields mentioned above as well as others. In conclusion therefore, one can say that the NL version of quantum mechanics based on a (complex) Riccati equation is able to cover all phenomena of standard quantum mechanics (including the superposition principle, due to its linearizablity). In addition, the sensitivity of NL differential equations to initial conditions provides further information that gets lost in the linearized form. The complex form of the Riccati equation (via its imaginary part) also supplies a new conservation law that resembles the conservation of angular momentum, but now for the motion in the complex plane (something closely related to the quantum property spin), expressing the connections between phase and amplitude of complex quantities that are essential in quantum mechanics, but not only there.

References

[1] Yang C N 1987 *Schrödinger – Centenary Celebration of a Polymath* ed C W Kilmister (Cambridge: Cambridge University Press) p 53
[2] Castaños O, Rosas-Ortiz O and Schuch D; in preparation
[3] Mielnik B and Rosas-Ortiz O 2004 *J. Phys. A: Math. Gen.* **37** 10007
[4] Andrianov A A and Cannata F 2004 *J. Phys. A: Math. Gen.* **37** 10297
[5] Cooper F, Khare A and Sukhatme U 2001 *Supersymmetry in Quantum Mechanics* (Singapore: World Scientific)
[6] Steen A 1874 *Overs. over d. K. Danske Vidensk. Selsk. Forh.* 1
[7] Redheffer R 1999 *Aequationes Math.* **58** 60
[8] Redheffer R and Redheffer I 2001 *Aequationes Math.* **61** 131
[9] Ermakov V P 1880 *Univ. Jzv. Kiev* **20** No 9, 1
[10] Milne W E 1930 *Phys. Rev.* **35** 863
[11] Pinney E 1950 *Proc. Amer. Math. Soc.* **1** 681
[12] Lewis H R 1967 *Phys. Rev. Lett.* **18** 510
[13] Schrödinger E 1926 *Naturwiss.* **14** 664
[14] Castaños O, Schuch D and Rosas-Ortiz O 2012 *J. Phys. A: Math. Theor.* **46** 075304
[15] Guerrero J, López-Ruiz F F, Aldaya V and Cossío F 2011 *J. Phys. A: Math. Theor.* **44** 445307
[16] Hartley J G and Ray J R 1981 *Phys. Rev.* A **24** 2873
[17] Kevrekidis P G and Drossinos Y 2007 *Math. Comp. Sim.* **74** 196
[18] Schuch D and Moshinsky M 2006 *Phys. Rev.* A **73** 062111
[19] Madelung E 1926 *Z. Physik* **40** 322

[20] Reinisch G 1994 *Physica* A **206** 229
[21] Reinisch G 1997 *Phys. Rev.* A **56** 3409
[22] Einstein A 1905 *Ann. d. Phys.* **17** 549
[23] Schuch D 2012 *J. Phys.: Conf. Series* **361** 012020
[24] Schuch D, Chung K-M and Hartmann H 1983 *J. Math. Phys.* **24** 1652; ibid 1984 *J. Math. Phys.* **25** 3086; Schuch D and Chung K-M 1986 *Int. J. Quantum Chem.* **29** 1561
[25] Madelung E 1950 *Die Mathematischen Hilfsmittel des Physikers* (Berlin: Springer) p 432
[26] Mrowka B 1951 *Z. Phys.* **130** 164
[27] Nassar A B 1986 *J. Math. Phys.* **27** 755
[28] Nassar A B 1986 *J. Math. Phys.* **27** 2949
[29] Schuch D 2007 *J. Math. Phys.* **48** 122701
[30] Feynman R P 1972 *Statistical Mechanics – A Set of Lectures* (Reading: Benjamin) 2.5
[31] Rosu H C and Aveces de la Cruz F 2002 *Phys. Scripta* **65** 377; Rosu H 2002 `arXiv:math-ph/020813v1`
[32] Goldstein H 1951 *Classical Mechanics* (Reading: Addison–Wesley) 9.8
[33] Garcia-Ripoll J J, Perez-Garcia V M and Torres P 1999 *Phys. Rev. Lett.* **83** 1715
[34] Schuch D and Kaushal R S 2011 *J. Phys.: Conf. Series* **306** 012032
[35] Sexl R U and Urbantke H K 1975 *Gravitation und Kosmologie* (Mannheim: Bibliographisches Institut)
[36] Lidsey J E 2004 *Class. Quantum Grav.* **21** 777
[37] Heisenberg W 1973 *Der Teil und das Ganze* (München: Deutscher Taschenbuch Verlag) p 17
[38] Friberg J 1981 *Historia Mathematica* **8** 277
[39] Bonella S, Monteferrante M, Pierleoni C and Ciccotti G 2010 *J. Chem. Phys.* **133** 164104

A classification of finite quantum kinematics

J Tolar [1]

[1] Department of Physics
Faculty of Nuclear Sciences and Physical Engineering
Czech Technical University in Prague
Břehová 7, 115 19 Prague 1, Czech Republic

E-mail: jiri.tolar@fjfi.cvut.cz

Abstract. Quantum mechanics in Hilbert spaces of finite dimension N is reviewed from the number theoretic point of view. For composite numbers N possible quantum kinematics are classified on the basis of Mackey's Imprimitivity Theorem for finite Abelian groups. This yields also a classification of finite Weyl-Heisenberg groups and the corresponding finite quantum kinematics. Simple number theory gets involved through the fundamental theorem describing all finite discrete Abelian groups of order N as direct products of cyclic groups, whose orders are powers of not necessarily distinct primes contained in the prime decomposition of N. The representation theoretic approach is further compared with the algebraic approach, where the basic object is the corresponding operator algebra. The consideration of fine gradings of this associative algebra then brings a fresh look on the relation between the mathematical formalism and physical realizations of finite quantum systems.

1. Introduction

Looking back on our papers [1, 2] on symmetries of finite Heisenberg groups, I find it appropriate — at this Symposium — to return to the foundations of our approach to quantum systems with finite-dimensional Hilbert spaces and supplement them with the algebraic treatment.

Non-relativistic quantum mechanics of particle systems can be divided into quantum kinematics and quantum dynamics. First the Hilbert space and the non-commuting operators of complementary observables, positions and momenta, are constructed. This kinematical structure then remains the same for all possible quantum dynamics which are determined by the respective Hamiltonians.

For systems with configuration space $\mathbb{R}^n$ one can define quantum kinematics according to H. Weyl [3] in terms of the Weyl system — a projective unitary representation of the Abelian group of translations of the corresponding classical phase space $\mathbb{R}^n \times \mathbb{R}^n$. Using Weyl's system of unitary operators, J. von Neumann was able to prove the uniqueness theorem stating that there is, up to unitary equivalence, unique irreducible quantum kinematics, commonly taken in the form of the Schrödinger representation. In the most general form the uniqueness theorem was proved on the basis of G.W. Mackey's Imprimitivity Theorem. In this mathematical generalization the configuration space $\mathbb{R}^n$ was replaced by an arbitrary locally compact second countable Abelian group G [4]. The direct product of G and its Pontryagin dual then plays the role of the phase space.

Quantum mechanics in finite-dimensional Hilbert spaces originally seemed to present only a nice and simple exercise in linear algebra. During the last decades it unexpectedly became the mathematical framework for the development of methods of quantum information processing with numerous applications to quantum cryptography, teleportation and quantum computing. For instance, the mathematical notion of complementary bases lies at the heart of quantum cryptography [5].

Historically, H. Weyl — not successful with the proof of the uniqueness theorem — wanted to present a simple example in $\mathbb{C}^n$ analogous to one-dimensional quantum particle and at the same time exhibiting the uniqueness feature [3]. This example was further developed by J. Schwinger [6] with the aim to approximate quantum mechanics of particles. He noticed the complementary nature of position and momentum observables which are here built in the finite Weyl-Heisenberg group generated by the generalized Pauli matrices. The elements of this group provide a useful basic set of quantum operators for finite quantum systems.

The special role of the finite Weyl-Heisenberg group has been recognized also in a distant domain of mathematics — the classification of fine gradings of classical Lie algebras [7, 8]. For classical Lie algebras of the type $\mathrm{sl}(n, \mathbb{C})$, $n = 2, 3, \ldots$, this classification contains — among others — the Pauli grading based on generalized Pauli matrices [9, 10]. The finite group generated by them has been called the Pauli group — a notion identical with the Weyl-Heisenberg group.

In our paper [11] we succeeded to prove the uniqueness of finite quantum kinematics using a simple version of Mackey's Imprimitivity Theorem. In this way also a geometric interpretation of finite quantum kinematics was obtained: it appears as quantum mechanics where a cyclic group serves as the configuration space. For composite dimensions N the fundamental theorem on finitely generated Abelian groups immediately leads to the classification of all finite quantum kinematics. This classification was also independently noted in 1995 by V.S. Varadarajan [12].

In this contribution our approach via representation theory is confronted with the algebraic formulation of quantum mechanics [13]. For finite quantum systems, the operator algebras are the associative complex matrix algebras $M_N(\mathbb{C})$ [15]. Inspired by Lie algebra gradings, we describe all fine gradings of $M_N(\mathbb{C})$ which are induced by inner automorphisms of $\mathrm{GL}(N, \mathbb{C})$.

In section 2 we reproduce the group theoretical approach based on representation theory of finite Abelian groups. Sections 3 and 4 are devoted to the classification of finite quantum kinematics and some historical remarks. In section 5 fine gradings of matrix algebras $M_N(\mathbb{C})$ are obtained and physical interpretation of the Pauli gradings is offered. Section 6 concludes the paper.

2. Simple quantum kinematics on cyclic groups

Ordinary quantum mechanics prescribes that the mathematical quantities representing the position and momentum should be self-adjoint operators on the Hilbert space of the system. Their algebraic properties constitute *quantum kinematics*, while *quantum dynamics* of the system is given by the unitary group generated by the Hamiltonian which is expressed as a function of the position and momentum operators.

According to Mackey, quantum kinematics of a system localized on a homogeneous space $M = G/H$ is determined by an irreducible transitive system of imprimitivity $(\mathcal{U}, \mathcal{E})$ for G based on M in a Hilbert space $\mathcal{H}$ [4]. Here $\mathcal{U} = \{U(g) | g \in G\}$ is a unitary representation of G in $\mathcal{H}$ and $\mathcal{E} = \{E(S) |\ S \text{ Borel subset of } M\}$ is a projection-valued measure in $\mathcal{H}$ satisfying the covariance condition

$$U(g)E(S)U(g)^{-1} = E(g^{-1}.S). \tag{1}$$

Given a positive integer $N \geq 2$, we assume that the configuration space M is the finite set

$$M = Z_N = \{\rho | \rho = 0, 1, \ldots, N-1\}$$

with additive group law modulo N. Since there is a natural transitive action of Z_N on itself, we may consider M as a homogeneous space of

$$G = Z_N = \{j | j = 0, 1, \ldots, N-1\},$$

realized as an additive group modulo N with the action

$$G \times M \to M : (j, \rho) \mapsto \rho + j (\text{mod } N)$$

and the isotropy subgroup $H = \{0\}$. For the finite set $M = Z_N$ the covariance condition simplifies to

$$U(j)E(\rho)U(j)^{-1} = E(\rho - j),$$

where $\rho \in M$, $j \in G$, $E(\rho) = E(\{\rho\})$ and $E(S) = \sum_{\rho \in S} E(\rho)$.

Complete classification of transitive systems of imprimitivity up to simultaneous unitary equivalence of both $\mathcal{U}$ and $\mathcal{E}$ is obtained from Mackey's Imprimitivity Theorem. Its application to our system yields

Theorem 2.1. *If $(\mathcal{U}, \mathcal{E})$ acts irreducibly in $\mathcal{H}$, then there is, up to unitary equivalence, only one system of imprimitivity, where:*

(i) $\mathcal{H}$ is the Hilbert space $\mathcal{H}_N = \ell^2(\mathbb{Z}_N)$, i.e. $\mathbb{C}^N$ with the inner product

$$(\varphi, \psi) = \sum_{\rho=0}^{N-1} \overline{\varphi_\rho} \psi_\rho,$$

where φ_ρ, ψ_ρ, $\rho = 0, 1, \ldots, N-1$, denote the components of φ, ψ in the standard basis.

(ii) $\mathcal{U}$ is the induced representation $\mathcal{U} = Ind_H^G\ I$ called the (right) regular representation

$$[U(j)\psi]_\rho = \psi_{\rho+j} \qquad (j \in G).$$

Its matrix form in the standard basis is

$$(U(j))_{\rho\sigma} = \delta_{\rho+j,\sigma}.$$

(iii) $\mathcal{E}$ *is given by*

$$[E(\rho)\psi]_\sigma = \delta_{\rho\sigma}\psi_\sigma.$$

This unique system of imprimitivity has a simple physical meaning. The localization operators $E(\rho)$ are projectors on the eigenvectors $e^{(\rho)} \in \mathcal{H}$ corresponding to the positions $\rho = 0, 1, \ldots, N-1$. Since the set $\{e^{(\rho)}\}$ forms the standard basis of $\mathcal{H}$ in the above matrix realization, this realization may be called position representation. Then in a normalized state $\psi = (\psi_0, \ldots, \psi_{N-1})$ the probability to measure the position ρ is equal to

$$(\psi, E(\rho)\psi) = |\psi_\rho|^2.$$

Unitary operators $U(j)$ act as displacement operators

$$U(j)e^{(\rho)} = e^{(\rho-j)}.$$

In the position representation they are given by unitary matrices equal to the powers $U(j) = P_N^j = U(1)^j$ of the one-step cyclic permutation matrix

$$P_N \equiv U(1) = \begin{pmatrix} 0 & 1 & 0 & \cdots 0 & 0 \\ 0 & 0 & 1 & \cdots 0 & 0 \\ \vdots & & & \ddots & \\ 0 & 0 & 0 & \ldots.0 & 1 \\ 1 & 0 & 0 & \ldots.0 & 0 \end{pmatrix}.$$

In this way finite-dimensional quantum mechanics can be viewed as quantum mechanics on configuration spaces given by finite sets equipped with the structure of a finite Abelian group [11]. In the above simplest case a single cyclic group $\mathbb{Z}_N$ was taken as the underlying configuration space. For given $N \in \mathbb{N}$ we set $\omega_N := e^{2\pi i/N} \in \mathbb{C}$. The *generalized Pauli matrices* of order N are given by

$$Q_N := \mathrm{diag}(1, \omega_N, \omega_N^2, \ldots, \omega_N^{N-1}), \quad (P_N)_{\rho,\sigma} = \delta_{\rho+1,\sigma}, \quad \rho, \sigma \in \mathbb{Z}_N.$$

The subgroup of unitary matrices in $\mathrm{GL}(N, \mathbb{C})$ generated by Q_N and P_N,

$$\Pi_N := \{\omega_N^j Q_N^k P_N^l | j, k, l \in \{0, 1, \ldots, N-1\}\}$$

is called the *finite Weyl-Heisenberg group*.

The special role of the generalized Pauli matrices has been confirmed as the cornerstone of finite-dimensional quantum mechanics and of quantum information science. The quantum mechanical operators, Q_N and P_N act in the N-dimensional Hilbert space $\mathcal{H}_N = \ell^2(\mathbb{Z}_N)$. Further properties of Π_N are as follows:

(1) The order of Π_N *is* N^3.

(2) The center of Π_N *is* $\{\omega_N^\rho I_N | \rho \in \{0, 1, \ldots, N-1\}\}$, *where* I_N *is the* $N \times N$ *unit matrix.*

(3) The commutation relation $P_N Q_N = \omega \ \ Q_N P_N$ *is equivalent with (1).*

In order to formulate finite quantum kinematics in terms of the equivalent *discrete Weyl system*, we need the dual system of unitary operators $\mathcal{V} = \{V(\rho)\}$ defined as powers of Q_N,

$$V(\rho) = Q_N^\rho.$$

Then the quantum kinematics $(\mathcal{U}, \mathcal{E})$ can be equivalently replaced by the discrete Weyl system $(\mathcal{U}, \mathcal{V})$ satisfying

$$U(j)V(\rho) = \omega^{j\rho}V(\rho)U(j).$$

The discrete Weyl displacement operators are defined by unitary operators[1]

$$W(\rho, j) = \omega^{j\rho/2}V(\rho)U(j) = \omega^{-j\rho/2}U(j)V(\rho).$$

They satisfy the composition law for a ray representation of $\mathbb{Z}_N \times \mathbb{Z}_N$

$$W(\rho, j)W(\rho', j') = \omega^{(\rho' j - \rho j')/2}W(\rho + \rho, j + j').$$

According to Schwinger, the discrete Weyl system $\mathcal{W}$ consisting of N^2 operators $W(\rho, j)$ provides an operator basis in the space of all linear operators in $\mathbb{C}^N$, i.e. in the full matrix algebra $M_N(\mathbb{C})$. The operators $W(\rho, j)$ are orthogonal with respect to the Hilbert-Schmidt inner product

$$\mathrm{Tr}(W(\rho, j)W(\rho', j')^*) = N\delta_{\rho\rho'}\delta_{jj'}$$

and satisfy the completeness relation

$$\sum_{\rho, j} W(\rho, j)W(\rho', j')^* = N^2 1.$$

This important result can be summarized as follows:

Theorem 2.2. *The set of* N^2 *matrices* $S(\rho, j) = Q_N^\rho P_N^j/\sqrt{N}$, $j, \rho = 0, 1, \ldots, N-1$, *constitutes a complete and orthonormal basis for the linear space* $M_N(\mathbb{C})$ *of* $N \times N$ *complex matrices. Any* $N \times N$ *complex matrix can thus be uniquely expanded in this basis. If* N *is odd, then the operator basis can be taken in the form* $\omega^{j\rho/2}S(\rho, j) = W(\rho, j)/\sqrt{N}$.

3. A classification of finite quantum kinematics

The cyclic group $\mathbb{Z}_N = \{0, 1, \ldots N-1\}$ is a configuration space for N-dimensional quantum kinematics of a single N-level system. However, the reasoning on the basis of Mackey's Imprimitivity Theorem allows direct extension to any finite Abelian group as configuration space because of the fundamental theorem describing the structure of finite Abelian groups [14].

Theorem 3.1. *Let* G *be a finite Abelian group. Then* G *is isomorphic with the direct product* $\mathbb{Z}_{N_1} \times \cdots \times \mathbb{Z}_{N_f}$ *of a finite number of cyclic groups for integers* $N_1, \ldots, N_f$ *greater than 1, each of which is a power of a prime, i.e.* $N_k = p_k^{r_k}$, *where the primes* p_k *need not be mutually different.*

[1] If N is odd, the factors $\omega_N^{j\rho/2}$ are well-defined on $\mathbb{Z}_N \times \mathbb{Z}_N$.

The integers $N_k = p_k^{r_k}$ are called the *elementary divisors* of G. Two finite Abelian groups are isomorphic if and only if they have the same *elementary divisor decomposition*. In the special case of $G = \mathbb{Z}_N$ with composite $N = p_1^{r_1} \dots p_f^{r_f}$ and distinct primes $p_k > 1$, the unique decomposition

$$\mathbb{Z}_N \cong \mathbb{Z}_{N_1} \times \cdots \times \mathbb{Z}_{N_f}, \qquad N_k = p_k^{r_k}$$

is obtained by the Chinese Remainder Theorem.

Now for a general finite Abelian group as configuration space, a transitive system of imprimitivity for $G = \mathbb{Z}_{N_1} \times \cdots \times \mathbb{Z}_{N_f}$ based on $M = \mathbb{Z}_{N_1} \times \cdots \times \mathbb{Z}_{N_f}$ is equivalent to the tensor product

$$\mathcal{U} = \mathcal{U}_1 \otimes \cdots \otimes \mathcal{U}_f, \quad \mathcal{E} = \mathcal{E}_1 \otimes \cdots \otimes \mathcal{E}_f \tag{2}$$

acting in the Hilbert space

$$\mathcal{H} = \mathcal{H}_1 \otimes \cdots \otimes \mathcal{H}_f,$$

where the dimensions are $\dim \mathcal{H}_k = N_k$, $\dim \mathcal{H} = N_1 \dots N_f$. Each such system of imprimitivity $(\mathcal{U}, \mathcal{E})$ is irreducible, if and only if each $(\mathcal{U}_k, \mathcal{E}_k)$ is irreducible, hence if irreducible, it is unique up to unitary equivalence by the Imprimitivity Theorem.

For a given composite dimension N, all irreducible quantum kinematics in the Hilbert space of dimension N can be obtained according to Theorem 3.1: just take all inequivalent choices of elementary divisors $N_k = p_k^{r_k}$ with not necessarily distinct primes $p_k > 1$. One can give physical interpretation to these tensor products in the sense that the factors correspond to the *elementary building blocks or constituents* forming a finite quantum system. Our results are summarized in

Theorem 3.2. *For a given finite Abelian group G there is a unique class of unitarily equivalent, irreducible imprimitivity systems $(\mathcal{U}, \mathcal{E})$ in a finite–dimensional Hilbert space $\mathcal{H}$. In the special case $G = \mathbb{Z}_N$ the irreducible imprimitivity system is unitarily equivalent to the tensor product (2) with $N_k = p_k^{r_k}$ and distinct primes $p_1, \dots, p_f > 1$.*

Corollary 3.3. *For a given finite Abelian group G there is a unique class of unitarily equivalent discrete Weyl systems $\mathcal{W}$ in a finite–dimensional Hilbert space $\mathcal{H}$,*

$$\mathcal{W} = \mathcal{W}_1 \otimes \cdots \otimes \mathcal{W}_f. \tag{3}$$

The above classification conclusively shows that finite quantum kinematics are mathematically composed of elementary constituents which are of standard types associated with finite Abelian groups $G = \mathbb{Z}_p, \mathbb{Z}_{p^2}, \dots, \mathbb{Z}_{p^r}, \dots$, where p runs through the set of prime numbers > 1.

As already mentioned, among possible factorizations into elementary constituents special role is played by the factorization associated with the 'best' prime decomposition of the dimension $N = p_1^{r_1} \dots p_f^{r_f}$ with mutually distinct primes $p_1, \dots, p_f$. Then the discrete Weyl system of composite dimension N is equivalent to a composite system where the constituent Weyl subsystems act in the Hilbert spaces of relatively prime dimensions $p_1^{r_1}, \dots, p_f^{r_f}$.

4. Quantum degrees of freedom taken seriously

A historical remark is in order: the structure of the finite–dimensional Weyl operators was thoroughly investigated by J. Schwinger. He noted in particular that, if the dimension N is a composite number which can be decomposed as a product $N = N_1 N_2$, where the positive integers N_1, $N_2 > 1$ are relatively prime, then all the Weyl operators in dimension N can be simultaneously factorized in tensor products of the Weyl operators in the dimensions N_1 and N_2. J. Schwinger then states ([6], pp. 578–579):

> *"The continuation of the factorization terminates in*
>
> $$N = \prod_{k=1}^{f} \nu_k,$$
>
> *where f is the total number of prime factors in N, including repetitions. We call this characteristic property of N the number of degrees of freedom for a system possessing N states."* And further, *"each degree of freedom is classified by the value of the prime integer $\nu = 2, 3, 5, \ldots \infty$."*

However, our application of Theorem 3.1 leads to the building blocks $\mathbb{Z}_{p^r}$ with $r \geq 1$ as constituent configuration spaces. This fact was independently noted by V.S. Varadarajan in his paper [12] devoted to the memory of J. Schwinger: *"Curiously, Schwinger missed the systems associated to the indecomposable groups Z_N where N is a prime power p^r, $r \geq 2$ being an integer."* And in another paper: *"In this way he arrived at the principle that the Weyl systems associated to $\mathbb{Z}_p$ where p runs over all the primes are the building blocks. Curiously this enumeration is incomplete and one has to include the cases with $\mathbb{Z}_{p^r}$ where p is as before a prime but r is any integer ≥ 1."*

Thus the elementary building blocks of finite quantum kinematics are given by the elementary divisor decomposition. In mathematics also exists equivalent *invariant factor decomposition* [14]. In that approach a finite Abelian group G is uniquely determined by an ordered finite list of integers $n_1 \geq n_2 \geq \ldots \geq n_s$ greater than 1 determining the invariant factors $\mathbb{Z}_{n_i}$ of G such that n_{i+1} divides n_i and $N = n_1 n_2 \ldots n_s$. For instance, if $N = 180$, the full list of non-isomorphic Abelian groups of order 180 consists of $\mathbb{Z}_{180}$, $\mathbb{Z}_2 \times \mathbb{Z}_{90}$, $\mathbb{Z}_3 \times \mathbb{Z}_{60}$, $\mathbb{Z}_6 \times \mathbb{Z}_{30}$. However, to exhibit the elementary building blocks of the corresponding composite system, the elementary divisor decomposition is indispensable. In the above example all inequivalent choices of elementary divisors $N_k = p_k^{r_k}$ with not necessarily distinct primes $p_k > 1$ are $2^2.3^2.5$, $2.2.3^2.5$, $2^2.3.3.5$, $2.2.3.3.5$.

The task of enumerating all finite quantum kinematics in dimension N then amounts to the determination of all finite Abelian groups of order N. It starts with the factorization of $N = p_1^{r_1} \ldots p_f^{r_f}$ with mutually distinct primes $p_1, \ldots, p_f$. First one finds all permissible lists for groups of orders $p_i^{r_i}$ for each i. For a prime power $p_i^{r_i}$ the problem of determining all permissible lists is equivalent to finding all *partitions of the exponent* r_i, and does not depend on p_i. Recall that the number of partitions of a natural number r is called *Bell's number* $B(r)$. Then the total number of groups of order N is equal to the product of Bell's numbers $B(r_1)B(r_2)\ldots B(r_f)$.

In physics, the dimensions of constituent Hilbert spaces are primarily fixed by the numbers of levels of physical subsystems. It follows that G is isomorphic to a direct product of cyclic groups of the respective orders. Whenever this is the case, in order to be able to work with the elementary building blocks of the corresponding quantum kinematics, we should find the elementary divisors. For example, if $G = \mathbb{Z}_6 \times \mathbb{Z}_{15}$, we factor $6 = 2.3$ and $15 = 3.5$. Then the elementary divisor decomposition of G is $\mathbb{Z}_2 \times \mathbb{Z}_3 \times \mathbb{Z}_3 \times \mathbb{Z}_5$.

5. Algebraic approach

In quantum information theory as in algebraic quantum theory, the operators of a finite quantum system belong to the full matrix algebra $M_N(\mathbb{C})$ of $N \times N$ complex matrices. In this contribution we started with finite Abelian groups and, via representation theory expressed by Mackey's systems of imprimitivity, arrived at the system of N^2 unitary operators forming a complete orthonormal basis of the linear space $M_N(\mathbb{C})$ with the Hilbert-Schmidt inner product.

Now we shall pass the opposite journey starting from the operator algebra $M_N(\mathbb{C})$ and equip it as associative algebra with additional structure of a fine grading induced by commuting inner automorphisms of $\mathrm{GL}(N, \mathbb{C})$.

A *grading* of an associative algebra $\mathcal{A}$ is defined as a direct sum decomposition of A as a vector space

$$\Gamma : \quad \mathcal{A} = \bigoplus_{\alpha} \mathcal{A}_\alpha \tag{4}$$

satisfying the property

$$x \in \mathcal{A}_\alpha, \quad y \in \mathcal{A}_\beta \quad \Rightarrow \quad xy \in \mathcal{A}_\gamma.$$

Similarly as for Lie algebras [7], the linear subspaces $\mathcal{A}_\alpha$ can be determined as eigenspaces of automorphisms of the associative algebra $\mathcal{A}$. For an automorphism g of $\mathcal{A}$, $g(xy) = g(x)g(y)$ holds for all $x, y \in \mathcal{A}$. Now if $g(x) = \lambda_\alpha x$ defines the subspace $\mathcal{A}_\alpha$ and $g(y) = \lambda_\beta y$ defines the subspace $\mathcal{A}_\beta$, then

$$g(xy) = g(x)g(y) = \lambda_\alpha \lambda_\beta xy$$

defines the subspace $\mathcal{A}_\gamma$ with $\lambda_\gamma = \lambda_\alpha \lambda_\beta$. In this way g induces the grading decomposition $\mathcal{A} = \bigoplus_\alpha \mathrm{Ker}(g - \lambda_\alpha)$. Further commuting automorphisms may refine the grading. If $gh = hg$, we have

$$g(h(x)) = h(g(x)) = h(\lambda_\alpha x) = \lambda_\alpha h(x)$$

implying that the grading decomposition using both g and h may lead to a refinement. By extending the set of commuting automorphisms one arrives at so called *fine gradings* where the grading subspaces have the lowest possible dimension.

For the associative algebra $M_N(\mathbb{C})$ we shall look for fine gradings induced by the inner automorphisms. For $M \in \mathrm{GL}(N, \mathbb{C})$ we denote $\mathrm{Ad}_M \in \mathrm{Int}(M_N(\mathbb{C}))$ be the *inner automorphism* of $M_N(\mathbb{C})$ induced by operator $M \in \mathrm{GL}(N, \mathbb{C})$, i.e.

$$\mathrm{Ad}_M(X) = MXM^{-1} \quad \text{for} \quad X \in M_N(\mathbb{C}).$$

The relevant properties of Ad_M are: for $M, N \in \mathrm{GL}(N, \mathbb{C})$

(i) $\mathrm{Ad}_M \mathrm{Ad}_N = \mathrm{Ad}_{MN}$.

(ii) $(\mathrm{Ad}_M)^{-1} = \mathrm{Ad}_{M^{-1}}$.

(iii) $\mathrm{Ad}_M = \mathrm{Ad}_N$ *if and only if there is a constant* $0 \neq \alpha \in \mathbb{C}$ *such that* $M = \alpha N$.

Since the commuting inner automorphisms form an Abelian subgroup of $\mathrm{Int}(M_N(\mathbb{C}))$, *fine gradings* of $M_N(\mathbb{C})$ can be obtained using the *maximal Abelian groups of diagonalizable automorphisms* – as subgroups of $\mathrm{Int}(M_N(\mathbb{C}))$ – which have been called the MAD-*groups* [8]. Thus we are looking for fine gradings which are induced via diagonalizable elements of maximal Abelian subgroups of $\mathrm{Int}(M_N(\mathbb{C}))$.

The MAD-groups of inner automorphisms of $M_N(\mathbb{C})$ can be derived by following [8]. The result is straightforward and is summarized in the following theorem:

Theorem 5.1. *Any MAD-group contained in* $\mathrm{Int}(M_N(\mathbb{C}))$ *is conjugated to one and only one of the groups of the form*

$$\mathcal{P}_{N_1} \otimes \mathcal{P}_{N_2} \otimes \ldots \otimes \mathcal{P}_{N_f} \otimes D(m), \tag{5}$$

where $N_i = p_i^{r_i}$ *are powers of primes,* $N = N_1 N_2 \ldots N_f\ m$ *and* $D(m)$ *is the image in* $\mathrm{Int}(M_N(\mathbb{C}))$ *of the group of* $m \times m$ *complex diagonal matrices under the adjoint action.*

Here $\mathcal{P}_N$ is defined as the group

$$\mathcal{P}_N = \{\mathrm{Ad}_{Q_N^i P_N^j} | (i,j) \in \mathbb{Z}_N \times \mathbb{Z}_N\}.$$

It is an Abelian subgroup of $\mathrm{Int}(M_N(\mathbb{C}))$ and is generated by two commuting automorphisms Ad_{Q_N}, Ad_{P_N}, each of order N. A geometric view is sometimes useful that $\mathcal{P}_N$ is isomorphic to the *quantum phase space* identified with the Abelian group $\mathbb{Z}_N \times \mathbb{Z}_N$.

For illustration, we give a list of MAD-groups in low dimensions:

- $n = 2$: $\mathcal{P}_2 \otimes D(1)$, $D(2)$
- $n = 3$: $\mathcal{P}_3 \otimes D(1)$, $D(3)$
- $n = 4$: $\mathcal{P}_4 \otimes D(1)$, $\mathcal{P}_2 \otimes \mathcal{P}_2 \otimes D(1)$, $\mathcal{P}_2 \otimes D(2)$, $D(4)$
- $n = 5$: $\mathcal{P}_5 \otimes D(1)$, $D(5)$
- $n = 6$: $\mathcal{P}_3 \otimes \mathcal{P}_2 \otimes D(1)$, $\mathcal{P}_3 \otimes D(2)$, $\mathcal{P}_2 \otimes D(3)$, $D(6)$
- $n = 7$: $\mathcal{P}_7 \otimes D(1)$, $D(7)$
- $n = 8$: $\mathcal{P}_8 \otimes D(1)$, $\mathcal{P}_4 \otimes \mathcal{P}_2 \otimes D(1)$, $\mathcal{P}_2 \otimes \mathcal{P}_2 \otimes \mathcal{P}_2 \otimes D(1)$, $\mathcal{P}_4 \otimes D(2)$, $\mathcal{P}_2 \otimes \mathcal{P}_2 \otimes D(2)$, $\mathcal{P}_2 \otimes D(4)$, $D(4)$

A part of the obtained MAD-groups containing the trivial diagonal subgroup $D(1)$ induce exactly all our Pauli decompositions (2), (3). However, there are still fine gradings induced by $D(m)$, $m = 2, 3 \ldots$. They include partial or complete decompositions which have the form of the Cartan root decompositions of Lie algebras $\mathrm{sl}(m, \mathbb{C})$ extended by the unit matrix. They contain the Abelian Cartan subalgebra of dimension $m - 1$, the unit matrix and one-dimensional root subspaces spanned by nilpotent matrices. Concerning physical interpretation of these Cartan parts of the decompositions, one can speculate that they may reflect extra degrees of freedom corresponding to some internal symmetries [16]. Leaving these decompositions aside, we are left with the *Pauli gradings* which decompose $M_N(\mathbb{C})$ in direct sums of N^2 one-dimensional subspaces. For these Pauli decompositions with given N one should realize that $M_N(\mathbb{C})$ encompasses all operators of any quantum system with N-dimensional Hilbert space.

Journal of Physics: Conference Series **538** (2014) 012020 doi:10.1088/1742-6596/538/1/012020

In this way we have got a new view on the relation between the general mathematical formalism and physical realizations of finite quantum systems. The apparent contradiction that $M_N(\mathbb{C})$ represents any N-dimensional quantum system and at the same time there is a multitude of inequivalent quantum kinematics for given N, is simply resolved: from the physical point of view the same algebra $M_N(\mathbb{C})$ is the operator algebra not only for a single N-level system, but also for all other members of the set of inequivalent quantum kinematics for this N.[2] They just correspond to different physical realizations of composite quantum systems. Of course, each such system has its preferred quantum operators (3).

6. Conclusions

In our study unexpectedly rich structures were obtained from number-theoretic properties connected with prime decompositions of numbers N. Our studies may also shed light on a long-standing unsolved problem related to complementary observables in finite-dimensional quantum mechanics. There the notion of complementarity of observables A, B with non-degenerate eigenvalues is equivalently reformulated in terms of their eigenvectors forming mutually unbiased bases: if the system is prepared in any eigenstate of A, then the transition probabilities to all eigenstates of the complementary observable B are the same (equal to $1/N$). It is known that the maximal set of mutually unbiased bases contains at most $N+1$ bases and that this maximal number is attained for N prime or a power of a prime. For composite numbers N the maximal number of mutually unbiased bases is unknown. The needed bases can be constructed as common eigenvectors of suitable subsets formed by commuting Pauli operators. In some cases the decomposition of the set of all Pauli operators into subsets of commuting operators can be reflected in the finite geometry [17]. The study of mutually unbiased bases may have implications for quantum information and communication science, since mutually unbiased bases are indispensable ingredients of quantum key distribution protocols [5].

There exist numerous studies of various aspects of the finite Weyl-Heisenberg group over finite fields for Hilbert spaces of prime or prime power dimensions, e.g. [18, 19, 20, 21, 22]. However, our main motivation to study finite quantum kinematics not in prime or prime power dimensions but for arbitrary dimensions stems from our previous research where we obtained results valid for arbitrary dimensions [10, 11, 23]. There are also other papers supporting our motivation [24, 25, 26].

Finite quantum systems are basic constituents of quantum information processing. Except single 2- and d-level quantum systems — qubits and qudits — many authors pay their attention to finitely composed systems, where the basic operators are formed by tensor products. Multiple qubits with Hilbert spaces $\mathbb{C}^2 \otimes \ldots \otimes \mathbb{C}^2$ are routinely employed in quantum algorithms, while multiple qudits with Hilbert spaces $\mathbb{C}^d \otimes \ldots \otimes \mathbb{C}^d$ may be interesting for quantum error-correction codes and for multipartite communication. What about general systems? They are sometimes called 'mixtures of multiple qudits'. We have shown the general classification of finite quantum kinematics and their physical interpretation: for given dimension there exists a broad variety of finitely composed distinct quantum kinematics involving inequivalent tensorial factorizations into elementary constituents. It is remarkable that in spite of the boom of quantum information

[2] Note that from the mathematical point of view M_{N_1} $) \otimes \cdots \otimes M_{N_f}(\mathbb{C})$ is isomorphic with $M_N(\mathbb{C})$.

science the community of quantum information and communication has not noticed this general classification up to now. Only the visible aspect of the finite-dimensional operator formalism is perceived in terms of examples constructed from generalized Pauli matrices in an *ad hoc* manner. In this way the underlying general structure remains hidden. In our approach we explicitly exhibit the elementary quantum degrees of freedom, since their relevance has not been emphasized in most of the literature on finite quantum systems.

It is clear that automorphisms or symmetries of the finite Weyl-Heisenberg group play very important role in the investigation of Lie algebras on the one hand [10, 27, 28] and of quantum mechanics in finite dimensions on the other [29, 21]. These symmetries find proper expression in the notion of the quotient group of a certain normalizer [7]. The groups of symmetries given by inner automorphisms were described in [10] as isomorphic to $\mathrm{SL}(2, \mathbb{Z}_N)$ for arbitrary $N \in \mathbb{N}$ and as $\mathrm{Sp}(4, \mathbb{Z}_p)$ for $n = p^2$, p prime in [28] (see also [30]). For the complete description of these symmetries — in quantum information conventionally called Clifford groups [31] — we refer to our papers [1, 2].

Acknowledgements

The author is grateful to M. Havlíček, M. Korbelář and P. Novotný for fruitful discussions on the subject of the paper. Support by the Ministry of Education of Czech Republic from the projects MSM6840770039, LC06002 and the research plan RVO:68407700 is acknowledged.

References

[1] Korbelář M and Tolar J 2010 Symmetries of the finite Heisenberg group for composite systems *J. Phys. A: Math. Theor.* **43** 375302
[2] Korbelář M and Tolar J 2012 Symmetries of the finite Heisenberg group for multipartite systems *J. Phys. A: Math. Theor.* **45** 285305
[3] Weyl H 1931 *The Theory of Groups and Quantum Mechanics* (New York: Dover) pp 272–280
[4] Mackey G W 1968 *Induced Representations and Quantum Mechanics* (Benjamin, New York)
[5] Gisin N, Ribordy G, Tittel W and Zbinden H 2002 Quantum cryptography *Rev. Mod. Phys.* **74** 145–195
[6] Schwinger J 1960 Unitary operator bases *Proc. Nat. Acad. Sci. U.S.A.* **46** 570–579, 1401–1415
[7] Patera J and Zassenhaus H 1989 On Lie gradings I *Lin. Alg. Appl.* **112** 87–159
[8] Havlíček M, Patera J, Pelantová E 1998 On Lie gradings II *Lin. Alg. Appl.* **277** 97–125
[9] Patera J and Zassenhaus H 1988 The Pauli matrices in n dimensions and finest gradings of simple Lie algebras of type A_{n-1} *J. Math. Phys.* **29** (3) 665–673
[10] Havlíček M, Patera J, Pelantová E and Tolar J 2002 Automorphisms of the fine grading of $sl(n, \mathbb{C})$ associated with the generalized Pauli matrices *J. Math. Phys.* **43** 1083-1094; arXiv: math-ph/0311015
[11] Šťovíček P and Tolar J 1984 Quantum mechanics in a discrete space-time *Rep. Math. Phys.* **20** 157–170
[12] Varadarajan V S 1995 Variations on a theme of Schwinger and Weyl *Lett. Math. Phys.* **34** 319–326
[13] Emch G G 1984 *Mathematical and Conceptual Foundations of 20th Century Physics* (Amsterdam: North-Holland)
[14] Dummit P S and Foote R M 2004 *Abstract Algebra* 3rd ed. (New York: Wiley)
[15] Petz D 2010 Algebraic complementarity in quantum theory *J. Math. Phys.* **51** 015215
[16] Kibler M R 2008 Variations on a theme of Heisenberg, Pauli and Weyl *J. Phys. A: Math. Theor.* **41** 375302
[17] Rosu H C, Planat M and Saniga M 2004 *Quantum Commun. Meas. Comput.* **734** 315–318
[18] Balian R and Itzykson C 1986 Observations sur la mécanique quantique finie *C. R. Acad. Sci. Paris* **303** Série I, n. 16, 773–777

[19] Fivel D I 1995 Remarkable phase oscillations appearing in the lattice dynamics of Einstein-Podolsky-Rosen states *Phys. Rev. Lett.* **74** 835–838

[20] Neuhauser M 2002 An explicit construction of the metaplectic representation over a finite field *Journal of Lie Theory* **12** 15–30

[21] Šulc P and Tolar J 2007 Group theoretical construction of mutually unbiased bases in Hilbert spaces of prime dimensions *J. Phys. A: Math. Theor.* **40** 15099–15111

[22] Vourdas A 2007 Quantum systems with finite Hilbert space: Galois fields in quantum mechanics *J. Phys. A: Math. Theor.* **40** R285–R331

[23] Tolar J and Chadzitaskos G 2009 Feynman's path integral and mutually unbiased bases *J. Phys. A: Math. Theor.* **42** 245306 (11pp)

[24] Vourdas A and Banderier C 2010 Symplectic transformations and quantum tomography in finite quantum systems *J. Phys. A: Math. Theor.* **43** 042001 (9pp)

[25] Digernes T, Husstad E and Varadarajan V S 1999 Finite approximation of Weyl systems *Math. Scand.* **84** 261–283

[26] Digernes T, Varadarajan V S and Varadhan S R S 1994 Finite approximations to quantum systems *Rev. Math. Phys.* **6** 621–648

[27] Hrivnák J, Novotný P, Patera J and Tolar J 2006 Graded contractions of the Pauli graded $\mathrm{sl}(3,\mathbb{C})$ *Lin. Alg. Appl.* **418** 498–550

[28] Pelantová E, Svobodová M and Tremblay J 2006 Fine grading of $\mathrm{sl}(p^2,\mathbb{C})$ generated by tensor product of generalized Pauli matrices and its symmetries *J. Math. Phys.* **47** 5341–5357

[29] Vourdas A 2004 Quantum systems with finite Hilbert space *Rep. Progr. Phys.* **67** 267–320

[30] Han G 2010 The symmetries of the fine gradings of $\mathrm{sl}(n^k,\mathbb{C})$ associated with direct product of Pauli groups *J. Math. Phys.* **51** 092104

[31] Gottesman D 1998 Theory of fault-tolerant quantum computation *Phys. Rev. A* **57** 127–137

Institute of Physics
Dirac House, Temple Back
Bristol BS1 6BE UK

CURRAN ASSOCIATES INC
proceedings
.com